A FIELD GUIDE TO

MICHIGAN WILDFLOWERS

HELEN VANDERVORT SMITH AND STEVE CHADDE

Dedicated to my parents
who encouraged my early interest in plants
and to my husband and daughter
whose aid and forbearance have made it possible
for me to devote myself to Michigan wildflowers
—H.V.S.

A FIELD GUIDE TO MICHIGAN WILDFLOWERS
Helen Vandervort Smith *and* Steve Chadde

ISBN 978-1951682880
Printed in the United States of America

A Pathfinder Field Guide
Published by Orchard Innovations, Mountain View, Arkansas

Author email: *steve@orchardinnovations.com*

TABLE OF CONTENTS

Preface (from 1961 edition) 5
Acknowledgments 6
Introduction 7
The Naming of Plants 7
The Parts of a Plant 8
Using this Book 9
Wildflower Families 11
Family and Species Descriptions
Dicotyledons (Dicots) 25
Monocotyledons (Monocots) 227
Family Keys 277
Glossary 287
Selected References 296
Index to Scientific (Latin) Names 297
Index to Common Names 307

DWARF LAKE IRIS (*Iris lacustris*), a regional-endemic species, found near the Great Lakes in Michigan, Wisconsin, and Ontario (see p. 234).

PREFACE (from 1961 Edition)

A book on the wildflowers of Michigan has been discussed by various groups for a number of years. There was obvious need for one which would include most of the wildflowers that one might encounter in the ordinary course of outdoor activities in the state. The larger publications are too technical and difficult for any but the trained botanist and are too large to carry about. The general handbooks include too many species not found in Michigan and too few species that are; in addition, the descriptions are often so sketchy that one cannot be sure of correct identification. It was to fill this need that Dr. Robert T. Hatt, Director of the Cranbrook Institute of Science, asked me to undertake a book on Michigan wildflowers. While I was extremely pleased, I felt considerable hestitation about accepting the task because of a realization of the difficulties which had to be overcome. However, I have had a lively interest in wildflowers most of my life. In pursuing this interest many manuals have been used as well as many of the small, limited guides designed to give quick identification of a few species. I have experienced much difficulty in trying to make identifications and consequently have developed a number of ideas as to just what is important to one who wants to learn the wildflowers. Hence writing such a book as this was a challenge too great to resist.

This work is offered with many misgivings. I find it much easier to criticize what others have done than to do a better job myself. If this publication enables the hobbyist to learn the names and something about the wildflowers of the state, my aim will have been accomplished.

Over 500 species of plants found in Michigan are included, most of which are herbaceous. Since there are excellent books on Michigan shrubs, trees, and weeds, few of these plants are treated here; even the showy Flowering Dogwood has been omitted. The plants to be included were selected by the Cranbrook Institute of Science, with the advice of members of the Herbarium Committee of the Institute. The majority of the descriptions were written up from living plants. The aim has been to give, through descriptions and illustrations, the picture of the plant as it appears in nature, coupled with such technical details as are essential in identifying it.

One species of each kind of plant is quite fully described; for closely similar kinds merely the identifying or distinguishing characters are given. The general aspect and most easily seen characteristics of the plant as a whole are given first, followed by a description of the flowers, the nature and arrangement of the leaves, the habitat, and the flowering time. In many cases interesting notes on the plant under discussion are added.

ACKNOWLEDGMENTS

No book of any size can be prepared without assistance; this one has been no exception. The late F. C. Gates, of Kansas State College and the University of Michigan Biological Station, gave me much interesting and helpful information based on his more than forty summers of work with Michigan plants. Professor Volney H. Jones of the University of Michigan made available to me the Museum of Anthropology file on the uses of native plants by Indians of the Great Lakes region. Professor Rogers McVaugh of the University of Michigan Herbarium has given advice on all phases of the work. Dr. E. B. Mains, Director of the University of Michigan Herbarium, offered access to the collection of Michigan flowering plants. Dr. Lloyd Shinners aided in some of the identifications, particularly in the Compositae. Throughout the course of the work Professor Alexander H. Smith of the University of Michigan Herbarium and Department of Botany has collected material for my use and has read the manuscript. Professor A. H. Stockard, Director of the University of Michigan Biological Station, provided me with facilities at the Station during the three summers I worked on the project there. Dr. Edward G. Voss of the University of Michigan Herbarium was most helpful in obtaining material for drawing and providing information on the distribution of species in the state; he also read the proofs. To all these I offer sincere gratitude.

—Helen V. Smith

Grateful acknowledgment is given to the numerous photographers who have made their photographs available for reuse on the **iNaturalist** platform (*www.inaturalist.org*). Their names (as provided on iNaturalist) are listed on each image. This book would not have been possible without their efforts to document our diverse flora.

Acknowledgment is also given to the **Biota of North America Program** (BONAP) for permission to use their data to generate the distribution maps (see *www.bonap.org*).

—Steve Chadde

INTRODUCTION

Michigan is unlike any other state in its combination of extensive forests of broad-leaved trees, vast areas of jack-pine plains, bits of prairie, and a shoreline unique in length. Its wilderness, lakes and streams, beaches, and sand dunes, its woods, meadows, and abundance of wildflowers and wildlife, bring pleasure each year both to its year-round residents and to hundreds of thousands of visitors. But residents and visitors alike who have gone afield in Michigan during the blooming season have often felt the lack of a good flower book especially adapted to the Michigan scene. ***Michigan Wildflowers*** has been prepared and published to meet this need. In it are described and illustrated almost every wildflower which attracts attention through charm, abundance, or rarity. This new (2024) edition updates scientific names to conform to the latest taxonomic research, adds Michigan county-level distribution maps, and provides color photographs for nearly all of the more than 500 plant species coered in the book.

THE NAMING OF PLANTS

When one discovers an unfamiliar object, the first question that comes to mind is: "What is its name?" Living organisms, plants and animals alike, have two types of names. The more familiar, everyday ones are called common names. These are often descriptive, as, for example, Blue-eyed Grass, Black-eyed Susan, or Milkweed. However, a single kind of plant may be called by different common names in different countries or even in different sections of the same country. Further, the same common name, in different regions, may be applied to entirely different plants.

To avoid the difficulties arising from conflicting common names, each kind of plant is given a scientific name. This consists of two Latin or Latinized words and is the same in all countries. The first word is capitalized and is the generic name, the name of the genus. A genus is a group of closely related kinds (species) of plants. The second word, called a species epithet, indicates which particular plant in the group is designated. *Viola* is the generic name for all our native violets. *Viola canadensis* is the name for the Canadian or Tall White Violet; *Viola pedata* (pedata meaning "footlike") is the Birdfoot Violet; and *Viola rostrata* ("having a beak") is the Long-Spur Violet. It is customary to italicize the scientific name and to indicate the name of the person who first named the species. Thus the full designation for Birdfoot Violet would be *Viola pedata* L., the L. standing for Linnaeus. Genera (plural of genus) are grouped in families. Family names end in the letters *aceae,* as in Violaceae. The scientific names used for the plants in this book conform (in nearly all cases), to those of the authoritative **Plants of the World Online** database maintained by the Royal Botanic Gardens, Kew (*https://powo.science.kew.org*).

THE PARTS OF A PLANT

In order to identify wildflowers it is necessary to understand a little about the plant as a whole, its parts, and the descriptive terms applied to them. The parts with which we are principally concerned are those that occur aboveground; the ***stem*** and its branches, the ***leaves,*** the ***flowers,*** and the ***fruits*** (see **Glossary**, pp. 287–295).

One of the first things to notice about a plant is whether it is ***herbaceous*** or ***woody,*** i.e., whether it dies down to the ground in the winter or has a persistent woody stem. Some plants are ***evergreen,*** i.e., they have green leaves the year around, but most of our species either shed their leaves (are ***deciduous***) or are ***herbaceous*** (non-woody). The habit of growth of plants, the direction in which the stem grows, is easily noted and is often characteristic (p. 291).

The **leaves** are extremely important as the photosynthetic or food manufacturing organs of the plant. All life is dependent upon the ability of green plants to combine carbon dioxide with water to form a simple sugar. This process can be carried on only by the green coloring material (chlorophyll) acting in the presence of sunlight or its equivalent. Leaves are also of importance in the classification of plants. Most of the common terms used in describing them are illustrated (see pp. 294–295). Definitions of the terms are given in the Glossary.

The **flowers** are the reproductive structures of the plant; their function is to produce seeds. A typical flower has four whorls or circles of parts. The outer whorl is called the ***calyx.*** This is composed of separate or united ***sepals*** and is usually green. The ***corolla*** is composed of separate or united ***petals,*** which are usually white or colored. The corolla is often lacking, and the calyx may be colored and corolla-like. The calyx and corolla (p. 292) function as floral envelopes to protect the essential organs and/or to attract insects. The ***stamens,*** each consisting of an ***anther*** and ***filament,*** and the ***pistil,*** or pistils, are essential for reproduction. The anther is borne on the filament and produces the ***pollen*** grains; these contain the male reproductive elements. The pistil consists of the (usually enlarged) ***ovary*** at the base, the ***style,*** and the ***stigma.*** The ovary contains the ovules which, when fertilized by the male element from the pollen grain, may develop into seeds. The stigma acts as the receptive organ for the pollen and usually has a moist or sticky surface. Flowers vary greatly in arrangement on the plants, number and shape of parts, etc. These variations are important in the classification of plants. Many of the common descriptive terms applied to flowers are illustrated (see pp. 292–293), and the terms are defined in the Glossary.

The **fruits** are the “ultimate purpose” of a plant; they are the structures that bear the seeds. Under proper conditions seeds germinate and produce new plants. Fruits are quite diverse and are an aid in identification. Some types of fruit are hard to classify; a few of the more easily distinguished kinds are illustrated (p. 293).

In addition to the names and shapes of the parts of a plant it is often necessary or helpful to know the sizes. In this book the size is expressed in the metric system. The measurements, with their abbreviations and approximate equivalents in inches, are as follows:

Millimeter (mm)	1/25 inch
Centimeter (cm = 10 mm)	2/5 inch
Decimeter (dm = 10 cm)	4 inches
Meter (m = 10 dm.)	39.4 inches

USING THIS BOOK

The more than 500 plant species described in ***Michigan Wildflowers*** are arranged in alphabetical order first by the scientific name of the family, with the dicotyledons ("dicots") presented first, followed by the monocotyledons ("monocots"). Within each family, plants are arranged (for the most part) in alphabetical order by their scientific name. In contrast to books arranged by flower color, this arrangement provides a deeper understanding of family characteristics, and, once family features are known, a more efficient way to identify unknown plant specimens.

MAPS

A county-level distribution map is provided for nearly all plants treated in this book. Maps indicate the verified presence of the plant in a county, based on data provided by the **Biota of North America Program** (BONAP, see *www.bonap.org*). Maps are color-coded as indicated below (the square icon – ■ – before the plant's common name is likewise color-coded):

NATIVE IN MICHIGAN
Considered part of the original flora of the state prior to European settlement

INTRODUCED
Not originally present in Michigan or North America but now well-established in the state

ADVENTIVE IN MICHIGAN
Native elsewhere in North America but not yet fully established in Michigan

CONSERVATION STATUS

If a plant is rare or uncommon in Michigan (or the United States), it may be given legally protected status as either:

Endangered: indicates the species is in danger of extinction throughout all or a significant portion of its range.

Threatened: indicates the species is likely to become endangered within the foreseeable future.

These plants (noted in their description) should never be picked or removed from their habitat. For a complete up-to-date listing of all protected plants in Michigan, see the website of the Michigan Natural Features Inventory (MNFI) at *https://mnfi.anr.msu.edu/species/plants.*

USES

For many species, notes on their traditional or cultural use by American Indians or by early settlers is given. Note that consumption of any plant should be made with caution unless absolutely certain of its identification and of its proven safety. Some plants, Poison-Hemlock, for example, can be fatal if eaten.

YELLOW TROUT-LILY (*Erythronium americanum*), one of Michigan's earliest spring wildflowers (see p. 245).

Wildflower plant families are listed below, first the **dicotyledons** ("dicots"), followed by the **monocotyledons** ("monocots"). For each family, a representative Michigan species is illustrated. A botanical key to the families of plants included in *Michigan Wildflowers* begins on page 277. Once family characteristics are known, identification of unknown plants becomes much easier!

DICOTS

25
AMARANTHACEAE
Amaranth Family

26
APIACEAE
Parsley Family

34
APOCYNACEAE
Dogbane Family

38
ARALIACEAE
Ginseng Family

41

ARISTOLOCHIACEAE
Birthwort Family

42

ASTERACEAE
Aster Family

80

BALSAMINACEAE
Touch-Me-Not Family

81

BERBERIDACEAE
Barberry Family

83

BORAGINACEAE
Borage Family

86

BRASSICACEAE
Mustard Family

93

CACTACEAE
Cactus Family

94

CAMPANULACEAE
Bluebell Family

97

CAPRIFOLIACEAE
Honeysuckle Family

100

CARYOPHYLLACEAE
Pink Family

105

CELASTRACEAE
Bittersweet Family

106

CISTACEAE
Rockrose Family

CONVOLVULACEAE
Morning-Glory Family

CORNACEAE
Dogwood Family

CRASSULACEAE
Stonecrop Family

CUCURBITACEAE
Gourd Family

DROSERACEAE
Sundew Family

ERICACEAE
Heath Family

EUPHORBIACEAE
Spurge Family

FABACEAE
Pea Family

GENTIANACEAE
Gentian Family

GERANIACEAE
Geranium Family

HYDROPHYLLACEAE
Waterleaf Family

HYPERICACEAE
St. John's-Wort Family

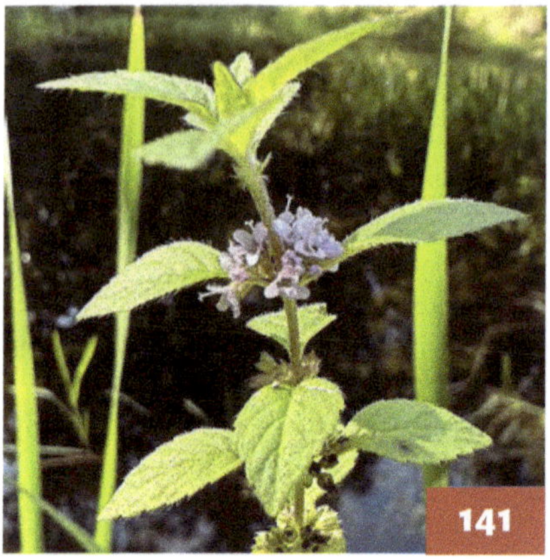

LAMIACEAE
Mint Family

LENTIBULARIACEAE
Bladderwort Family

LINACEAE
Flax Family

LYTHRACEAE
Loosestrife Family

MALVACEAE
Mallow Family

MENYANTHACEAE
Buck-Bean Family

MONTIACEAE
Candy-Flower Family

NELUMBONACEAE
Lotus-Lily Family

NYCTAGINACEAE
Four-O'clock Family

NYMPHAEACEAE
Water-Lily Family

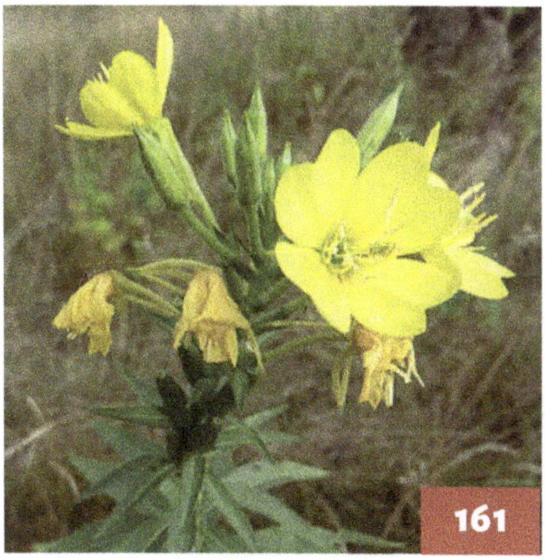

ONAGRACEAE
Evening-Primrose Family

OROBANCHACEAE
Broom-Rape Family

OXALIDACEAE
Wood-Sorrel Family

PAPAVERACEAE
Poppy Family

PHRYMACEAE
Lopseed Family

PHYTOLACCACEAE
Pokeweed Family

PLANTAGINACEAE
Plantain Family

POLEMONIACEAE
Phlox Family

POLYGALACEAE
Milkwort Family

POLYGONACEAE
Buckwheat Family

PRIMULACEAE
Primrose Family

RANUNCULACEAE
Buttercup Family

ROSACEAE
Rose Family

RUBIACEAE
Madder Family

SANTALACEAE
Sandalwood Family

SARRACENIACEAE
Pitcher-Plant Family

SAXIFRAGACEAE
Saxifrage Family

SCROPHULARIACEAE
Figwort Family

SOLANACEAE
Nightshade Family

VERBENACEAE
Vervain Family

VIOLACEAE
Violet Family

MONOCOTS

ACORACEAE
Calamus Family

ALISMATACEAE
Water-Plantain Family

AMARYLLIDACEAE
Daffodil Family

ARACEAE
Arum Family

ASPARAGACEAE
Asparagus Family

ASPHODELACEAE
Onionweed Family

BUTOMACEAE
Flowering-Rush Family

COLCHICACEAE
Autumn-Crocus Family

COMMELINACEAE
Spiderwort Family

HYPOXIDACEAE
Yellow Star-Grass Family

IRIDACEAE
Iris Family

JUNCAGINACEAE
Arrow-Grass Family

LILIACEAE
Lily Family

MELANTHIACEAE
False Hellebore Family

NARTHECIACEAE
Asphodel Family

ORCHIDACEAE
Orchid Family

PONTEDERIACEAE
Pickerelweed Family

SMILACACEAE
Greenbrier Family

TOFIELDIACEAE
Featherling Family

TYPHACEAE
Cat-Tail Family

XYRIDACEAE
Yellow-Eyed-Grass Family

DICOTYLEDONS

AMARANTHACEAE *Amaranth Family*

This is a family of wide distribution, with 183 genera and 2,040 species; 16 genera occur in Michigan. The family is of some economic importance: garden and sugar beets, spinach, and Swiss chard are common foods; pigweed and Russian thistle are troublesome weeds; Kochia is grown as an ornamental.

The family, which now includes the former **Goosefoot Family** (Chenopodiaceae), is characterized by its minute flowers. The perianth is bractlike and lacks petals; there are 5 sepals, 5 stamens, and a single pistil.

Strawberry-Blite

Blitum capitatum L.

SYNONYM *Chenopodium capitatum* (L.) Aschers.

Glabrous, erect annual up to 6 dm tall, simple or branching. **Flowers** very small, borne in globose clusters in the axils of the leaves (or the clusters confluent, without intervening leaves, near top of stem); in fruit the clusters become large, juicy, and bright red, and are very conspicuous; seed minute, black. **Leaves** alternate, bright green, triangular to somewhat arrow-shaped, long-petioled, the margin wavy or with large irregular teeth.

HABITAT In woods, clearings, burns, usually in light soil.

FLOWERING June to August.

STRAWBERRY-BLITE

Lamb's-Quarters, Goosefoot

Chenopodium album L.

Mealy, erect annual up to 2 m tall. **Flowers** small, greenish, often becoming reddish in fall, borne in small clusters in terminal or axillary panicles. **Leaves** alternate, green or (often) covered with a white meal-like powder, lanceolate or somewhat rhombic, wedge-shaped at base, 3–10 cm long, the larger leaves usually coarsely toothed.

HABITAT In waste ground and dry woods, along roadsides, in fields and gardens. Introduced from Europe.

FLOWERING June to October.

NOTES This is a common and troublesome weed, but it is sometimes cooked and served as "greens." Some quite similar very common weeds (called Pigweed or Tumbleweed) belong to the genus *Amaranthus*, with tiny flowers with dry bracts at the base.

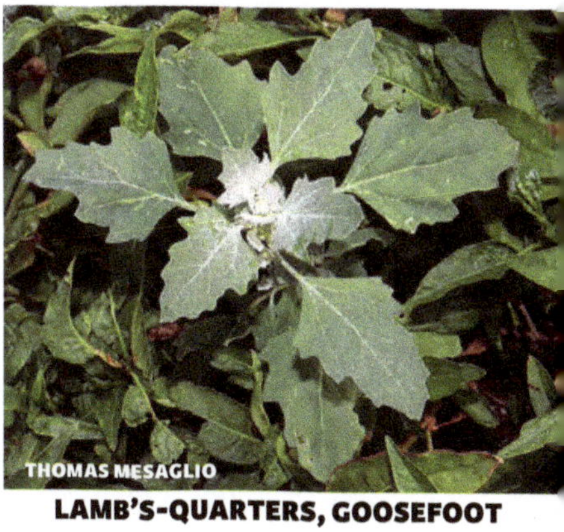

LAMB'S-QUARTERS, GOOSEFOOT

APIACEAE *Parsley Family*

This is a large, cosmopolitan family of 448 genera and with more than 3,800 species; 37 genera occur in Michigan. The family is quite important economically. Carrots, parsnips, celery, and parsley are important foods. Anise, caraway, chervil, dill, and fennel are used for flavoring. Blue Laceflower, Angelica, Sea Holly, and Cow-Parsnip are grown as ornamentals. Water-Hemlock and Poison-Hemlock are very poisonous and several other species slightly so.

The Apiaceae are characterized by having small flowers usually borne in umbels. The flower parts are in 5's except for the single pistil with its inferior, 2-celled ovary. The petals are usually incurved, and the fruit consists of 2 seed-like, dry carpels which separate from each other when ripe. The foliage is aromatic, and the leaves are usually compound or deeply lobed. The petioles are enlarged to form a sheathing base.

KEY TO APIACEAE (PARSLEY FAMILY) SPECIES

1 Ovary and fruit covered with bristles or hooked or barbed prickles 12

1 Ovary and fruit lacking bristles or prickles 2

2 Flowers yellow 3

2 Flowers white 6

3 Leaflets having entire margins .... ***Taenidia integerrima***

3 Leaflets with toothed margins 4

4 Leaves pinnately compound, having 5 or more leaflets .... ***Pastinaca sativa***

4 Leaves compound or twice compound in 3's 5

5 Central flower of each umbel on a stalk; terminal leaflets longer than lateral leaflets .... ***Thaspium trifoliatum***

5 Central flower of each umbel not stalked; terminal leaflet shorter than lateral leaflets .... ***Zizia aurea***

6 Leaves 2–3 times compound 7

6 Leaves once compound, leaflets distinct 10

7 Flowering in early spring; leaves mostly from base of plant .... ***Erigenia bulbosa***

7 Flowering in summer; leaves principally on stem 8

8 Margin of leaflets merely toothed 9

8 Margin of leaflets deeply lobed or divided into segments to coarsely incised, so that leaf appears to be greatly dissected .... ***Conium maculatum***

9 Umbel densely hairy; blades of upper leaves reduced, their sheaths greatly enlarged .... ***Angelica venenosa***

9 Umbel not hairy; upper leaves having distinct blades, sheaths less than 1 cm wide .... ***Cicuta maculata*** or ***C. bulbifera***

10 Leaves pinnately compound, leaflets linear-lanceolate (submerged leaves often more than once compound) .... ***Sium suave***

10 Leaves trifoliolate, leaflets ovate or cordate 11

11 Umbels irregular, few-flowered; plants not hairy; not especially robust .... ***Cryptotaenia canadensis***

11 Umbels large, regular; leaflets cordate at base, palmately veined; plants robust .... ***Heracleum maximum***

12 Fruit several times longer than wide .... ***Osmorhiza claytoni*** or ***O. longistylis***

12 Fruit not several times longer than wide 13

13 Plants pubescent, leaves finely dissected .... ***Daucus carota***

13 Plants smooth, leaves palmately compound with 3–7 leaflets .... ***Sanicula*, 14**

14 Styles longer than bristles of fruit 15

14 Styles shorter than bristles of fruit and hidden by them .... *Sanicula marilandica*
15 Fruits sessile, 5–8 mm long; sepals of stamnate flowers rigid .... *Sanicula canadensis*
15 Fruits on a stalk, less than 5 mm long; sepals of staminate flowers soft . *Sanicula odorata*

■ Hairy Angelica

Angelica venenosa (Greenway) Fern.

Rather slender perennial up to 1.8 m tall. **Flowers** white or greenish, borne in large, terminal, densely hairy, compound umbels, the central umbel up to 1.5 dm broad. **Fruit** tan, flattened, ribbed, and wing-margined, with several oil tubes between the ribs. **Basal and lower leaves** compound, with 3 primary divisions, odd-pinnate, the leaflets thick, lanceolate to oblong, toothed, 2–7 cm long; **upper leaves** reduced to tubular sheaths with or without small blades; blades, if present, shorter than the sheaths: sheaths greatly dilated, more than 1 cm wide.

HABITAT In dry woods, thickets, and openings.

FLOWERING July to September.

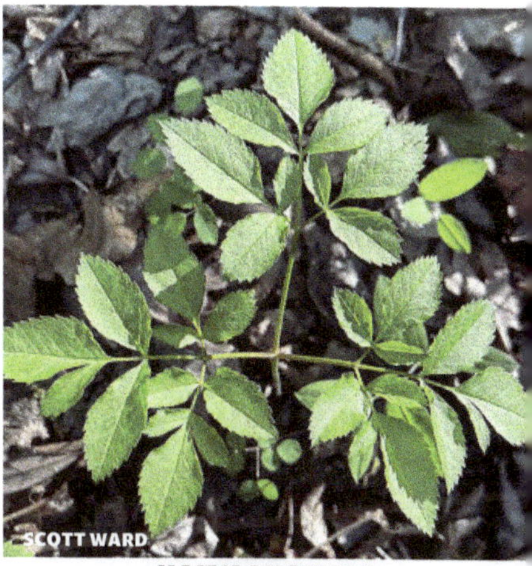

HAIRY ANGELICA

SIMILAR SPECIES Purple-Stem Angelica, *Angelica atropurpurea* L., is more widespread in Michigan, and has a very stout stem which is glabrous or nearly so.

This species may be distinguished from the closely similar *Cicuta* by the leaf veins which end in the tips of the teeth; in *Cicuta* they terminate in or near the notches, and the sheaths of the upper leaves are usually less than 1 cm wide.

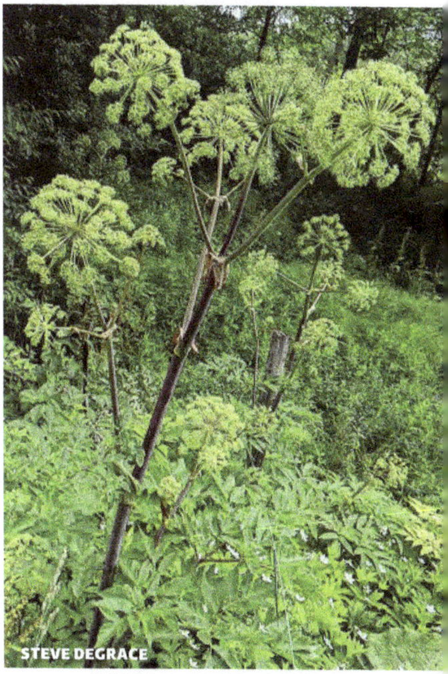

PURPLE-STEM ANGELICA

■ Bulblet-Bearing Water-Hemlock

Cicuta bulbifera L.

This species is characterized by the small clustered **bulblets** in the axils of the upper leaves; the leaflets are linear or narrowly lanceolate and have only a few teeth.

HABITAT In swamps and marshes.

FLOWERING July to September.

■ Water-Hemlock

Cicuta maculata L.

Stout, erect, many-branched perennial up to 2.2 m tall from a cluster of fleshy, finger-like roots, the stem often mottled with purple below. **Flowers** white, tiny, borne in flat-topped terminal or axillary umbels which are usually above the leaves; involucres lacking but the small umbels having several slender, drooping bracts. **Fruit** ovoid or ellipsoid, having alternate ribs and

BULBLET-BEARING WATER-HEMLOCK

WATER-HEMLOCK

furrows. **Leaves** 2–3 times pinnately compound, the lower leaves long-petioled; leaflets linear to lanceolate, bright green, sharply toothed, with conspicuous pinnate veins; upper leaves having distinct blades, the sheaths less than 1 cm wide.

HABITAT In marshy places, along streams, in wet meadows and swales.

FLOWERING June to September.

CAUTION The clustered roots resemble small sweet potatoes. They smell like parsnips and are deadly poisonous.

■ Poison-Hemlock

Conium maculatum L.

Smooth, purple-spotted, hollow-stemmed biennial up to 2 m tall. **Flowers** white, tiny, in showy umbels, the central or first umbel being quickly overtopped by later ones; involucre of narrow bracts. **Fruit** ovoid, somewhat flattened, smooth, having prominent wavy ribs, no oil tubes. **Leaves** alternate, petioled, 2–3 times compound, the leaflets lanceolate, pinnately cut and deeply crenate-dentate, the fresh leaves having a nauseous taste and parsnip-like odor when bruised.

HABITAT In waste places, woodlots, roadsides, around farm buildings. Naturalized from Europe.

FLOWERING late June to September.

CAUTION All parts of this species are poisonous and should be avoided. It is said that the poison the Greeks gave to Socrates was derived from this species. One can be poisoned by just blowing a whistle made from the hollow stems. The plant is most poisonous at the time the fruits are ripe.

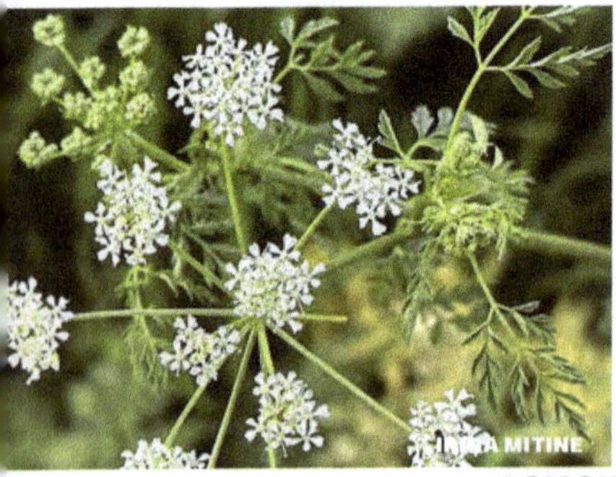

POISON-HEMLOCK

Wild Chervil, Honewort

Cryptotaenia canadensis (L.) DC.

Smooth perennial up to 1 m tall. **Flowers** white, small, in irregular, few-rayed umbels which are not subtended by an involucre. **Fruits** dark brown, stiffly erect, slenderly ellipsoid, beaked, ribbed, about 5–7 mm long. **Leaves** alternate, 3-foliolate (sometimes appearing 5-foliolate), thin, the leaflets large, ovate, often deeply lobed.

HABITAT In rich woods and thickets, on flood plains, and along roads.

FLOWERING June to September.

Wild Carrot, Queen Anne's Lace

Daucus carota L.

Coarse, stiff, freely branching, bristly hairy biennial up to 1 m tall from a deep, fleshy tap root. **Flowers** very small, white or rarely roseate, borne in compact, compound, many-rayed umbels which usually have a purple flower or flowers in the center, the umbels flattish at first, but becoming concave, the marginal flowers often enlarged and irregular; **fruits** flattened, ribbed, armed with barbed prickles; involucre of numerous large, pinnate bracts with long, very narrow, threadlike or linear, pointed lobes. **Leaves** finely pinnately 2–3 times compound, the segments narrow, lanceolate, tipped with a firm sharp point.

HABITAT In dry fields and sunny waste places. Introduced from Europe; now a pernicious weed.

FLOWERING June to September.

NOTES The cultivated carrot was derived from this wild species. Some people are sensitive to this plant and get a severe skin irritation from handling the foliage.

Harbinger-Of-Spring, Pepper-And-Salt

Erigenia bulbosa (Michx.) Nutt.

Delicate, low perennial not over 4 dm tall at flowering time, from a deep-seated globose tuber. **Flowers** white, small, borne in a single, compound, leafy-bracted umbel, the peduncle at first very short so that the flowers often appear to be half buried, later raised a little. Petals flat; anthers blackish-red; seeds flattened, kidney-

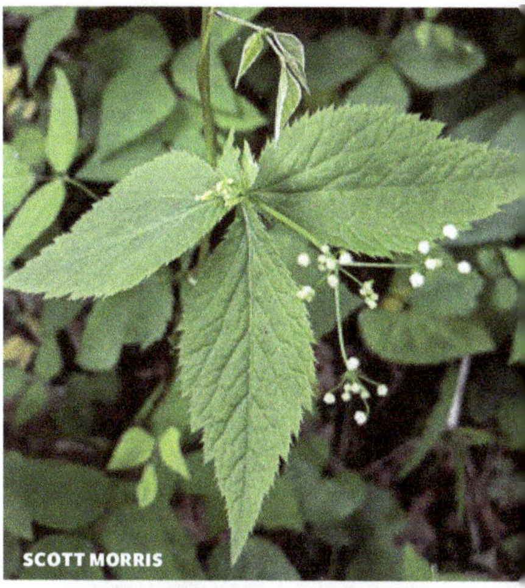

WILD CHERVIL, HONEWORT

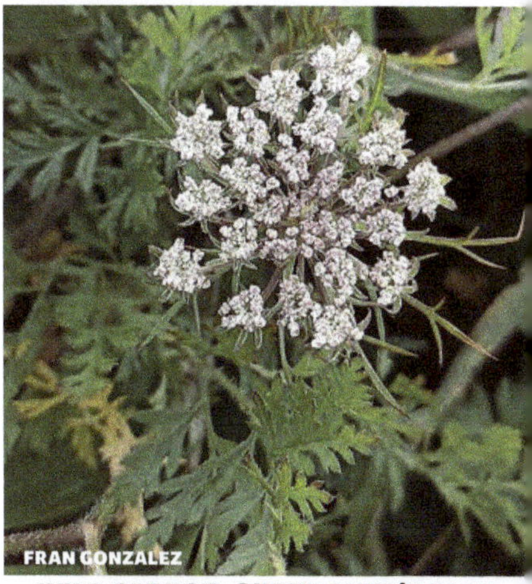

WILD CARROT, QUEEN ANNE'S LACE

HARBINGER-OF-SPRING

COW-PARSNIP

HAIRY SWEET CICELY

shaped. **Leaves** 1–2, 1–3 times compound, the divisions in 3's, the leaf segments linear to oblong.

HABITAT In deciduous woods.

FLOWERING March (rarely February) and April.

NOTES Although not particularly attractive, this species is of interest because of its early flowering. It is also one of the most easily identified species of this rather difficult family. The contrast between the blackish-red anthers and the white petals is the basis for another common name, Pepper-and-Salt.

Cow-Parsnip

Heracleum maximum Bartr.

Very stout perennial up to 3 m tall, the stems hairy, strongly ridged, up to 5–6 cm thick at the base. **Flowers** white, borne in very large, flattish umbels, composed of 15–35 smaller umbels; involucre soon falling. Corolla irregular, largest on the outer flowers. **Fruit** broadly oval or obovate, flattened. **Leaves** 3-foliolate, up to 6 dm long, the leaflets broadly ovate, circular, or heart-shaped, deeply lobed and sharply toothed.

HABITAT In low, moist places, usually in rich soil.

FLOWERING June to August.

USES American Indians used this species for medicine and food. The young stalks were roasted over hot coals. The leaf stalks were peeled and eaten raw like celery. The young roots when cooked taste like rutabaga.

Hairy Sweet Cicely

Osmorhiza claytonii (Michx.) C. B. Clarke

Slender, downy to long-hairy perennial up to 9 dm tall. **Flowers** white, tiny, borne in compound umbels, the umbels usually held above the leaves, subtended by an involucre. **Fruit** linear to narrowly club-shaped, upright, blackish, ribbed, the slender ribs covered with stiff, upward-pointing appressed bristles. **Leaves** thin, basal or alternate, compound, the main divisions in 3's or 5's; lower leaves long-petioled, upper ones sessile, the leaflets ovate to lanceolate, pointed at apex, often cleft and having coarse rounded teeth, 4–7 cm long, green on upper surface, whitened beneath with stiff hairs.

HABITAT In low, or rich deciduous woods, on wooded slopes and river banks, in thickets and moist, open fields.

FLOWERING May to June.

USES American Indians ate the roots and branches as an aid in gaining weight, and used decoctions made from the root in treating sore throat, ulcers, and running sores.

Anise-Root

Osmorhiza longistylis (Torr.) DC.

Coarser than the preceding species, up to 1.2 m tall, the leaflets larger, the roots sweet and aromatic with an anise odor. The styles are longer than the petals at flowering time and become 3.5–4 mm long in fruit.

HABITAT In rich, often alluvial, woods and thickets.

FLOWERING May to June.

NOTES These two species of *Osmorhiza* seldom grow together. Two other species also occur in the northern part of the state.

ANISE-ROOT

Wild Parsnip

Pastinaca sativa L.

Smooth, coarse, stout biennial up to 1.5 m tall. **Flowers** small, yellow, borne in very broad, flat-topped, compound umbels, the terminal umbels soon overtopped by the others; primary rays 15–25, unequal; involucres lacking. **Fruits** broadly oval, flattened, ribbed. **Leaves** pinnately compound, the lower and basal leaves petioled, up to 5 dm long, thin, the leaflets 5–15, ovate to oblong, 5–10 cm long, sessile, lobed or cut, sharply toothed; upper leaves similar to lower leaves but smaller.

HABITAT In old fields, sunny waste places, along roads. Native of Eurasia; long cultivated.

FLOWERING May to October.

NOTES According to some authorities two varieties may be distinguished. The edible cultivated form is var. *hortensis* Ehrh.; the weedy, reputedly poisonous form, with a more slender root, is var. *pratensis* Pers.

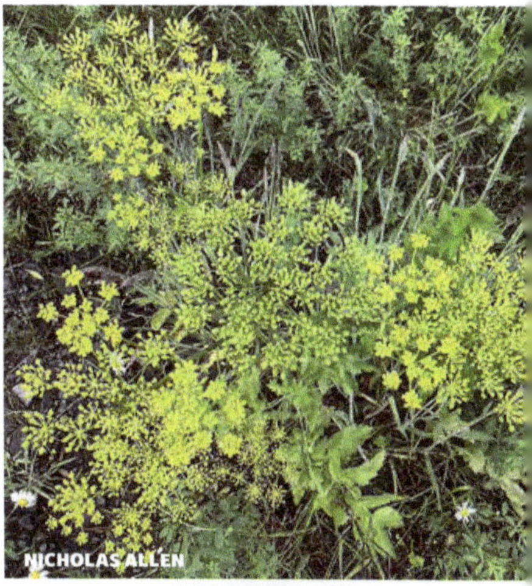

WILD PARSNIP

Canadian Black Snakeroot

Sanicula canadensis L.

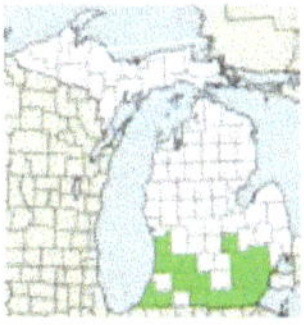

More spreading than the following two species, the branches forking 2 or 3 times, the **lower leaves** mostly 3-foliolate, the **upper leaves** greatly reduced, those in the flower clusters very small and usually in pairs; **flowers** white, anthers white, **fruit** subglobose.

HABITAT In dry open woods.

FLOWERING May to July.

CANADIAN BLACK SNAKEROOT

MARYLAND BLACK SNAKEROOT

CLUSTERED BLACK SNAKEROOT

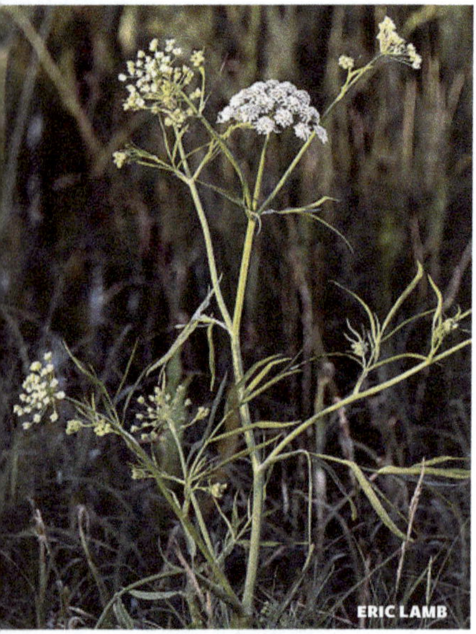

WATER-PARSNIP

Maryland Black Snakeroot

Sanicula marilandica L.

Perennial with strongly ascending branches up to 1.2 m tall from a thick rootstock. **Flowers** greenish-white or yellowish, tiny, borne in rather stiff, head-like umbels which have 2 or more stiff, ascending, 3-forked rays and a shorter, unbranched central ray; staminate flowers mixed with the perfect flowers or in separate small umbels. **Fruits** 3–8, ovoid, 5–8 mm long, sessile, brownish, 2-celled, not readily separating, covered with stiff, hooked bristles, the styles longer than the bristles, curved and spreading, conspicuous. **Basal leaves** long-petioled, 5-parted or appearing 7-parted, the leaflets obovate, unequally cut, toothed; **stem leaves** alternate, similar to the basal or with elliptic leaflets, the upper leaves smaller and sessile; greatly reduced leaves forming the involucre at the base of the umbel.

HABITAT In thickets and open woods, on ridges, in wet swampy places and marshes. Common.

FLOWERING May to July.

Clustered Black Snakeroot

Sanicula odorata (Raf.) Pryer & Phillippe

SYNONYM *Sanicula gregaria* Bickn.

Quite similar in aspect to *S. marilandica,* but the **fruits** globose, 2.5–4 mm long, and are borne on a stalk. This species tends to be more slender and shorter (up to 8 dm tall), the **leaves** mostly smaller, with sharply toothed leaflets; **flowers** greenish-yellow; anthers bright yellow.

HABITAT In rich woods and thickets.

FLOWERING late April to July.

Water-Parsnip

Sium suave Walt.

Erect, branching perennial up to 1.5 2 m tall, branched mostly above the middle, the stems corrugated or angled. **Flowers** white, borne in large compound umbels; involucre of narrow bracts. **Fruit** stalked, prominently corky-ribbed and having 1–3 oil tubes between the ribs. **Leaves** alternate, petioled, compound, the leaflets 5–17, linear or lanceolate, finely toothed, the earliest submersed rosette leaves very thin, 2–3 times pinnately dissected into linear segments.

HABITAT In low swampy ground, along streams, on muddy banks, in wet thickets, most commonly in land that is covered with water part of the time.

FLOWERING July to September.

NOTES This species is easily distinguished from Poison Hemlock by the once-pinnate leaves and the corrugations of the stems.

SIMILAR SPECIES A different plant of swampy places, **Cowbane**, *Oxypolis rigidior* (L.) Raf., is superficially similar to Water Parsnip, but the leaflets are entire or have only a few teeth.

YELLOW PIMPERNEL

Yellow Pimpernel

Taenidia integerrima (L.) Drude

Smooth, often glaucous perennial up to 8 dm tall. **Flowers** tiny, yellow, borne in compound umbels which do not have subtending involucres. **Fruit** flattened, wingless. **Leaves** compound or more usually 2–3 times compound, the divisions in 3's, leaflets smooth, sessile or short-stalked, lanceolate to ovate, entire, having a celery-like but somewhat disagreeable odor when crushed.

HABITAT In dry, rocky or gravelly woods and thickets and along roads.

FLOWERING late May to July.

YELLOW MEADOW-PARSNIP

Yellow Meadow-Parsnip

Thaspium trifoliatum (L.) Gray

Erect, branching perennial up to 1.5 m tall. **Flowers** tiny, yellow, all pedicelled, including the central flower in each umbel, the large, compound terminal umbel flattish at first but becoming globular. **Fruit** ovoid to ellipsoid, slightly flattened, the ribs usually winged. **Leaves** mostly 3-foliolate or with 5 leaflets, the lateral leaflets sometimes 2–3-lobed; leaflets lanceolate to ovate, sharply toothed; basal leaves simple, and heart-shaped at base, rarely 3-foliolate.

HABITAT In rich woods and thickets and in marshy places.

FLOWERING May to July.

SIMILAR SPECIES A quite similar species, *Thaspium chapmanii* (Coult. & Rose) Small, also grows in southern Michigan. It is slightly smaller, the stems are hairy at the nodes, the stem leaves are twice pinnate, and the flowers are pale yellow. Yellow Meadow-Parsnip can be distinguished from Golden Alexanders by its pedicelled flowers and fruit.

GOLDEN ALEXANDERS

Golden Alexanders

Zizia aurea (L.) W.D.J. Koch

Erect, often red-tinged, perennial up to 1 m tall, having a strong parsley odor when crushed. **Flowers** yellow, tiny, borne in a flattish compound umbel, the central flower of each umbel sessile, the others on pedicels; primary rays of the umbels 10–18, the outer rays of the terminal umbel becoming 3–5 cm long and stiffly ascending in fruit; involucre lacking. **Fruit** oblong, glabrous, the ribs threadlike. **Leaves** membranous; stem leaves mostly 2–3 times compound, the leaflets ovate to lanceolate, sharply toothed, the terminal leaflet no longer than the lateral leaflets; upper leaves merely 3-foliolate.

HABITAT In fields, along roadsides and shores, in swamps and damp thickets.

FLOWERING April to June.

APOCYNACEAE *Dogbane Family*

This primarily tropical family of 378 genera and about 5,100 species; 5 genera occur in Michigan. The family includes such ornamentals as Amsonia, Oleander, and Periwinkle, and now incorporates the former Milkweed Family (Asclepiadaceae).

Apocynum have a milky juice, the leaves are usually opposite, the flower parts are in 5's, the stamens are free from the stigma, the pollen grains are separate, and the 2 superior ovaries have a single style and stigma.

Michigan *Asclepias* are easily recognized by the peculiar extra whorl (the crown) of parts of the flower, the usually opposite or whorled leaves, and the milky juice. (Butterfly-weed has alternate leaves, at least on the lower part of the stem, and the juice is not milky.) The crown is composed of 5 petal-like hoods which stick up between the reflexed corolla and the stamens; this crown is usually more conspicuous than the corolla, and the hoods may or may not have each an incurved horn. The 5 anthers are more or less united to each other and to the stigma, forming a peculiar columnar structure in the center of the flower. The pollen is borne in pear-shaped masses (pollinia). The ovaries and styles are 2, with a single shield-shaped, 5-lobed stigma.

A few species, including the well-known Butterfly-weed, are grown for ornamentals. Rubber Vine, *Cryptostegia grandiflora,* has been grown commercially for rubber. The down from the seeds of some species was used as a substitute for kapok during World War II.

KEY TO APOCYNUM SPECIES

1 Corolla 5–9 mm long when mature, with spreading to recurved lobes, pink to rosy (at least in lines within) when fresh; leaves ovate to broadly elliptic, widely spreading or drooping .... ***A. androsaemifolium***

1 Corolla 3–4 mm long, with erect lobes when mature, white to greenish when fresh; leaves oblong to broadly lanceolate, ascending .... ***A. cannabinum***

Spreading Dogbane

Apocynum androsaemifolium L.

More or less inclined, often reddish-stemmed, milky-juiced perennial up to 1.3 m tall, usually unbranched below but having wide-spreading, arching branches above. **Flowers** white or pink and white, 6–9 mm long, fragrant, spreading or nodding in rather loose terminal and axillary cymes. Corolla bell-shaped, the 5 lobes turned back, the tube (at least in young flowers) pink-striped within; calyx much shorter than the corolla; **pods** very slender, hanging down, 7–20 cm long. **Leaves** opposite, petioled, ovate to ovate-oblong, tapered to both ends, dark green and smooth on upper surface, paler beneath.

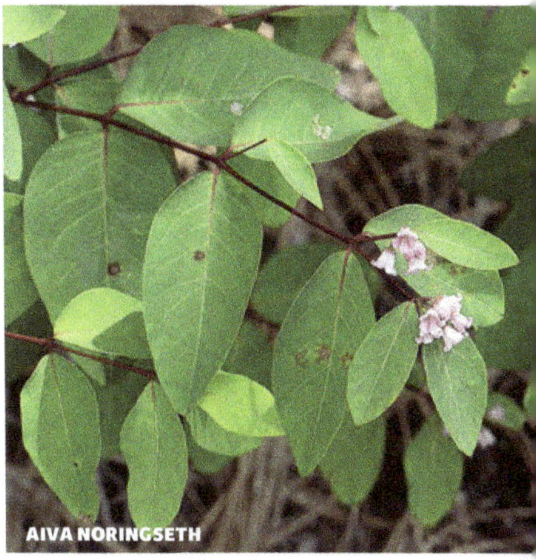

SPREADING DOGBANE

HABITAT In fields, in thickets, and along the borders of woods.

FLOWERING June to August.

USES American Indians used the pulverized roots to treat headache, babies' colds, dropsy, and heart palpitations. They also derived a fiber from this species which they made into a very strong cord. This fiber is said to be stronger than that of the related commercial hemp. Fibers from the outer rind of the stem were used for fine sewing.

Indian-Hemp

Apocynum cannabinum L.

Glabrous, erect perennial with milky juice, simple or with ascending to erect, often reddish branches, up to 1.5 m tall. **Flowers** small, white or greenish-white, 3–6 mm long, borne in terminal cymes. Corolla tubular to ovoid, the 5 lobes erect, not veined with red within; calyx 5-lobed, the lobes about as long as the corolla tube; stamens 5, attached to base of the corolla tube, the anthers converging; ovaries 2, readily separating, having a single style; **pods** 2, elongate, slender, 5–20 cm long, the seeds with a tuft of silky white hair 2–2.5 cm long. **Leaves** opposite, mostly ascending, ovate to lanceolate, 5–25 cm long, sessile to petioled, prominently veined.

HABITAT In open ground along shores, in thickets, borders of woods, and low places, in damp or dry soil.

FLOWERING June to August.

USES American Indians used fibers from both these species for making rope, twine, and thread. After the stalks of the previous year's growth had become naturally retted the fibers were carded

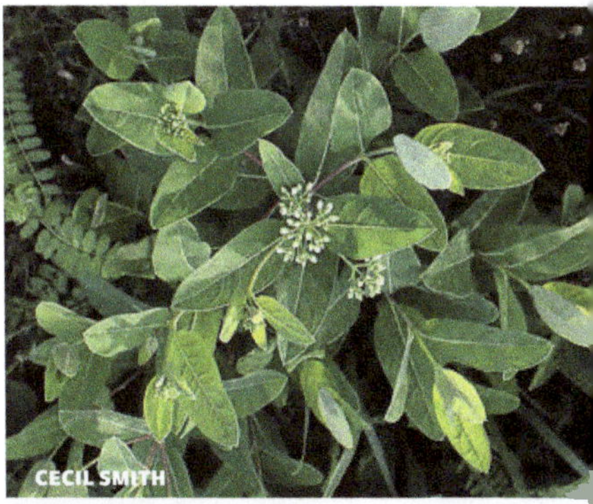

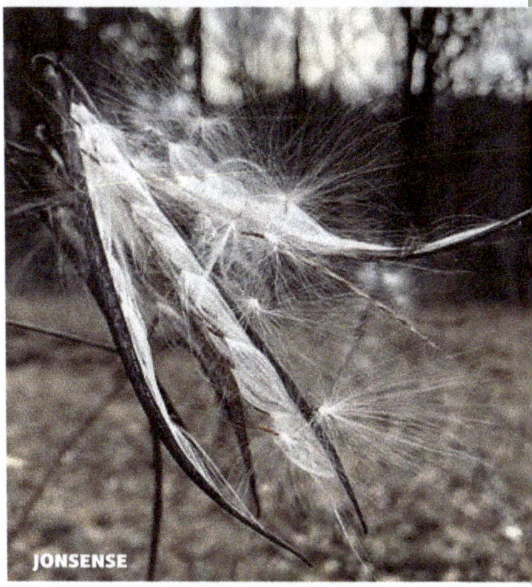

INDIAN HEMP

out. They were spun by twining or by rolling the strands on the leg under the palm of the hand. Fabric for diapers was sometimes woven from this thread. A laxative drink was also made from the plant. Horses, cattle, and sheep have been poisoned by eating dried or fresh Indian Hemp; 15–30 grams of the green leaves are said to be fatal to a horse or cow.

KEY TO ASCLEPIAS (MILKWEED) SPECIES

1 Flowers orange; juice not milky . . . . *A. tuberosa*

1 Flowers not orange; juice milky 2

  2 Flowers greenish; no pointed horns present in the hoods; stems trailing or leaning . . . . *A. viridiflora*

  2 Flowers pink, rose, or purplish, sometimes suffused with green; pointed horns present in the hoods; stems erect 3

3 Stem and leaves glabrous or nearly so; leaves thin, tapering to a long, slender apex; flowers pink to rose-purple . . . . *A. incarnata*

3 Stem and leaves finely pubescent; leaves thick, rounded at the apex; flowers dull purple and greenish . . . . *A. syriaca*

MICHAEL HINCZEWSKI

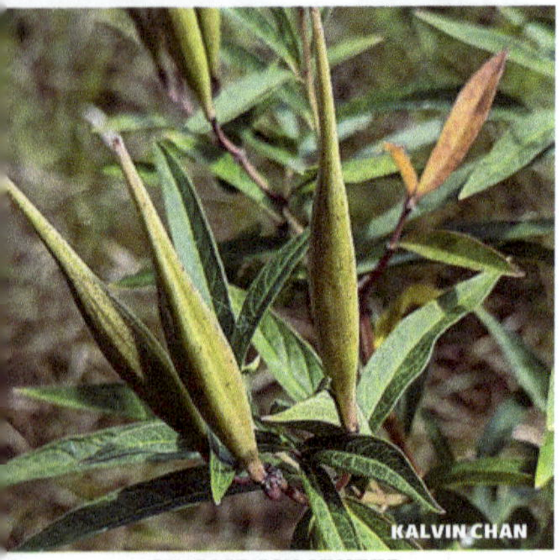

KALVIN CHAN

SWAMP MILKWEED

Swamp Milkweed

Asclepias incarnata L.

Glabrous or hairy, up to 1.5 m tall, branched above, the stems solitary or clustered. **Flowers** purplish-red to pink, rarely whitish, borne in umbels at the top of the plant. Corolla 6–7 mm long; the hoods about 3 mm long, the horns longer; **pods** 5–10 cm long, spindle-shaped, erect. **Leaves** opposite, lanceolate to oblong-lanceolate, the veins prominent.

HABITAT In swamps, wet meadows, and thickets and on shores. The dried pods are a familiar sight in the fall in marshes and wet places.

FLOWERING July and August.

USES American Indians made the buds into soup with deer broth; they also dried the pods for winter use. The liquid from the boiled roots was used for a gargle. Twine and thread were made from the fibers of the naturally retted stalks by rolling the fibers on the leg with the palm of the hand.

SIMILAR SPECIES **Tall Milkweed**, *Asclepias exaltata* L., has duller, purplish, greenish, or whitish flowers; it grows in moist upland woods statewide.

Common Milkweed, Silkweed

Asclepias syriaca L.

Stout, finely downy, grayish-green, usually unbranched perennial up to 2 m tall, spreading rapidly by underground stems; juice copious, milky. **Flowers** greenish-purple to dull purple, buff, or whitish, often fragrant, borne in globose umbels in the axils of the leaves. **Flowers** typical of the family, the horns short and incurved; **pods** ovoid, usually covered with short, soft warts, splitting along one side; seeds flat, brownish, the hairs long and silky.

HABITAT In fields and meadows, along roadsides.

FLOWERING June to August.

USES The young plants make a satisfactory cooked vegetable but are said to require long cooking. American Indians ate the young sprouts, buds, and pods. A crude brown sugar was made from the flowers by some tribes. The silky down of the seeds was used in making pads for babies. It was believed that taking the juice of this plant would increase the milk production of new mothers.

Butterfly-Weed, Pleurisy-Root

Asclepias tuberosa L.

Rough-hairy, erect or ascending perennial up to 1 m tall, branching only at the top; juice not milky. **Flowers** variable from yellow to deep red but usually bright orange, borne in broad terminal umbels or in smaller umbels in the axils of the leaves. Corolla deeply 5-lobed, the lobes bent downward and covering the 5-toothed calyx; the 5 hoods orange, upstanding, each with a slender horn; stamens and pistils united into a column; **pods** slender, elongate, 8–12 cm long, erect; seeds with a tuft of white hairs. **Leaves** alternate (sometimes opposite on the branches), lanceolate to oblong.

HABITAT In open sunny places and upland woods, especially in sandy soil. Common in southern Michigan; rare to absent northward.

FLOWERING late June to September.

NOTES This is the only one of our milkweeds that does not have milky juice. It is a most attractive garden plant but does not stand transplanting well. The deep roots are easily broken and are susceptible to attacks by fungi.

USES American Indians used the young seed pods and the tubers or roots for food; some tribes made a crude sugar from the flowers, and some used the plant medicinally. This species is reputedly poisonous to livestock, but apparently they seldom eat it.

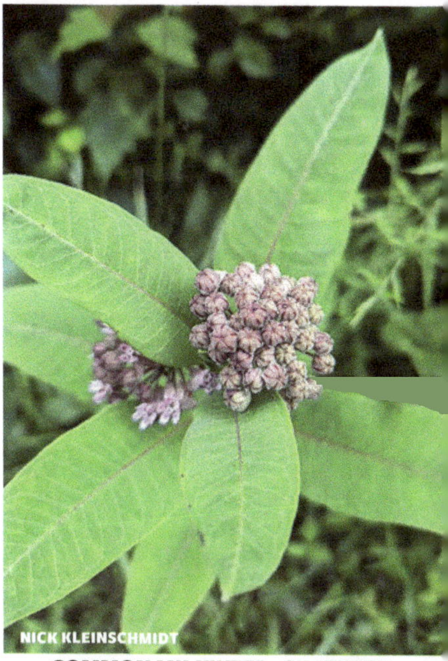

NICK KLEINSCHMIDT

COMMON MILKWEED, SILKWEED

PETER KISNER

BUTTERFLY-WEED, PLEURISY-ROOT

GREEN COMET MILKWEED

Green Comet Milkweed

Asclepias viridiflora Raf.

Minutely hairy to nearly smooth, prostrate to ascending perennial up to 8 dm tall; juice milky. **Flowers** greenish, borne in dense, globose, nearly sessile umbels in the axils of the leaves. Hoods lacking horns; **pods** slender, 8–11 cm long. **Leaves** mostly opposite, very thick, variable in shape, oval, oblong, or linear-lanceolate.

HABITAT In sandy, open ground, on lakeshores and dunes.

FLOWERING mid-June to August.

ARALIACEAE *Ginseng Family*

This primarily tropical family includes about 46 genera and about 1,500 species; 8 genera occur in Michigan. English Ivy (*Hedera*), its variants, and some related species which climb by aerial roots are cultivated evergreen vines, and several species of shrubs and small trees are also cultivated. The pith used in making Chinese rice paper is derived from a member of this family. Ginseng roots much used by the Chinese are obtained from a Manchurian species and from ***Panax quinquefolius*** L. **Devil's-Club** (***Oplopanax horridus*** (Small) Miq.), found in western United States and on Isle Royale, is the bane of hikers because of its numerous prickles cause festering sores.

The flowers are borne in umbels, the flower parts are mostly in 5's, the petals are not incurved. The ovary is inferior and is topped by 2 or more styles; the fruit is a berry. The leaves are usually alternate and compound, the petioles are usually adnate to the stipule and are not sheathing at the base.

KEY TO ARALIACEAE (GINSENG FAMILY) SPECIES

1 Umbels solitary; leaves once compound, borne in whorls .... ***Panax trifolius*** (common) or ***P. quinquefolius*** (rare)

1 Umbels 2 or more; leaves 2–3 times compound, alternate or basal .... ***Aralia*, 2**

2 Plants having sharp bristles on stem, at least at base .... ***Aralia hispida***

2 Stems not bristly **3**

3 Leaves and peduncles coming from base; umbels usually 3 .... ***Aralia nudicaulis***

3 Leaves borne on stems; umbels usually numerous .... ***Aralia racemosa***

Bristly Sarsaparilla

Aralia hispida Vent.

Bristly, strong-scented, erect perennial up to 1 m tall, the slender spines especially abundant on lower part of the stem. **Flowers** small, greenish, borne in simple, somewhat globose umbels, the umbels several, usually above the leaves. **Berry** globose, black. **Leaves** pinnately compound or twice. compound, the leaflets oblong-ovate, toothed, sessile, smooth or hairy on the veins beneath.

HABITAT In clearings and in rocky or sandy, open woods.

FLOWERING June to August.

USES Jelly is sometimes made from the berries of this plant; it has a rather strong flavor.

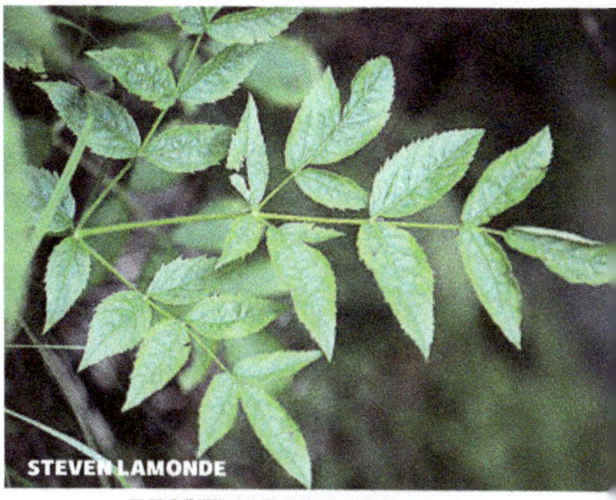

BRISTLY SARSAPARILLA

Wild Sarsaparilla

Aralia nudicaulis L.

Stemless perennial up to 5 dm tall, the solitary leaf and peduncle arising from a thick brown rootstock. **Flowers** whitish, small, borne usually in 3 (2–7) umbels, on a scape which is shorter than the leaf. Petals 5, often tinged with green or purple outside, inserted at the top of the calyx tube, often deflexed but not incurved; calyx teeth 5, small, alternate with the petals; stamens 5, white, inserted on the calyx tube; pistil 1, the ovary 5-celled, inferior; styles 5, distinct to the base; **berry** dark bluish-black. **Leaf** with 3 divisions, each division pinnately 3–5-foliolate, the leaflets oblong-ovate to ovate, up to 15 cm long, finely and regularly toothed, the upper surface varnished-looking at least when young, the under surface downy.

HABITAT In moist or dry woods.

FLOWERING late May and June.

USES The long, horizontal rootstocks are aromatic. American Indians used the pounded roots for treating nosebleed and to make poultices for infections. The root steeped with other herbs was applied to the chest and legs of horses suffering from exhaustion. The fruits and seeds are eaten by birds and chipmunks.

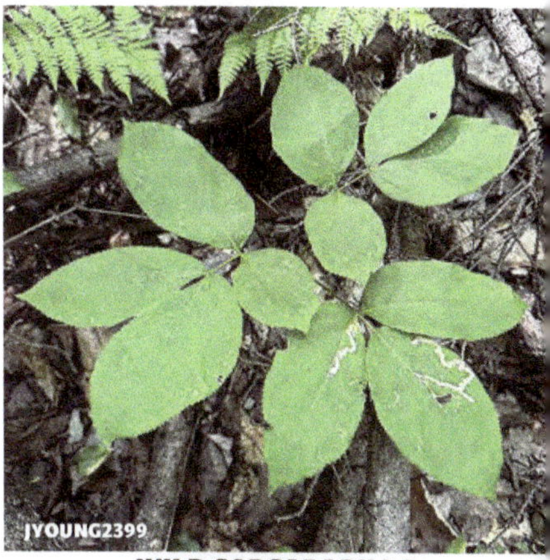

WILD SARSAPARILLA

Spikenard

Aralia racemosa L.

Stout, spreading, smooth or slightly hairy perennial up to 2 m tall from large, thick, spicy-aromatic roots. **Flowers** small, greenish, borne in numerous umbels which are usually at least partly hidden by the leaves. Styles united; **fruits** nearly globular, reddish-brown to dark purple. **Leaves** few, large, widely spreading, the 3 primary divisions pinnately compound, the leaflets ovate, variable in size, up to 15 cm long, sharply toothed, obliquely heart-shaped at base.

SPIKENARD

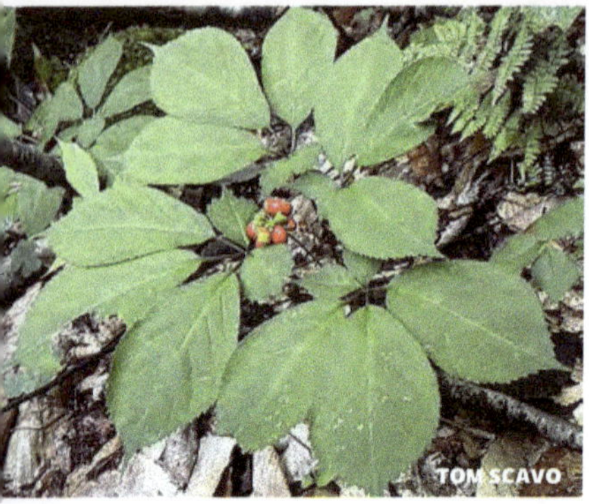

HABITAT In rich woods and thickets.

FLOWERING June to August.

USES This species was used medicinally by American Indians, as a poultice and blood purifier and for treating coughs. The roots were also prepared with wild onion, gooseberries, and maple sugar as a food.

American Ginseng

Panax quinquefolius L.

Unbranched perennial 2–6 dm tall, bearing a single long-peduncled umbel. **Flowers** greenish-white, mostly perfect, the styles usually 3; **berry** bright red. **Leaves** usually 3, borne in a whorl, palmately compound, the leaflets usually 5 (3–7), oblong-ovate to ovate, sharply toothed, petioled.

HABITAT In rich woods.

FLOWERING June and July.

STATUS Threatened in Michigan.

NOTES This species was formerly very common. Quantities of the roots were shipped to China for medicinal uses, and it was collected so extensively that it is now quite rare.

AMERICAN GINSENG

Dwarf Ginseng, Ground-Nut

Panax trifolius L.

SYNONYM *Nanopanax trifolius* (L.) A. Haines

Small, delicate, smooth, unbranched perennial 1–2 dm tall from a deepseated, globose root. **Flowers** small, white, often unisexual, 15 or more in a single terminal, nearly globose umbel, the peduncle 2–8 cm long, the pedicels and upper part of the peduncle white. Petals 5, purplish or rose in bud, attached at the top of the ovary; calyx white, tubular, adherent to the ovary, the 5 lobes scarcely noticeable; stamens 5, alternate with the petals; ovary inferior, the styles usually 3; **berry** small, yellowish, 3-angled. **Leaves** 3, in a whorl near the summit of the stem, palmately compound, the 3–5 leaflets sessile, narrowly oblong, 4–8 cm long.

HABITAT In rich woods and thickets and in clearings, usually in moist soil.

FLOWERING April to June.

DWARF GINSENG, GROUND-NUT

ARISTOLOCHIACEAE *Birthwort Family*

This primarily tropical family includes 8 genera and about 400 known species; 3 genera occur in Michigan. Several species of Dutchman's-pipe (*Aristolochia*) and Wild Ginger are cultivated as ornamentals.

Wild Ginger

Asarum canadense L.

Low, creeping hairy perennial up to 24 cm tall at flowering time, the rootstock thick, elongate, brown to green, on or just below the surface, and giving rise to the pairs of leaves, spicy and aromatic. **Flowers** solitary, borne on short stout peduncles from the leaf axils and usually hidden under the leaves. Petals lacking; calyx bell-shaped, with 3 spreading, pointed lobes, purplish-green and woolly outside, rich brownish-purple, chocolate, or dark maroon inside; stamens 12, rich purple, attached to the top of the ovary, at first curved outward and down, becoming erect; pistil solitary, the ovary inferior, the styles united into a purple column, spreading somewhat at the top. **Leaves** more or less in pairs, kidney-shaped, sometimes with a short point at the apex, glabrous on the upper surface, finely hairy beneath, 7–16 cm broad, the petioles stout, densely long-woolly.

HABITAT In rich woods; often abundant.

FLOWERING late April to early June.

NOTES Studies have shown that this species is self-pollinated by an interesting mechanism. The stamens are at first curved downward and outward, well away from the stigmas. As the stigmas mature and become bristly and sticky, the stamens become erect and thus can drop the pollen on the sticky surface.

USES The rootstocks have a strong aromatic flavor and have sometimes been dried and pulverized as a substitute for ginger. A perfume was made by dampening the roasted rootstocks and roots. Medicinal uses of the plant have included treatment of whooping cough and other coughs, as well as treatment of bruises, contusions, and inflammations by application of the leaves.

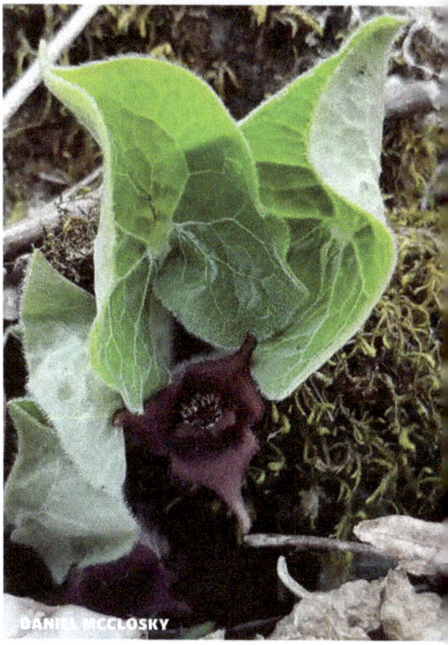

WILD GINGER

ASTERACEAE *Aster Family*

This, the largest family of flowering plants, includes 1,706 genera and over 32,000 species; 108 genera occur in Michigan. A number of our most common species, including dandelions and burdock, have been naturalized from Europe and South America. The family is important for both food and ornamentals. Lettuce and endive, salsify and chicory are important foods. Some 200 genera used for ornamentals include Aster, Coreopsis, Cosmos, Dahlia, Cineraria, Marigold, Shasta Daisy, and Zinnia. The insecticide pyrethrum is derived from a species of *Leucanthemum,* and sunflower seeds are used for chicken and bird feed. Among the many noxious weeds in the family are hawkweed, yarrow, sow-thistle, fleabane, everlasting, and ragweed. The pollen of ragweed is one of the principal causes of hay fever.

Members of this family are easily recognized. Several to many sessile flowers in a head, itself resembling a flower, are borne on a common receptacle; each head is surrounded by an involucre of many bracts (sometimes called *phyllaries*) which functions as a calyx in relation to the head. The five petals are united into a tube at least at the base and the five stamens are inserted on the corolla tube and their anthers are united into a tube. There is a single pistil, the style is usually 2-cleft at the apex in fertile flowers and entire in sterile flowers, the ovary is inferior, united with the calyx tube, and usually has a crown of scales, awns, and teeth, called a *pappus,* and the fruit is seed-like (an *achene*). The flowers may be unisexual or bisexual. In addition to the flowers, the receptacle may bear bracts or scales called *chaff*. When there is no chaff the receptacle is said to be naked.

The corolla of Michigan species of the Asteraceae may be either of two types, *tubular* or *ligulate*. Flowers which have a tube-shaped or trumpet-shaped corolla are said to be tubular, and heads composed solely of this type of flower are said to be *discoid*. The second type of corolla is tubular only at the base; above this it is flat and strap-like, and is usually toothed or lobed at the apex. This type of corolla is referred to as *ligulate*. A head composed solely of this type of flowers, called ray flowers, is also said to be *ligulate*. A head in which the center (disk) is occupied by tubular flowers and the edge is provided with ray flowers (sometimes mistakenly called "petals," as in sunflowers and daisies) is said to be *radiate*. Some discoid heads have flowers of two lengths, the shorter ones being in the center; these are said to be *falsely radiate* (e.g., knapweed).

NOTE In order to identify the composites one must determine whether or not the plant has a milky juice; the type or types of flowers present; and the nature of the pappus (the "crown" of the ovary). The pappus is best seen in fruit, but if none is available the oldest flowers should be studied.

KEY TO ASTERACEAE (ASTER FAMILY) SPECIES

1 Some or all of flowers having tubular corollas; juice watery .... **TUBULIFLORAE, 2**

1 Flowers all having straplike corollas; juice of plant (at least in older stems or roots) milky .... **LIGULIFLORAE, Key C**

  2 Heads discoid, having only tubular (disk) flowers .... **Key A**

  2 Heads radiate, having both tubular and ray flowers .... **Key B**

KEY A. TUBULIFLORAE (Heads discoid, with disk flowers only)

1 Leaves prickly .... ***Cirsium***

1 Leaves not prickly **2**

2 Fruit a spiny bur 3
2 Fruit not a spiny bur 4
3 Flowers greenish; heads of 2 kinds . . . . . *Xanthium strumarium*
3 Flowers lavender; heads alike . . . . . *Arctium minus*
4 Flowers some shade of purple or lavender 5
4 Flowers not purple or lavender 8
5 Leaves deeply pinnately cut into narrow segments . . . . . *Centaurea stoebe*
5 Leaves entire or somewhat toothed 6
6 Leaves mostly in whorls . . . . . *Eutrochium purpureum*
6 Leaves alternate 7
7 Leaf margins entire . . . . . *Liatris*
7 Leaf margins toothed . . . . . *Vernonia gigantea*
8 Flowers yellow 9
8 Flowers not yellow 12
9 Plants grayish, leaves deeply cut and dissected . . . . . *Artemisia caudata*
9 Plants green 10
10 Heads fairly large and conspicuous; leaves deeply cut or lobed; aromatic . . . . . *Tanacetum vulgare* (introduced, common) or *T. bipinnatum* (native, rare)
10 Heads small, borne in large clusters, leaves merely toothed or entire 11
11 Flower clusters flattish on top; leaves fairly uniform in size, linear to linear-lanceolate . . . . . *Euthamia graminifolia*
11 Flower clusters not flat-topped; leaves variable in size . . . . . *Solidago*
12 Flowers green or greenish 13
12 Flowers white, whitish, sordid, or very pale dirty yellow. 15
13 Plants having a distinct pineapple odor when crushed . . . . . *Matricaria discoidea*
13 Plants not having a pineapple odor 14
14 Leaves entire or toothed . . . . . *Erigeron canadensis*
14 Leaves deeply pinnately cut . . . . . *Ambrosia artemisiifolia*
15 Heads wilting on drying 16
15 Heads not wilting on drying, retaining virtually their original appearance (everlastings) 17
16 Leaves deeply pinnately cut . . . . . *Centaurea diffusa*
16 Leaves toothed or shallowly lobed 17
17 Leaves alternate, lower leaves largest . . . . . *Arnoglossum atriplicifolium*
17 Leaves opposite, lower leaves smallest . . *Eupatorium perfoliatum* or *Ageratina altissima*
18 Plants usually having a basal rosette of well-developed leaves, stem leaves small, bract like . . . . . *Antennaria parlinii*
18 Plants usually lacking a basal rosette, stem leaves well developed 19
19 Heads pearly white; plants white-woolly, essentially odorless . . *Anaphalis margaritacea*
19 Heads sordid white to yellowish; plants often sticky to touch, spicy-fragrant when crushed . . . . . *Pseudognaphalium macounii* or *P. obtusifolium*

KEY B. TUBULIFLORAE (Heads radiate, with both disk and ray flowers)

1 Ray and disk flowers same or nearly same color 2
1 Ray and disk flowers different colors. 13
2 Heads white . . . . . *Achillea millefolium*
2 Heads yellow. 3
3 Involucre very sticky . . . . . *Grindelia squarrosa*
3 Involucre not sticky 4

4 Plants having a very strong odor when crushed; leaves deeply pinnately cut; rays very small ...... ***Tanacetum vulgare*** (introduced, common) or ***T. bipinnatum*** (native, rare)
4 Not entirely as in alternate choice 5
5 Heads small, numerous, borne in compact clusters; pappus of bristles; stem leaves sessile or nearly so .... ***Euthamia, Solidago***
5 Not as in alternate choice 6
6 Disk globose or conic 7
6 Disk flat to slightly convex 8
7 Disk conic; rays persistent, becoming papery, not drooping ...... ***Heliopsis helianthoides***
7 Disk globose; rays soon drooping .... ***Helenium autumnale*** or ***H. flexuosum***
8 Leaf margin entire (may be lobed but not toothed) ***Coreopsis lanceolata*** or ***C. tripteris***
8 Leaf margin toothed or wavy 9
9 Fruits with sharp, barbed, awl-like teeth .... ***Bidens cernua*** or ***B. trichosperma***
9 Fruits lacking awl-like teeth 10
10 Heads rather small, disk not over 1.2 cm wide; pappus 10 of soft, threadlike bristles .... ***Packera aurea*** or ***P. paupercula***
10 Heads generally larger; if pappus is of bristles, disk is 3–5 cm wide 11
11 Disk flowers fertile, styles split 12
11 Disk flowers sterile, style not split .... ***Silphium perfoliatum*** or ***S. terebinthinaceum***
12 Receptacle bearing chaff .... ***Helianthus***
12 Receptacle not bearing chaff .... ***Inula helenium***
13 Ray flowers yellow 14
13 Ray flowers not yellow 16
14 Disk flat to broadly rounded .... ***Helianthus***
14 Disk conic, columnar, or ellipsoid 15
15 Ray flowers having bracts at base .... ***Ratibida pinnata***
15 Ray flowers lacking bracts at base .... ***Rudbeckia***
16 Heads mostly solitary, relatively large and showy, flat or nearly flat, rays white, disk yellow; leaves with prominent lobes or finely dissected 17
16 Heads smaller, mostly clustered; leaf margins entire or merely toothed 18
17 Leaves finely dissected, heads 1.5–3 cm broad .... ***Anthemis cotula***
17 Leaves pinnately lobed, heads 2–5 cm broad .... ***Leucanthemum vulgare***
18 Heads clustered on leafy branches; bracts of involucre of unequal length, successively shorter and overlapping (formerly genus *Aster*) .... ***Eurybia, Symphyotrichum***
18 Heads solitary or in clusters on mostly leafless branches; bracts of involucre nearly equal in length or with one long series surrounded by a short basal series; rays usually narrower and more numerous than in alternate choice .... ***Erigeron canadensis*** or ***E. philadelphicus***

KEY C. LIGULIFLORAE (Heads with ray flowers only)

1 Leaves grasslike with parallel veins .... ***Tragopogon*, 2**
1 Not as in alternate choice 4
2 Heads yellow 3
2 Heads purple .... ***Tragopogon porrifolius***
3 Bracts of involucre longer than flowers; peduncles enlarged upward . ***Tragopogon dubius***
3 Not as in alternate choice .... ***Tragopogon pratensis***
4 Heads bright blue (rarely white) .... ***Cichorium intybus***
4 Not as in alternate choice 5
5 Leaves all or principally at base of plant 6
5 Leaves borne along stem 10

6 Heads solitary on hollow, unbranched scapes; leaves usually deeply pinnately cut or lobed; achenes having a ring of bristles attached to end of a long beak . . . . . . ***Taraxacum officinale*** or ***T. erythrospermum***

6 Not as in alternate choice **7**

7 Leaf margins entire; heads usually several in a cluster; achenes topped by bristles . . . . . . ***Hieracium, Pilosella***, **8**

7 At least the later leaves toothed on margin; heads solitary or few; achenes topped with a row of scales and a row of bristles . . . . . . ***Krigia virginica***

8 Heads deep orange to orange-red . . . . . . ***Pilosella aurantiaca***

8 Heads yellow. **9**

9 Leaves narrowly oblanceolate, pale green . . . . . . ***Pilosella piloselloides***

9 Leaves elliptic-oblong, marked with reddish-purple along the veins . ***Hieracium venosum***

10 Heads white or whitish, often tinged with rose or green ***Nabalus albus*** or ***N. altissimus***

10 Heads yellow. **11**

11 Heads 3–5 cm broad, becoming swollen at base, including 50 or more flowers . . . . . . ***Sonchus arvensis***

11 Heads smaller, not becoming swollen at base, including 12–20 flowers . . . . . . ***Lactuca canadensis***

Common Yarrow, Milfoil

Achillea millefolium L.

Strong-scented perennial up to 1 m tall from a creeping underground rootstock, the flowering stems usually simple below the inflorescence, glabrous to cobwebby (covered with entangled hairs). **Heads** small, in compound, stiffly branched flattish-topped corymbs, usually white but sometimes pink to rose-purple. Rays few, small, the heads resembling small 5-petaled flowers; involucral bracts in 3–4 series. **Basal leaves** very finely 2-pinnately cut and finely dissected; **upper leaves** similar, but gradually decreasing in size upward.

HABITAT Along roads and railways, in fields, open woods, dryish hillsides, and waste places. A common weed introduced from Europe.

FLOWERING June to October.

NOTES During the Middle Ages this plant was held in high esteem for its supposed curative powers.

COMMON YARROW, MILFOIL

White Snakeroot

Ageratina altissima (L.) King & H.E. Robins.

SYNONYM *Eupatorium rugosum* Houtt.

Erect, glabrous to hairy perennial with solitary or clustered stems up to 1.5 m tall. **Heads** white, 15–30-flowered, borne in loose corymbs from the upper axils (corymbs rather compact and terminal in the smaller

WHITE SNAKEROOT

plants). **Leaves** opposite, petioled, ovate, coarsely and often sharply toothed, 2–8 cm long.

HABITAT In rich woods, thickets, and clearings.

FLOWERING late July to October.

NOTES This species is responsible for trembles in cattle and milk sickness in human beings. The poison, tremulol, is soluble in milk fat and may be transferred to other animals or human beings through milk. Dried plants in hay are less dangerous than fresh plants.

Common Ragweed

Ambrosia artemisiifolia L.

Extremely variable, simple to many-branched, smooth to hairy annual up to 3 m tall. **Heads** small, inconspicuous, greenish, of 2 kinds, the pistillate heads 1 to a few together, sessile in the axils of the leaves or bracts at the bases of the slender staminate racemes or spikes; staminate flowers 5–20 in a head, bearing copious quantities of powdery yellow pollen which is readily wind-borne. **Leaves** both opposite and alternate, petioled, merely pinnately lobed to 2 or 3 times pinnately cut into small segments, thin, smooth on upper surface.

HABITAT In vacant lots, along roads, in cultivated and waste ground of all kinds; frequently in great abundance.

FLOWERING July to October.

NOTES The genus name means "food of the gods," but the pollen of these plants is the main cause of fall hay fever for millions of people. This plant causes so much distress that many communities conduct eradication drives, and many newspapers give the ragweed-pollen count daily during the hay fever season. Its virtual absence is a great boon to the resort regions of northern Michigan, where traditionally many go in August and September to escape the pollen.

RELATED SPECIES **Great Ragweed**, *Ambrosia trifida* L., also occurs in Michigan, but to a more limited extent. It is much more important farther south, where, in low ground along streams and in rich openings, it may make a dense forest of plants up to 15 or 18 feet tall. The **Perennial Ragweed**, *Ambrosia psilostachya* DC. (synonym *Ambrosia coronopifolia*), is found in drier areas.

WHITE SNAKEROOT

COMMON RAGWEED

Pearly-Everlasting

Anaphalis margaritacea (L.) Benth. & Hook. f.

White-woolly, erect, many-stemmed perennial up to 1 m tall, branching at the summit, having a basal rosette only when young. **Flower heads** numerous, white with a yellowish center, globose, up to 1 cm wide, not changing appreciably on drying, borne in somewhat leafy corymbs at the summit of the stem, the branches of the corymb white and covered with long, appressed, white hairs. Corolla tubular, threadlike; pappus of long bristles; involucre dry, the bracts papery, white, ovate-lanceolate, spreading in age. **Leaves** alternate, linear-lanceolate, sessile, bright green with some hairs on upper surface, densely white-hairy or woolly beneath.

HABITAT In dry, usually open, gravelly or sandy soil, on road cuts, or sometimes along streams or in woods.

FLOWERING July to September.

NOTES Several varieties, distinguishable mostly on leaf characters, have been recognized. The flowers are often dried and used in winter bouquets, either in their natural color or dyed. *Anaphalis* lacks the characteristic, somewhat spicy, odor which is detectable even in long-dried specimens of *Pseudognaphalium*.

PEARLY-EVERLASTING

Parlin's Pussytoes

Antennaria parlinii Fern.

SYNONYM *Antennaria fallax* Greene

Densely white-woolly, stoloniferous, mat-forming, dioecious perennial, the pistillate plants up to 5 dm tall, the staminate less frequent and about half as tall. **Heads** small, whitish, several in a compact cyme at the top of the flowering stalk. **Flowers** tiny, tubular, the corolla threadlike; styles yellowish or brown; receptacle naked; involucre of several series of dry, white-tipped bracts. **Stem leaves** alternate, the lower ones oblong lanceolate, sessile, the upper ones narrower; **basal leaves** petioled, ovate, obovate, or nearly round, the larger 2–8 cm long, having 4 prominent veins, often woolly, sometimes becoming glabrous,

HABITAT In dry ground, open woods, clearings.

FLOWERING April to June.

PARLIN'S PUSSYTOES

NOTES The various species of *Antennaria* are easily recognized as everlastings. They are frequently used in winter bouquets. They may be grown in a dry rock garden where a soil cover is more important than the beauty of the flower. Several to numerous species are recognized in this genus. They are extremely difficult to distinguish, and botanists are not agreed upon them.

MAYWEED, STINKING CHAMOMILE

Mayweed, Stinking Chamomile

Anthemis cotula L.

Ill-scented, simple or freely branching annual up to 9 dm tall. **Heads** solitary and terminal, the ray flowers white, sterile, 6–10 mm long, the disk yellow, convex; pappus none, receptacle chaffy. **Leaves** very finely pinnately dissected into narrow, pointed segments.

HABITAT Along roadsides and in waste places.

FLOWERING June to October.

NOTES **Corn Chamomile**, *Anthemis arvensis* L., is similar, but the heads tend to be slightly larger and the leaves less finely dissected; ray flowers are fertile, and the plants are not ill-scented.

LESSER BURDOCK

Lesser Burdock

Arctium minus (Hill) Bernh.

Coarse, weedy biennial up to 1.5 m tall. **Heads** lavender to rose-purple, occasionally white, sessile or borne on short peduncles, solitary or in racemes; flowers all tubular; involucre subglobose, the bracts green, numerous, closely appressed at the base but with long, stiff, pointed, hooked ends, forming a spiny bur in fruit. **Leaves** broadly ovate, mostly heart-shaped at base, veiny, thin, serrate to entire, green above, gray below, large, up to 7.5 dm, long, hollow-petioled.

HABITAT In waste land. Introduced from Europe; now a widespread weed.

FLOWERING July to October.

NOTES Almost everyone knows the tenacity with which the dry burs stick to clothing. The bristles are so strong that small birds lighting on a cluster of burs may be held fast by the strong, sharply curved hooks, and die of exhaustion before they can free themselves.

Pale Indian-Plantain

Arnoglossum atriplicifolium (L.) H.Rob.

SYNONYM *Cacalia atriplicifolia* L.

Smooth, glaucous perennial up to 2 m tall. **Heads** numerous, without rays, whitish, rather large, borne in short, broad, more or less flat corymbs. Flowers 5 or more to a head, the corolla deeply 5-lobed; pappus of numerous soft hairlike bristles; receptacle usually with thickish hairs in the center; involucre narrowly cylindric, the bracts usually 5, erect in a single row, 7–12 mm long.

Leaves alternate, irregularly toothed and shallowly lobed, the lower leaves very large, kidney-shaped or slightly heart-shaped, palmately veined, pale beneath, the upper leaves smaller and rhombic to ovate.

HABITAT In dry soil in open woods and thickets.

FLOWERING late June to September.

PALE INDIAN-PLANTAIN

Beach Wormwood

Artemisia caudata Michx.

SYNONYM *Artemisia campestris* subsp. *caudata* (Michx.) H.M.Hall & Clem.

Slightly aromatic, grayish biennial, usually with a single stem up to 1 m tall rising from the first year's rosette. **Heads** small, very numerous, crowded in narrow, elongate, green to bronze, terminal leafy panicles 1.5–7.5 dm long. Flowers all tubular, 14–25 in a head, yellowish; involucre of thin, green, dry-margined bracts. **Basal leaves** numerous, grayish-downy, 2–3 times pinnately divided into very long, narrow, almost threadlike segments; **stem leaves** similar but smaller, numerous.

HABITAT On dry plains, prairies, and sandy beaches. Common on the sandy beaches of the Great Lakes.

FLOWERING August and September.

NOTES Like the ragweeds, which they greatly resemble, the wormwoods have wind-borne pollen.

BEACH WORMWOOD

KEY TO FORMER ASTER SPECIES (EURYBIA, SYMPHYOTRICHUM)

1 Rays purple to violet or blue-violet 2

1 Rays white, heads borne on one-sided racemes .... *Symphyotrichum ericoides*

  2 Leaves sessile, usually clasping at base 3

  2 Leaves petioled, heart-shaped at base .... *Eurybia macrophylla*

3 Involucre covered with glandular hairs and sticky to touch; stem hairy .... *Symphyotrichum novae-angliae*

3 Involucre not sticky; stem glabrous .... *Symphyotrichum laeve*

Large-Leaved Aster

Eurybia macrophylla (L.) Cass.

SYNONYM *Aster macrophyllus* L.

Coarse perennial frequently occurring in colonies of many clusters of leaves, the flowering stems when present up to 1.5 m tall, sticky at least in the inflorescence. **Heads** numerous, borne in broad, irregular, somewhat flat-topped corymbs; rays about 16, violet or pale blue, about 1 cm, long; disk becoming reddish-brown in age; involucre slender, bell-shaped, of 3 or 4 se-

LARGE-LEAVED ASTER

WHITE ASTER

ries of greenish bracts, the inner ones often with roseate margins. **Basal leaves** large, thick, firm, heart-shaped, coarsely toothed, rough to touch, 420 cm long, petioled; **stem leaves** smaller and narrower, the upper ones ovate or oblong, and sessile.

HABITAT In dry to moist shady places, open woods, and thickets.

FLOWERING August and September.

SIMILAR SPECIES ***Symphyotrichum cordifolium*** (L.) G.L. Nesom (synonym *Aster cordifolius* L.), also has heart-shaped basal leaves, but it does not usually occur in colonies. It has a more elongate inflorescence, with smaller heads; the leaves are relatively thin and up to 12 cm long.

White Aster

Symphyotrichum ericoides (L.) G.L. Nesom
SYNONYM *Aster ericoides* L.

Stiffly erect to partially prostrate perennial up to 2 m tall, densely covered with curved or divergent hairs. **Heads** small, about 1 cm wide, borne in slender panicles of elongate, mostly one-sided racemes (or solitary and terminal on the branches); rays 8–20, narrow, usually white but sometimes blue, violet, or roseate, 3–5 mm long; disk yellow; involucral bracts with spreading tips. **Leaves** numerous and crowded, somewhat rigid, linear, scarcely narrowing to the sessile base, the uppermost leaves much reduced.

HABITAT In dry open soil in fields, along roads, and in thickets.

FLOWERING July to October. The arrangement of the small heads on one side of the stems, rather like that in most goldenrods, makes this white aster fairly easy to recognize.

SIMILAR SPECIES **Calico Aster**, ***Symphyotrichum lateriflorum*** (L.) Á. & D. Löve (synonym *Aster lateriflorus* (L.) Britt.), has 9–15 white (rarely pinkish) rays 4.5–7.5 mm long; the disk corollas are purplish; the stems usually have arching or wide-spreading branches from the middle or below, and the thin, lanceolate leaves taper to apex and base. ***Symphyotrichum lanceolatum*** (Willd.) Nesom has larger heads, with 20–40 white rays 4.5–12 mm long.

Smooth Blue Aster

Symphyotrichum laeve (L.) Á. & D. Löve
SYNONYM *Aster laevis* L.

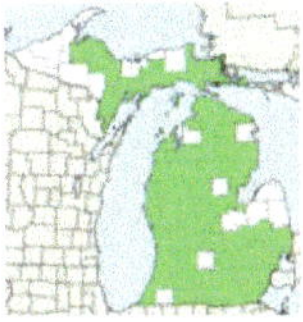

Smooth and usually glaucous, erect, simple or branched perennial up to 1.3 m tall. **Heads** numerous, borne in stiff panicles which have greatly reduced, bractlike leaves; rays 15–20, blue, lavender, violet, or rarely white, 8–15 mm long; disk yellow at first, becoming dark; involucre bell-shaped, the bracts obtuse to acute, rigid, green, appressed in several rows. **Leaves** thick, obscurely veined, the basal leaves tapered to winged petioles, the upper and middle leaves narrowly lanceolate, oblanceolate, or elliptic, sessile and more or less clasping, usually entire.

HABITAT In dry open habitats, in thin woods, borders of woods, and in thickets.

FLOWERING August to October.

SIMILAR SPECIES **Red-Stemmed Aster**, ***Symphyotrichum puniceum*** (L.) Á. & D. Löve (synonym *Aster puniceus* L.), also usually has blue rays. It is somewhat taller (up to 2.5 m); stems are reddish and usually pubescent; rays 30–60, which are 7–18 mm long; involucral bracts loose and spreading.

SMOOTH BLUE ASTER

New England Aster

Symphyotrichum novae-angliae (L.) G.L. Nesom
SYNONYM *Aster novae-angliae* L.

Stout, many-stemmed, corymbosely branched, long-hairy perennial up to 2.6 m tall. **Heads** numerous, showy, about 3 cm wide, borne on short peduncles in leafy corymbs; rays violet-purple, sometimes roseate or white, 40–50 in a single series; disk flowers orange at first; achenes more or less flattened, the pappus a single series of bristles; receptacle flat. Involucre broadly hemispheric, 8–10 mm high, of thin, long-tapering, often purple-tinged, recurving bracts which are sticky owing to glandular hairs. **Leaves** numerous, crowded, usually hairy on both surfaces, lanceolate, cordate-auriculate and clasping at the base, entire, the principal ones 5–10 cm long.

NEW ENGLAND ASTER

HABITAT In damp thickets, meadows, open woods, and on shores.

FLOWERING August to October.

NOTES More than 2 dozen species of aster occur in Michigan. Since it is often difficult to identify species, only a few of the more distinctive kinds are treated here. Various forms of the New England Aster are cultivated under the name 'Michaelmas Daisy.'

NODDING BEGGAR-TICKS

Nodding Beggarticks

Bidens cernua L.

Annual up to 1.8 cm tall, sometimes decumbent and rooting. **Heads** erect at flowering time, nodding in fruit; discoid or radiate, the rays bright yellow. Achenes curved, olivaceous, the margins somewhat winged, awns with backward pointing barbs. **Leaves** linear, lanceolate, or oblong, sessile or united at the base (rarely petioled). Highly variable species.

HABITAT In wet places, springs, and pools.

FLOWERING August to October.

MARSH TICKSEED

Marsh Tickseed

Bidens trichosperma (Michx.) Britt.

SYNONYM *Bidens coronata* (L.) Britt.

Glabrous, slender annual or biennial 3–15 dm tall. **Heads** yellow, borne on long, slender peduncles; rays golden-yellow, spreading, longer than the outer bracts of the involucre; disk flowers yellow, with purple-black anthers; achenes wedge-shaped at base, blunt at apex and tipped with 2 stiff, sharp, barbed awns, the barbs pointing backward; involucres of 2 series of bracts, the outer series of usually 8 (6–11) linear, blunt, leaflike bracts. **Leaves** pinnately divided into 3–7 lanceolate or linear, pinnately cut to coarsely toothed divisions.

HABITAT In bogs, swales, and other wet places.

FLOWERING August to October.

NOTES Several other species of Beggarticks occur in Michigan. They are all notable for the tenacity with which the fruits stick to clothing.

DIFFUSE KNAPWEED

Diffuse Knapweed

Centaurea diffusa Lam.

Shorter (1–6 dm), stiffer, and more erect than the following species. The heads are smaller, very numerous, white, creamy, roseate, or sometimes rose-purple; all flowers with corollas the same length; involucral bracts pinnately cut and spiny-tipped.

HABITAT In fields and along roadsides. Introduced from Europe.

FLOWERING July to September.

Spotted Knapweed

Centaurea stoebe L.

SYNONYM *Centaurea maculosa* Lam.

SPOTTED KNAPWEED

Tough, wiry, grayish-green, usually bushy, many-branched, spreading biennial or short-lived perennial up to 15 dm tall. **Heads** numerous, terminal on the branches, rose-purple, rose, or at times white, slightly fragrant, 2–3 cm wide, falsely radiate; flowers all tubular with long narrow lobes, the outer ones, the largest, sterile, the inner flowers shorter and perfect; involucre ovoid, of numerous rows of strongly ribbed, overlapping bracts, the bracts with short, dark, pinnately cut tips, the inner bracts the longest. **Leaves** alternate, grayish-green, sessile; principal stem leaves deeply pinnately cut into long, linear segments, the upper leaves smaller; leaves of the short flowering branches linear.

HABITAT In fields, along roadsides, and in waste places. Introduced from Europe.

FLOWERING July to August.

NOTES This weed has spread rapidly across the U.S., especially in the drier western states. The familiar garden **Cornflower** or **Bachelor's Button** is *Centaurea cyanus* L.

Chicory

Cichorium intybus L.

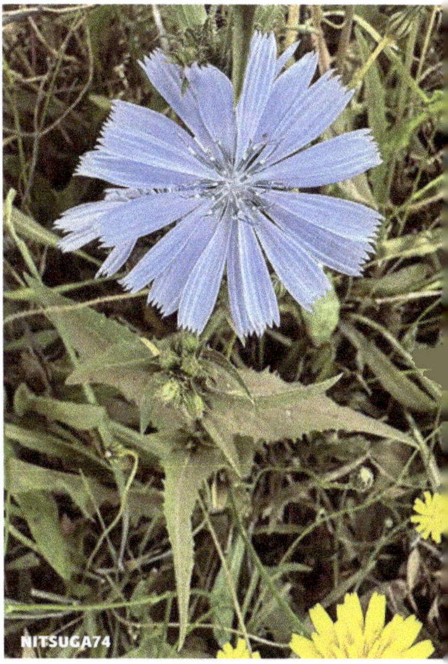

CHICORY

Erect, stiffly branching biennial or perennial up to 1.2 m tall from a long, stout, fleshy root; juice milky. **Heads** showy, blue (rarely pink or white), 2–5 cm in diameter, opening in the morning, closing in the afternoon or in cloudy weather, solitary or 2–3, sessile in the axils and terminal on the stem and short, thick branches. **Flowers** all ligulate, the outer flowers with corollas about 2.5 cm long, the inner flowers shorter (falsely radiate); involucre double, the outer bracts 5, broad and spreading, the inner bracts 8–10, narrower, erect. **Basal leaves** spreading, spatulate, pinnately cut to merely sharply toothed, the divisions directed backward; **stem leaves** greatly reduced, toothed or entire.

HABITAT Along roads, in fields and waste places. Introduced from Europe.

FLOWERING June to October.

NOTES This species is most attractive when the heads are expanded; it often grows in large patches and makes a striking display. It can be a troublesome weed, however, and if eaten by cows gives the milk an unpleasant taste.

USES The ground roots are roasted and used as an adulterant in, or substitute for, coffee. **Garden Endive**, *Cichorium endivia* L., is closely related.

KEY TO CIRSIUM (THISTLE) SPECIES

1 Plants grayish, owing to white-woolly hairs; growing on shores of the Great Lakes .... *C. pitcheri*

1 Plants green; not usually growing on shores of the Great Lakes 2

2 Upper stems and branches having broad, prickly-margined wings extending from leaf bases .... *C. vulgare*

2 Stems and branches not having wings 3

3 Flowering stems rising from a basal rosette; involucre 2–3.5 cm high .... *C. muticum*

3 Flowering stems rising from a deep rootstock, no rosette present; involucre 1–2 cm high .... *C. arvense*

CANADA THISTLE

SWAMP THISTLE

Canada Thistle

Cirsium arvense (L.) Scop.

Prickly, erect perennial up to 1 m tall from creeping, freely branching, and sprouting roots. **Heads** unisexual, pinkish-purple or white, numerous, about 2.5 cm in diameter, the staminate heads globose and with projecting corollas, the pistillate heads more oblong, the corollas shorter, and the pappus long and conspicuous; involucral bracts appressed, short-pointed but not spiny-tipped. **Leaves** alternate, oblong to lanceolate, and deeply pinnately cut, toothed, very prickly, at first often woolly beneath but finally green on both sides, the lower leaves up to 3 dm long, petioled, the upper leaves shorter and sessile; no rosette formed.

HABITAT In cultivated and waste ground. This common weed was introduced from Europe; not, in spite of its name, from Canada.

FLOWERING June to August.

Swamp Thistle

Cirsium muticum Michx.

Stout, erect, prickly biennial up to 3 m tall. **Heads** reddish-purple to rose-purple, several to many, medium-sized; involucre sticky and cobwebby (covered with entangled hairs), the bracts not tipped with spines. **Leaves** deeply pinnately cut into lanceolate segments, spiny, densely covered with matted wool beneath when young, becoming smooth and green on both sides.

HABITAT In swamps, low woods, thickets, and meadows.

FLOWERING July to September.

Pitcher's Thistle

Cirsium pitcheri (Torr. ex Eat.) Torr. & Gray

Stout, erect, white-woolly, prickly biennial up to 1 m tall from a basal rosette. **Heads** cream-color, solitary or several in a raceme; flowers all alike, the corolla tubes deeply 5-lobed; involucre ovoid, about 2.5 cm high, the bracts overlapping in several series, bristle-tipped. **Basal leaves** white-woolly, long-petioled, cleft to the midrib into a few, remote, elongate linear segments which may be spiny at the tip or merely rounded, up to 3 dm long; **stem leaves** similar but smaller and partly clasping or slightly decurrent at base, the segments about 5 mm wide, the margins inrolled, sparingly prickly.

HABITAT On sandy shores and dunes of Lakes Michigan, Superior, and Huron.

FLOWERING May to September.

STATUS Threatened in Michigan.

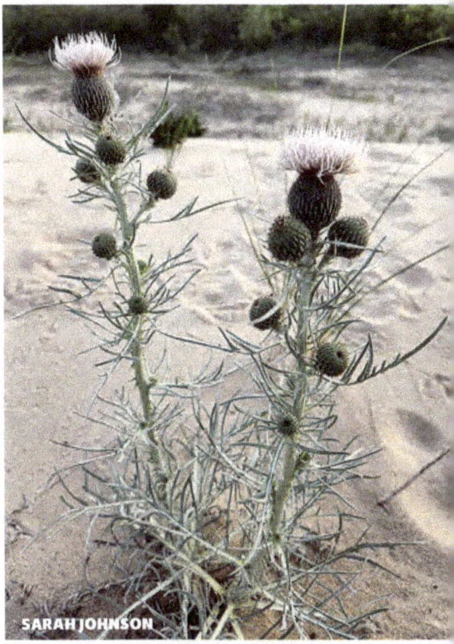

PITCHER'S THISTLE

Bull Thistle

Cirsium vulgare (Savi) Tenore

Coarse, prickly biennial with conspicuously spiny-winged stems up to 2 m tall. **Heads** discoid, purple, several or solitary at the tips of short prickly winged branches; involucre ovoid to subglobose, 3–4 cm high, the bracts lanceolate to linear, tipped with rigid spines. **Leaves** of the first year pale, woolly or webbed beneath, green above, forming basal rosettes, pinnately cut, the larger ones with the lobes again lobed or toothed, the lobes and teeth with long, strong spines, the stem leaves of the second year similar.

HABITAT In fields and pastures, along roadsides, and in waste places. Naturalized from Europe.

FLOWERING June to September.

BULL THISTLE

Lance-Leaf Coreopsis

Coreopsis lanceolata L.

Glabrous, erect or sprawling, many-stemmed, sparsely branched perennial up to 1 m tall, leafy only near the base. **Heads** showy, bright golden-yellow, 3–5 cm wide, upright on long slender peduncles 2–4 dm long; ray flowers usually 8, in a single marginal row, long, conspic-

LANCE-LEAF COREOPSIS

uous, 4–7 lobed at apex, narrowed to the base; disk flat, the flowers numerous, yellow; **achenes** flattened, oblong; receptacle with some long thin chaff which falls off with the achenes; involucre of 2 series of about 8 bracts, the inner ones broad, closely appressed, the outer ones spreading, leaflike.

HABITAT In dry, sandy or rocky soils. It is common in the upper part of the Lower Peninsula and along the sand dunes on the Lake Michigan shore of the Upper Peninsula.

FLOWERING May to July.

USES This attractive plant is frequently used in the garden.

TALL COREOPSIS, TALL TICKSEED

Tall Coreopsis, Tall Tickseed

Coreopsis tripteris L.

Smooth perennial with solitary leafy stems up to 3 m tall, freely branched above. **Heads** several or numerous, about 3.5 cm broad, having an anise-like odor; rays usually 8, rounded at tip or with low round teeth; disk flat, the flowers yellow, becoming brown; involucre of 2 series of 8 bracts, the outer ones leaflike and spreading, the inner ones broader, membranous, appressed. **Leaves** opposite, petioled, mostly 3- or 5-foliolate, the leaflets oblong-lanceolate to linear, stalked, the margin entire; upper leaves often simple and sessile.

HABITAT In open woods, borders of woods, and thickets, on prairies.

FLOWERING July to September.

HORSEWEED, BUTTERWEED

Horseweed, Butterweed

Erigeron canadensis L.

Erect, hairy or downy annual up to 2 m tall, simple below, branched into very slender panicles above. **Heads** usually very numerous, small, few-flowered, and inconspicuous; rays white, usually shorter than the diameter of the disk or lacking; involucre narrowly bell-shaped at first, spreading in fruit, often persisting. **Leaves** numerous, oblanceolate to linear, toothed to entire.

HABITAT In waste places, along roadsides, in cultivated fields.

FLOWERING July to November.

NOTES This is an unattractive but common, semi-cosmopolitan weed. It is easily recognized, even from a distance, by its narrow, small-headed panicles.

Common Fleabane, Daisy Fleabane

Erigeron philadelphicus L.

Soft-stemmed, softly downy, short-lived perennial up to 1 m tall. Heads fairly numerous, (about 2.5 cm wide,) nodding in bud but becoming upright; rays more than 100, very narrow, pale pink, pinkish-lavender, or whitish; disk yellow; achenes flattened, covered with scattered stiff hairs; pappus a single row of bristles with smaller bristles between the long ones; involucre saucer-shaped, the bracts narrow, equal, greenish, the receptacle slightly rounded, naked. **Leaves** long-hairy, the basal and lower stem leaves obovate or spatulate, toothed, 2–7 cm long, narrowed to short petioles, the upper leaves smaller, oblong to lanceolate, heart-shaped below and clasping.

HABITAT In fields and rich thickets, on shores and springy places.

FLOWERING May to August.

SIMILAR SPECIES **Robin's-Plantain**, *Erigeron pulchellus* Michx., is quite similar to Common Fleabane, but the heads are fewer and larger (2–3.5 cm broad) and have about 50 rather broad rays. Several species of *Erigeron* with white rays are commonly known as White-Top, or as Daisy Fleabane. They may be distinguished from the White Aster by their earlier blooming (chiefly in spring and early summer), the typically narrower and more numerous rays, and the heads borne on essentially leafless stems.

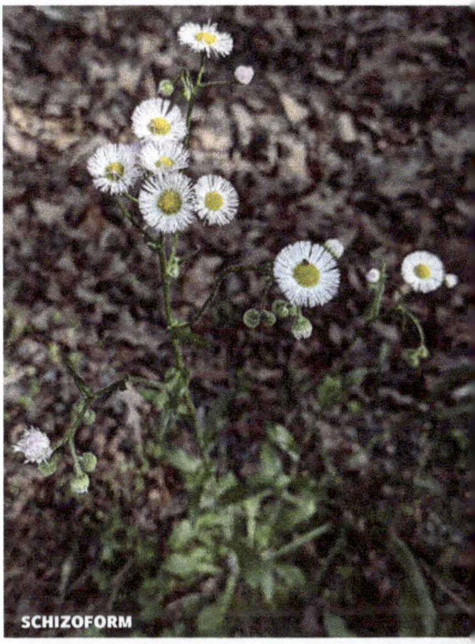

COMMON FLEABANE, DAISY FLEABANE

Boneset, Thoroughwort

Eupatorium perfoliatum L.

Coarse, erect, long-hairy perennial up to 1.5 m tall, branched above. **Heads** small, dull white, 10–16-flowered, crowded in a flattish-topped terminal corymb; involucre bell-shaped to cylindric, green, the hairy lanceolate bracts in 2 or 3 overlapping series; pappus a row of whitish bristles. **Leaves** opposite, usually united by their bases and completely encircling the stem, tough, wrinkled, veiny, lanceolate, crenate, the larger blades 7–20 cm long and 1.5–4.5 cm wide.

BONESET, THOROUGHWORT

HABITAT In low woods or thickets, in swamps, on wet shores, along streams, and in swales.

FLOWERING late July to October.

USES This is one of our best-known medicinal plants. It was used by American Indians and early settlers as a general tonic and for treating influenza, rheumatism, and fevers. American

FLAT-TOPPED GOLDENROD

SPOTTED JOE-PYE WEED

Indians made a cathartic from the steeped foliage, and dressings for snake bites were made by chewing any part of the plant. Used with milkweed, the root fibers were supposed to act as a charm when applied to a whistle for calling deer.

I have seen two explanations for the name boneset. According to one, this species was used for the treatment of breakbone fever; according to the other, it was believed by the doctrine of signatures that the union of the leaves at the base signified that the plant would promote the knitting of bones.

Flat-Topped Goldenrod

Euthamia graminifolia (L.) Nutt.

SYNONYM *Solidago graminifolia* (L.) Salisb.

Smooth, hairy, or minutely hairy perennial 2.5–6 dm tall, branched only above. **Heads** small, yellow, 12–45-flowered, borne in a flat-topped many-branched corymb up to 3 dm broad; ray flowers pistillate, disk flowers perfect; achenes round, ribbed involucre cylindric. **Leaves** linear to lanceolate, sessile, grasslike, with 2 or 4 lateral veins parallel to the midrib.

HABITAT On lakeshores, beaches, and rocky places, in thickets.

FLOWERING August and September. This is a variable and very common species.

Spotted Joe-Pye Weed

Eutrochium maculatum (L.) E.E. Lamont

Quite similar to the following species, but the **stem** is deep purple or purple-spotted, not glaucous; the **inflorescence** flat-topped; the **flowers** mostly 9–22 in a head and usually more deeply colored.

HABITAT In low wet places, swamps, and meadows, along streams and shores, usually in rich or calcareous soil.

FLOWERING mid-July to early September.

USES American Indians used a decoction made from this species to bathe weak or paralyzed children, believing that it strengthened the legs and feet. This species is suitable for a large garden of wildflowers.

■ Sweet-Scented Joe-Pye Weed

Eutrochium purpureum (L.) E. Lamont

SYNONYM *Eupatorium purpureum* L.

Erect perennial up to 2 m tall, the stems slightly glaucous, usually green (purple at the nodes), the drying foliage having a sweet vanilla odor when bruised or crushed. **Heads** numerous, small, usually 4–7-flowered, pale pink to pinkish-purple, in an open, somewhat domed, corymb. **Flowers** all tubular and perfect, 4.5–7.5 mm long; involucre cylindric, the bracts closely overlapping; pappus of slender bristles in single row. **Leaves** mostly in whorls of 3 or 4, lanceolate, ovate, or elliptic, narrowed to the short petiole, sharply and coarsely toothed, 8–30 cm long, up to 15 cm wide.

HABITAT In rich, dry or dryish, usually calcareous soil in thickets and open woods.

FLOWERING July to mid-September.

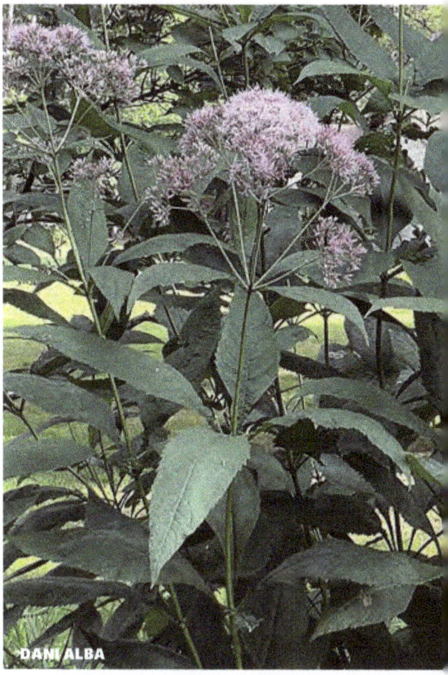

SWEET-SCENTED JOE-PYE WEED

■ Curly-Cup Gumweed

Grindelia squarrosa (Pursh) Dunal

Stout, erect, smooth biennial or perennial up to 1 m tall. **Heads** several to many, yellow, 3–4 cm wide, terminal on leafy branches, having both tubular and ray flowers. Rays up to 1 cm long or sometimes lacking, narrowly oblong, entire at apex, pistillate; disk flowers perfect; achenes flattened, those of the ray flowers the largest; pappus of 2–8 awns; involucre hemispheric, sticky, the bracts in several rows, linear and with strongly recurving pointed tips. **Leaves** clasping, thickish, alternate, narrowly oblong to oblanceolate or ovate, 3–7 cm long, abundantly punctate, finely serrate, entire, or coarsely toothed, obscurely veined.

HABITAT In open or waste places, usually in dry soil. Native on the prairies and plains but now found locally as far east as the Atlantic states.

FLOWERING July to September.

USES American Indians used the species medicinally. They ground the seeds to make a flavoring and used the resin to hold women's hair in place.

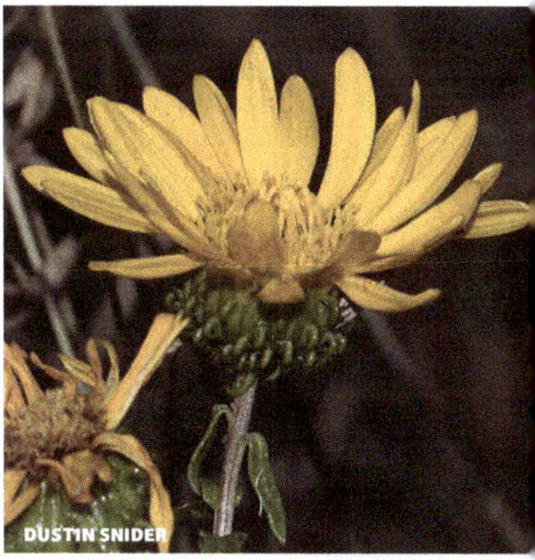

CURLY-CUP GUMWEED

■ Sneezeweed, Swamp Sunflower

Helenium autumnale L.

Aromatic, resinous perennial up to 2 m tall. **Heads** yellow, borne on long peduncles; ray flowers 10–20, deep yellow, soon drooping, 3-notched at apex, usually fertile; disk a darker yellow, depressed-globose, achenes ribbed; involucre small, the bracts reflexed. **Leaves** thin,

SNEEZEWEED, SWAMP SUNFLOWER

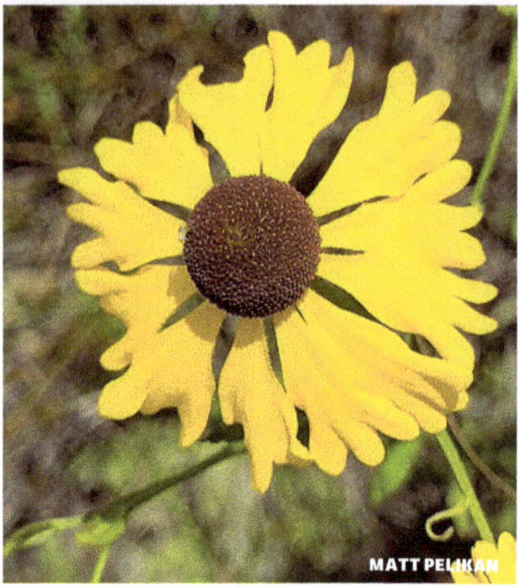

PURPLE-HEAD SNEEZEWEED

alternate, oblong-lanceolate, usually toothed, 5–15 cm long, up to 4 cm wide, the decurrent base continued as wings on the stem.

HABITAT In swamps, wet meadows, along streams, on shores and in rich thickets.

FLOWERING August to November.

Purple-Head Sneezeweed

Helenium flexuosum Raf.

SYNONYM *Helenium nudiflorum* Nutt.

Quite similar to the preceding species, but the plants tending to be shorter, the ray flowers usually sterile, golden yellow overall or purplish at the base; the disk brown or purplish, globose; leaves smaller, firm, less numerous and more erect, usually entire or nearly so.

HABITAT In moist ground along roads and in waste places; native to the southeastern United States.

FLOWERING June to October.

KEY TO HELIANTHUS (SUNFLOWER) SPECIES

1 Annuals; flowers large, disk red-purple to dark brown, flat or nearly so, large, usually over 4 cm broad . . . . *H. annuus*

1 Perennials; disk yellow (sometimes browish yellow), convex or obtusely conic; usually not over 3.5 cm broad 2

2 Leaves mostly basal; flowering stems having only 1–3 pairs of small, opposite leaves; . bracts of involucre tightly appressed . . . . *H. occidentalis*

2 Leaves borne along flowering stem; outer bracts of involucre spreading at tips 3

3 Leaves mostly folded together and curving back; peduncles short . . . . *H. maximiliani*

3 Leaves flat or nearly so; peduncles moderately long 4

4 Leaves sessile or on very short petioles (not over 5 mm long), opposite, spreading horizontally . . . . *H. divaricatus*

4 Leaves having definite petioles 5

5 Leaves broadly lanceolate to ovate, at least 1/3 as wide as long 6

5 Leaves lanceolate or narrower, more than 3 times as long as wide . . . . *H. giganteus*

6 Plants not producing tubers; leaves thin, veins thin and inconspicuous; bracts of involucre green .... *H. decapetalus*

6 Plants producing tubers; leaves thick and hard, veins prominent; bracts of involucre dark at base .... *H. tuberosus*

Common Sunflower

Helianthus annuus L.

Erect, simple to many-branched annual up to 5 m or more tall. **Heads** large, 5–7 cm wide or wider, the rays orange-yellow, the disk rather flat, dark brown to purple; achene obovate, thick, slightly compressed, whitish to grayish with dark lines; involucre of wide-spreading, rather broad bracts with tapering tips. **Leaves** mostly alternate, heart-shaped to elliptic-ovate, long-petioled, 3-nerved, toothed, rough on both surfaces.

HABITAT In old fields, along fences, and in waste places, mostly as an escape from cultivation; native to the western United States.

FLOWERING August and September.

USES As is true of most of the sunflowers, nearly all parts of this plant have some use. It is said that the plants make good fodder, the silage having about 90 per cent of the food value of corn silage. The young heads may be boiled and eaten; the seeds contain considerable oil which may be expressed and used in cooking, the residue making a good food for cattle and poultry; the seeds themselves are extensively used for poultry and bird feeding, and are sometimes prepared to be eaten like peanuts. The Huron Indians used the oil obtained by boiling the ground-up meal for cooking and as a hairdressing. A floss may be obtained from the stalks; the pith is light and has been used in life preservers.

COMMON SUNFLOWER

Thin-Leaf Sunflower, Pale Sunflower

Helianthus decapetalus L.

Smooth-stemmed perennial, simple or branched above. **Heads** small, the rays 8–12, 2–2.5 cm long, the disk about 1.5 cm broad; involucre of spreading bracts. **Leaves** opposite (upper stem leaves sometimes alternate), ovate, coarsely toothed to entire, 3-nerved, thin, green both sides, glabrous or nearly so.

HABITAT In open woods or thickets.

FLOWERING late July to September.

THIN-LEAF SUNFLOWER

WOODLAND SUNFLOWER

GIANT SUNFLOWER

MAXIMILIAN SUNFLOWER

Woodland Sunflower

Helianthus divaricatus L.

Glabrous, often glaucous perennial up to 2 m tall, simple or branching above. **Heads** few, bright yellow, sessile or on short peduncles, 3–6 cm broad, the rays narrow, about 2.5 cm long, the disk about 1 cm broad; involucre of somewhat recurving bracts. **Leaves** opposite, sessile or nearly so, spreading at right angles to the stem, ovate to lanceolate, rounded at the base and tapering to a long point at the apex, 3-nerved, green and rough above, pale beneath.

HABITAT In dry woodlands, thickets, openings and on roadsides.

FLOWERING July to September.

Giant Sunflower

Helianthus giganteus L.

Rough or hairy-stemmed perennial up to 3 m tall. **Heads** 1 to a few, the rays 1.5–2 cm long, the disk about 2 cm broad. **Leaves** alternate (the lower ones sometimes opposite), lanceolate, shallowly toothed or nearly entire, pinnately veined, 1–3 cm broad.

HABITAT In damp or rich thickets, swamps, and other moist places.

FLOWERING July to October.

Maximilian Sunflower

Helianthus maximiliani Schrad.

Stout, rough perennial up to 3 m tall, unbranched up to the inflorescence, the lower part often leafless or nearly so. **Heads** yellow, up to 10 cm wide, borne on short peduncles in slender racemes, the rays 15–30, 2–3 cm long; involucral bracts loosely spreading, narrowly lanceolate, pointed. **Leaves** numerous, alternate, lanceolate, pointed at apex, somewhat folded together upward (trough-like) and curled backward (somewhat like peach leaves), sessile or on short petioles.

HABITAT Along roads and railways, on rich prairies and in waste ground, mostly in sandy or dry soil; native to the Great Plains.

FLOWERING July to October.

Few-Leaf Sunflower

Helianthus occidentalis Riddell

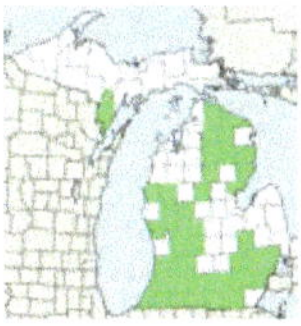

Slender, erect, smooth to hairy perennial up to 1.5 m tall, branching only at the inflorescence, almost leafless above. **Heads** generally few, yellow, 2–4 cm wide, the rays 10–16; achenes 4-sided, pappus soon falling, of 2 thin, chaffy scales. **Leaves** mostly at the base of the plant, opposite, ovate to oblong-lanceolate, tapering to the winged petiole, entire, wavy, or crenate; stem leaves 1–3 pairs, small, lanceolate or ovate, sessile or short-petioled.

HABITAT In dry soil, sandhills, abandoned fields, pine plains, and on rocky banks.

FLOWERING August and September.

FEW-LEAF SUNFLOWER

Jerusalem Artichoke

Helianthus tuberosus L.

Stout, somewhat hairy perennial up to 3 m tall, the elongate rootstocks producing fleshy edible tubers. **Heads** several to numerous, 1.5–2.5 cm wide, rays 10–20, 2–4 cm long; bracts of the involucre usually rather dark, at least at the base. **Lower leaves** opposite, ovate or broadly lanceolate, 3-veined at the base, 10–25 cm long, 4–12 cm wide, petioles mostly 2–8 cm long, winged.

HABITAT Along streams, in damp thickets, borders of woods, and moist open ground.

FLOWERING August to mid-October.

USES The edible tubers are formed only as the days become shorter. They were eaten extensively by American Indians. Introduced into England early in the seventeenth century and is still cultivated in parts of this country.

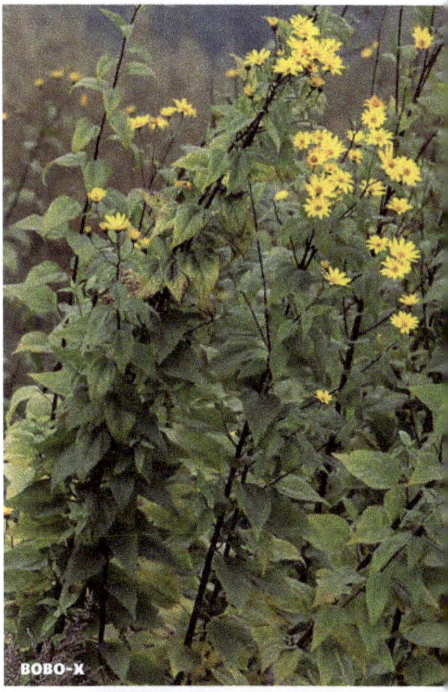

JERUSALEM ARTICHOKE

Ox-Eye, False Sunflower

Heliopsis helianthoides (L.) Sweet

Nearly smooth perennial up to 1.5 m tall. **Heads** showy, yellow, few to several; rays 8–15, 1.5–4 cm long, pistillate, persisting on the achenes and becoming papery; disk conical, 1–2.5 cm wide, the flowers fertile; pappus of 2–4 obscure teeth or lacking; involucral bracts in 2 or 3 rows, the outer ones leaflike and spreading **Leaves** opposite, ovate-lanceolate to ovate, toothed, petioled, 3-ribbed, rough.

HABITAT In open woods and thickets and on banks.

OX-EYE, FALSE SUNFLOWER

RATTLESNAKE-WEED

ELECAMPANE

FLOWERING July to September.

NOTES Ox-eye differs from the true sunflowers by its more conical disk; the rays are persistent on the achenes, and the disk and ray flowers are fertile. In the sunflowers the disk is quite flat, the rays drop off, and only the disk flowers are fertile.

Rattlesnake-Weed

Hieracium venosum L.

Erect perennial with a solitary stem up to 1 m tall from a basal rosette, the stem many-branched above; juice milky. **Heads** numerous, yellow 15–40-flowered, in open panicles; involucre cylindric with a single series of long, narrow bracts and a few short outer ones. **Leaves** all basal, oblong-spatulate, 2.5–12 cm long, distinctly marked with reddish-purple along the veins.

HABITAT In open woods and clearings, usually in sandy or poor soil.

FLOWERING May to September.

Elecampane

Inula helenium L.

Stout, woolly, usually unbranched perennial up to 2 m tall from a thick mucilaginous root. **Heads** few, large (5–10 cm broad), terminal on stout peduncles, many-flowered; rays yellow, narrow; disk flowers dingy yellow or brownish; achenes 4–5 ribbed; pappus a single row of bristles; involucre hemispherical, 2–2.5 cm high, of overlapping bracts, the outer ones leaflike. **Leaves** large, woolly beneath, the basal ones ovate, 2.5–5 dm long and 2 dm wide, petioled, the upper leaves becoming sessile and cordate-clasping.

HABITAT Along roadsides and fences, in rich clearings, fields, and waste places. Introduced from Europe; escaped from cultivation,

FLOWERING May to August.

Dwarf Dandelion

Krigia virginica (L.) Willd.

Annual with several flowering stems up to 3 dm tall from fibrous roots; unbranched, leafy at or near the base only; with milky juice. **Heads** yellow, solitary at the end of the slender

flowering stem. **Flowers** all ligulate and perfect; bell-shaped involucre of 9–18 bracts which become reflexed in age. Pappus double, of 5–7 short roundish scales alternating with long bristles. **Earliest leaves** roundish and entire; **later leaves** linear to obovate, often pinnately cut, 1.5–12 cm, long, 1–12 mm, wide.

HABITAT In sterile, dry or sandy soil.

FLOWERING April to August.

Wild Lettuce

Lactuca canadensis L.

Glaucous, usually glabrous, leafy-stemmed biennial up to 3.5 m tall, with basal rosette from a tap root; juice milky. **Heads** yellow, sometimes becoming purplish in age, borne in open, elongate panicles; flowers 12–20 in a head, all ligulate; achenes flat, 1–3 nerved on each face, contracted to a long, threadlike beak which bears a short-lived, very soft, copious white pappus of hairlike bristles; involucre green or tinged with purple, urn-shaped, the bracts of unequal length, closely appressed, in 2 or more series. **Leaves** glabrous, or slightly hairy beneath, numerous, alternate, variable, unlobed to wavy-pinnate with lanceolate to oblong segments that are broader at the base, the upper leaves usually lanceolate and entire, sessile, clasping.

HABITAT In thickets, open woods, borders of woods, and waste places.

FLOWERING July to September.

SIMILAR SPECIES Prickly Lettuce, *Lactuca serriola* L., is an annual or biennial. It has only 5–7 flowers in a head, the leaves are spiny-toothed, and the achenes have 5–7 nerves on each face. **Garden Lettuce**, *Lactuca sativa* L., may escape from cultivation and be found along roads and in waste places, but does not usually persist.

USES American Indians used the fresh white latex of Wild Lettuce to treat warts.

Ox-Eye Daisy

Leucanthemum vulgare Lam.

SYNONYM *Chrysanthemum leucanthemum* L.

Smooth, erect, simple or sparingly branched, many-stemmed perennial up to 1 m tall, often growing in colonies. **Heads** solitary at the ends of branches, 2–

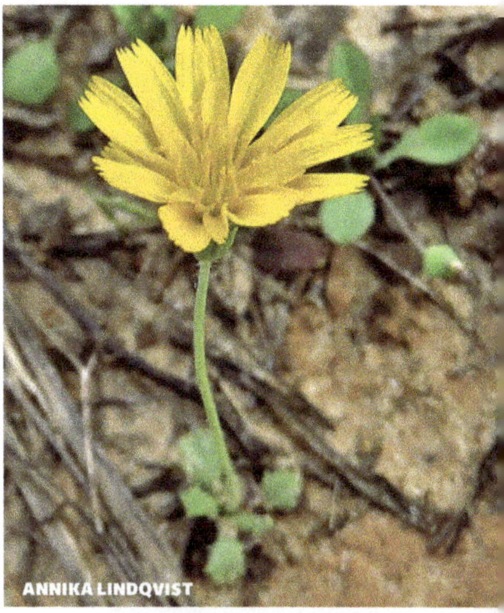

ANNIKA LINDQVIST

DWARF DANDELION

DAVID GEORGE

WILD LETTUCE

DIEGO DOPICO

OX-EYE DAISY

5 cm wide, flattish; rays 14–20, white, long, spreading, 2–3-toothed at apex; disk yellow, the receptacle low, convex, naked; achenes striate; involucre of small, overlapping, compact bracts with dry margins. **Basal leaves** often forming compact rosettes, oblanceolate or spatulate, usually pinnately cleft or coarsely and irregularly toothed, petioled; middle and upper **stem leaves** becoming smaller, less deeply cut, sessile and usually somewhat clasping at base.

HABITAT In fields and meadows, along roadsides. A common weed, introduced from Europe.

FLOWERING June and July.

NOTES This attractive weed often virtually takes over run-down meadows and pastures. Cattle seldom browse these plants, but when they do the milk has an unpleasant taste. The Ox-eye Daisy is a very attractive cut flower, and one does not have to worry about diminishing the supply.

KEY TO LIATRIS (GAYFEATHER, BLAZING-STAR) SPECIES

1 Heads solitary or few; bristles of pappus having fine, elongate hairs ...... *L. cylindracea*
1 Heads numerous; bristles of pappus covered with fine barbs 2
  2 Heads sessile or nearly so, having 7–9 flowers .... *L. spicata*
  2 Heads borne on peduncles or nearly sessile, having 16–80 flowers 3
3 Heads containing 16–35 flowers .... *L. aspera*
3 Heads containing 30–80 flowers .... *L. scariosa*

TALL GAYFEATHER

ONTARIO GAYFEATHER

Tall Gayfeather, Rough Blazing Star

Liatris aspera Michx.

Distinguishable from *Liatris scariosa* (below) only with difficulty. It may have more leaves below the inflorescence, fewer flowers in a head (16–35), and generally shorter peduncles. The involucre is usually glabrous, and the middle involucral bracts have broad, dry, and papery slashed borders.

HABITAT In dry, often sandy, soil.

FLOWERING August and September.

Ontario Gayfeather

Liatris cylindracea Michx.

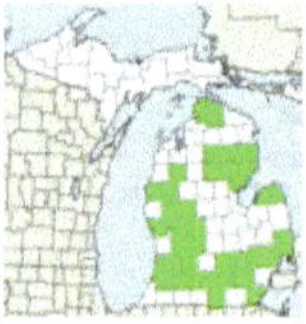

Slender, stiffly erect, smooth to sparsely hairy, leafy, unbranched perennial, up to 7 dm tall from a bulb-like base. **Heads** 30–60-flowered, rose-purple to light purple, few or solitary on short peduncles from the upper axils, or sessile. **Flowers** all tubular, perfect; corolla slender; stigmas 2, colored, extending beyond the corolla tube, giving the heads a ragged appearance; involucre of leathery, closely overlapping bracts; pappus of bristles covered with elongate fine hairs; receptacle naked. **Leaves** alternately sessile, linear and grasslike in appearance, rigid, punctate, entire, becoming progressively smaller above, parallel-veined.

HABITAT In dry open soil, on gravelly banks, pine plains.

FLOWERING July to September.

Devil's-Bite

Liatris scariosa (L.) Willd.

Smooth or hairy perennial up to 10 dm tall. **Heads** 30–80-flowered, pinkish-purple, 6–50 on peduncles 5–50 mm long. Corolla lobes long and slender, the stigmas extending beyond them; pappus of bristles covered with small barbs (scarcely visible to the naked eye); receptacle naked; involucre 12–17 mm high; achenes slender, tapering to the base, ribbed. **Leaves** crowded, linear-elliptic, parallel-veined, 25–60 below the inflorescence, the upper leaves small and sessile, the lowest petioled and 2–5 cm wide.

HABITAT In dry or sandy soil, often along railways and roads.

FLOWERING August and September.

DEVIL'S-BITE

Dense Gayfeather, Spiked Blazing Star

Liatris spicata (L.) Willd.

Stiffly erect, smooth or rarely hairy perennial up to 20 dm tall, often with a strong turpentine like odor when crushed. **Heads** 10–18-flowered, cylindric bell-shaped, purplish-pink, in a dense spike 1–7 dm long, the upper heads over-topping the bractlike leaves, the lower heads usually shorter than the leaves, mostly sessile. **Flowers** 7–9 (sometimes up to 18); involucre 8–10 mm long and about half as thick, the bracts green, or purple-tinged. **Leaves** numerous, linear to lanceolate or the lower ones oblanceolate, tapering to long, narrow-winged petioles, upper leaves smaller.

HABITAT In meadows, damp slopes, marshes, and other moist open places.

FLOWERING July to September.

DENSE GAYFEATHER

Pineapple-Weed

Matricaria discoidea DC.

SYNONYM *Matricaria matricarioides* (Less.) Porter

Low-growing, weedy annual with the odor of fresh pineapple, at least when bruised. **Heads** without rays, erect, terminal, greenish to yellowish, somewhat resembling acorns in

PINEAPPLE-WEED

WHITE RATTLESNAKE-ROOT

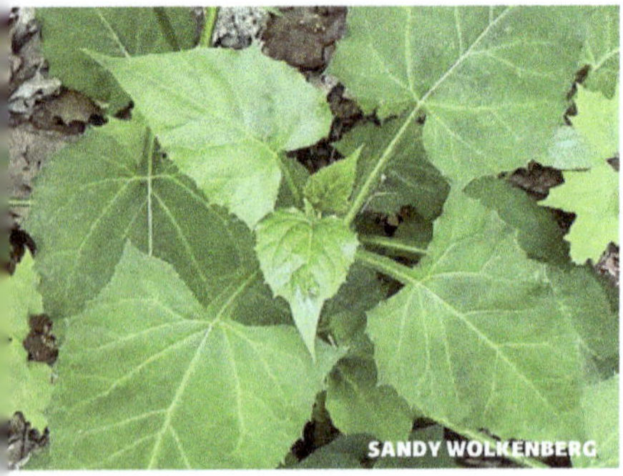

WHITE RATTLESNAKE-ROOT

their cups, the disk being conical and the involucre of short, overlapping papery bracts. **Leaves** 2–3 times pinnately divided into very narrow segments.

HABITAT Along roads and railways, in fields and dry waste land.

FLOWERING all summer.

White Rattlesnake-Root

Nabalus albus (L.) Hook.

SYNONYM *Prenanthes alba* L.

Glaucous, smooth perennial up to 1.5 m tall; juice milky, copious. **Heads** nodding, slenderly cylindric, whitish or greenish, tinged with rose or pale purple, closing at night and in cloudy weather, borne in loose terminal panicles; **flowers** 8–15 in a head, all ligulate and perfect, fragrant; the styles with the 2-cleft stigmas extending well beyond the rays to produce a ragged appearance; involucre cylindrical, of 6–8 long, purplish-tinged bracts and many much smaller triangular bracts; pappus of rough, brown to deep reddish-brown bristles on short, columnar, beakless achenes. **Leaves** alternate, variable, the lower ones more or less triangular, often 3–5 lobed, the base heart-shaped or hastate, the margin wavy-toothed, the petioles long, wing-margined; upper leaves much smaller, more or less oblong, often with angular teeth or small lobes, the petioles short.

HABITAT In rich woodlands and thickets.

FLOWERING August and September.

Tall Rattlesnake-Root

Nabalus altissimus (L.) Hook.

SYNONYM *Prenanthes altissima* L.

Somewhat taller than the preceding species (up to 2 m) the **heads** usually 5–6-flowered and borne in a long, leafy, open panicle; the involucre of 5 principal bracts; the pappus creamy-white to bright yellow-brown; the **leaves** often unlobed, the petioles winged, or not.

HABITAT In moist woods.

FLOWERING July to October.

Golden Ragwort

Packera aurea (L.) Á. & D.Löve

SYNONYM *Senecio aureus* L.

Erect perennial up to 16 dm tall, freely branched above, the rootstocks and basal offshoots horizontal, creeping, usually bearing distinct tufts of leaves. **Heads** golden-yellow, 2–3 cm wide; ray flowers 8–15, pistillate; disk flowers perfect; achenes smooth, the pappus white, of numerous, soft hairlike bristles; receptacle flat, naked; involucre of linear, pointed greenish bracts. **Basal leaves** heart-shaped to nearly round, toothed, long-petioled, often purplish below; **lower stem leaves** more or less lyrate, short-petioled; **upper leaves** pinnately cut and sessile.

HABITAT In wet meadows, swampy ground, and moist thickets.

FLOWERING May to August.

GOLDEN RAGWORT

Balsam Ragwort

Packera paupercula (Michx.) Á. & D. Löve

SYNONYM *Senecio pauperculus* Michx.

Slender perennial up to 6 dm tall. **Heads** usually 1–8, rarely more than 20, deep yellow; rays few, usually conspicuous; involucre bell-shaped, the bracts often purple-tipped. **Basal leaves** oblanceolate to oblong-elliptic, tapering to the petiole, crenate or toothed, up to 10 cm long; **stem leaves** usually pinnately cut, but sometimes undivided, the upper leaves reduced and often sessile.

HABITAT In gravel, rocky places, on sandy lakeshores and dunes, on cliffs, or in bogs and peaty places.

FLOWERING late May to September.

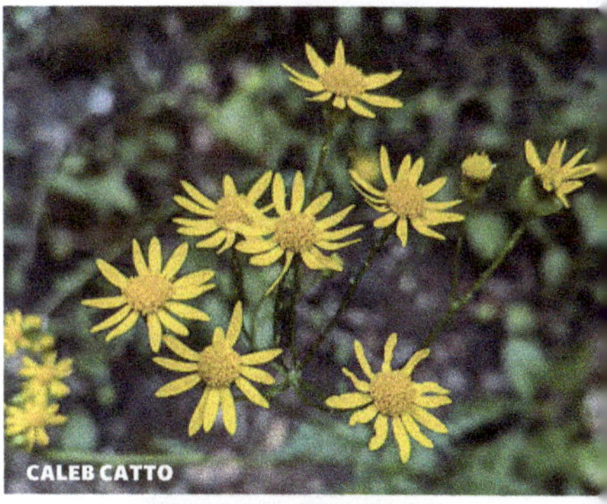

BALSAM RAGWORT

Orange Hawkweed

Pilosella aurantiaca (L.) F.W. Schultz & Sch.Bip.

SYNONYM *Hieracium aurantiacum* L.

Erect, stiffly-hairy perennial up to 6 dm tall from a basal rosette; juice milky. **Heads** orange-red to deep orange, few, in loose, more or less flat-topped terminal clusters, the pedicels and involucres thickly clothed with black glandular dots. **Flowers** all ligulate and perfect, the outer row or two with longer rays; involucre of 2 or 3 rows of nearly equal, pointed bracts; achenes black, columnar, tapering to the base; pappus a single series of nearly equal, brittle, rough bristles. **Leaves** mostly in a basal rosette (oc-

ORANGE HAWKWEED

SMOOTH HAWKWEED, KINGDEVIL

MACOUN'S EVERLASTING

casionally 1 or 2 on a scape), oblong or oblanceolate, 5–13 cm long, green and long-hairy on both sides; vigorous, leafy stolons are frequent.

HABITAT In fields and open woods, along roads, and in waste places. Introduced from Europe; now a troublesome weed.

FLOWERING June to August.

Smooth Hawkweed, Kingdevil

Pilosella piloselloides (Vill.) Soják

SYNONYM *Hieracium florentinum* All.

Slender, smooth or hairy, many-stemmed perennial up to 1 m tall from basal rosettes, the leafless stems many branched above. **Heads** yellow, 3–75, on glandular-hairy peduncles in a corymb. **Flowers** all perfect and ligulate; pappus of many bristles; achenes columnar, narrowed at the base; involucre with black glandular hairs. **Leaves** all basal, narrowly oblong or spatulate, thick, entire, pale green, narrowed to a winged petiole.

HABITAT In fields, meadows, and clearings, along roadsides. Naturalized from Europe.

FLOWERING May to August.

NOTES This aggressive weed has become very common in the tip of the Lower Peninsula and in the Upper Peninsula within the last few years. Flowering at the same time as Orange Hawkweed, Ox-eye Daisy, and some of the ragworts, it adds color to wasteland, particularly in late June.

Macoun's Everlasting

Pseudognaphalium macounii (Greene) Kartesz

SYNONYM *Gnaphalium macounii* Greene

Strongly fragrant, white-woolly, somewhat viscid biennial up to 1.5 m tall. **Heads** numerous, many-flowered, dirty-yellowish, densely clustered in corymbs; **flowers** all tubular, the outer ones very slender, pistillate, the central flowers perfect; pappus a single row of entirely distinct, rough bristles; involucre yellowish-white, woolly only at the base, the bracts dry, overlapping in several rows. **Stem leaves** alternate, linear-lanceolate, tapering to apex, clasping at the base and having decurrent wings extending down the stem, greenish on upper surface, densely white-woolly with a felt-like covering beneath, the margin entire, somewhat inrolled; **basal leaves** oblanceolate.

HABITAT In clearings, along roadsides, in pastures, waste places, and the borders of woods.

FLOWERING July to October.

Catfoot *(not illustrated)*

Pseudognaphalium obtusifolium (L.) Hilliard & B.L. Burtt

SYNONYM *Gnaphalium obtusifolium* L.

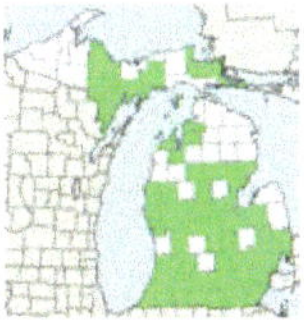

Quite similar to the preceding species, but usually more slender, the inflorescence more open and spreading, the leaves less woolly, narrowing somewhat to the sessile base, and the stem not winged.

HABITAT In open, often in sandy, places.

FLOWERING July to October.

GRAY-HEAD CONEFLOWER

Gray-Head Coneflower

Ratibida pinnata (Vent.) Barnh.

Hoary perennial up to 1.5 m tall. **Heads** several or solitary; showy; rays 5–10, pale yellow, drooping, 2.5–6 cm long; disk elliptic, grayish, the receptacle having an anise-like odor when crushed, the chaffy bracts subtending the ray and disk flowers and enclosing the 4-sided achenes; involucre a single series of a few small, spreading bracts. **Leaves** at or near the base of the plant, alternate, pinnately divided into 3–7 lanceolate segments.

HABITAT In dry soil.

FLOWERING June to August.

KEY TO RUDBECKIA SPECIES

1 Leaves entire; plants coarsely hairy . . . . *R. hirta*

1 At least some leaves 3-lobed or pinnately cut; plants hairy to glabrous 2

2 Leaves usually 3-lobed; disk dark purple; heads on short peduncles . . . . *R. triloba*

2 Lower leaves usually pinnately cut; disk greenish-yellow to grayish; heads on long peduncles . . . . *R. laciniata*

BLACK-EYED SUSAN

Black-Eyed Susan

Rudbeckia hirta L.

Rough-hairy, erect biennial or short-lived perennial up to 1 m tall, simple or sparsely branched. **Heads** solitary or few on hairy peduncles; rays 8–26, usually orange-yellow or orange, but variable and sometimes red or red-tipped, or with a purple or brown spot at base, 2–4 cm long; disk hemispheric to ovoid, purple or brown, rarely yellow, 12–20 mm wide; achenes 4-angled, smooth, flat on top; bracts of the involucre leaflike, spreading or reflexed, and much shorter than the rays. **Leaves** alternate, thick, variable in shape, the basal ones mostly oblanceolate, 1–3 cm broad, entire, petioled, the upper leaves linear-lanceolate to oblong or ovate, mostly sessile.

CUTLEAF CONEFLOWER

BROWN-EYED SUSAN

HABITAT In fields, clearings, meadows, along roadsides, mostly in disturbed or waste ground.

FLOWERING July to September.

Cutleaf Coneflower

Rudbeckia laciniata L.

Smooth, coarse, glaucous, erect perennial up to 3 m tall. **Heads** solitary on long peduncles, showy; rays yellow, soon reflexed; disk dull greenish-yellow, up to 2.5 cm in diameter, at first hemispherical to globular, but elongating and becoming columnar in fruit. **Lower leaves** pinnate with 5–7-cut or 3-lobed leaflets; petioled; **upper leaves** similar or 3–5 parted or uncut, sessile.

HABITAT In rich, low ground.

FLOWERING July to September.

NOTES One cultivated variety, *hortensis* Bailey, is a well-known garden flower under the name 'Golden-glow.' It has many more ray flowers.

Brown-Eyed Susan

Rudbeckia triloba L.

Smooth to hairy, leafy, branched annual or biennial (rarely short-lived perennial) up to 1.6 m tall. **Heads** on short peduncles, showy, about 5 cm wide; rays yellow or the base orange to brown; disk blackish-purple, soon ovoid; involucre of leaflike, soon drooping bracts. **Lower stem leaves** usually 3-lobed, pointed at apex, coarsely toothed.

HABITAT In fields, thickets, and open woods.

FLOWERING June to October.

Cup-Plant

Silphium perfoliatum L.

Stout, coarse, rough perennial up to 2.5 m tall, the 4-angled stems leafy, branched above, the juice resinous, **Heads** numerous, large (5–7 cm broad), yellow; rays 20–30, pistillate and fertile; disk about 1 cm broad, the flowers sterile, the style not cleft; achenes (of the rays) flat, broad-obovate, surrounded by a wing; involucre broad and rather flat, the outer bracts smooth, spreading, leaflike, the inner bracts similar to the chaff of the receptacle. **Leaves** rough, coarsely toothed, opposite and (at least the upper leaves) joined by their bases or petioles, often forming a cuplike structure.

HABITAT In moist soil in prairies, rich woods, and thickets, along river banks, and in low ground.

FLOWERING July to September.

STATUS Threatened in Michigan.

CUP-PLANT

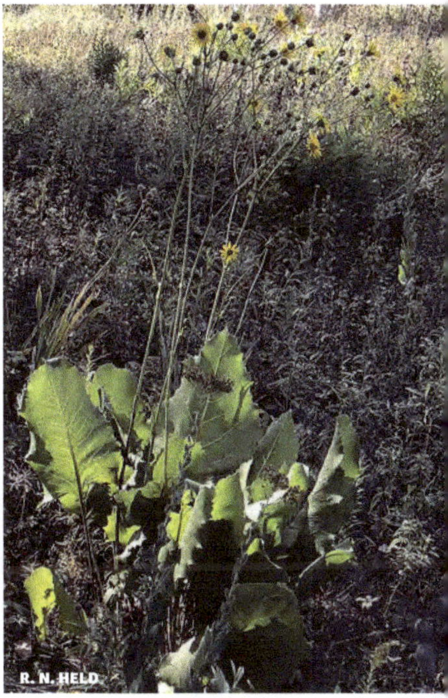

PRAIRIE-DOCK

Prairie-Dock

Silphium terebinthinaceum Jacq.

Differing from the preceding chiefly by having most of the leaves near the base of the plant; neither the leaf bases nor the petioles are joined.

HABITAT In fields and waste places, along roads, mostly in dry open ground.

FLOWERING July to September.

KEY TO SOLIDAGO (GOLDENROD) SPECIES

NOTE About two dozen species of goldenrod occur in Michigan, several of them distinguishable only with difficulty. Only a few of the state's species are described and illustrated here.

1 Flower cluster long and slender; heads spirally arranged on branches 2

1 Flower cluster pyramidal; heads borne mostly on upper side of branches 3

2 Leaves hairy to bristly .... *S. hispida*

2 Leaves glabrous to minutely hairy .... *S. gillmanii*

3 Lower leaves much larger than upper; basal rosettes usually formed 4

3 Basal leaves mostly smaller than upper leaves; basal rosettes rarely formed S. canadensis

4 Flower cluster slender, usually arched or leaning, and somewhat one-sided; stems and leaves covered with minute grayish hairs .... *S. nemoralis*

4 Flower cluster as broad as tall, or broader; plants not hairy except for margins of leaves .... *S. juncea*

CANADA GOLDENROD

Canada Goldenrod

Solidago canadensis L.

Perennial with solitary or clustered stems up to 11 dm tall, glabrous below, becoming hairy above. **Heads** yellow, borne in a broad, pyramidal panicle 5–40 cm high, the dense, 1-sided divergent racemes recurved at the tip. **Leaves** numerous, crowded, lanceolate, long-tapering to the tip, mostly sharply toothed, 3-nerved, the basal and lower leaves mostly smaller than the middle and upper.

HABITAT In open woods, thickets, and clearings and along dry roadsides.

FLOWERING mid-July to September.

USES Both the flowers and roots of this species were used medicinally by American Indians. The flowers of various species of goldenrod were boiled to make a yellow dye.

DUNE GOLDENROD

Dune Goldenrod

Solidago gillmanii (A. Gray) E.S. Steele

SYNONYM *Solidago racemosa* Greene

Sticky, glabrous to slightly hairy perennial up to 1 m, tall; stolons lacking at the base. **Heads** yellow (white in one form), borne on upright pedicels mostly 5–15 mm long and forming a slender, loose terminal raceme or panicle of upright racemes; rays about 10; involucres slenderly bell-shaped, 5–8 mm high. **Leaves** numerous, the basal and lower leaves mostly oblanceolate, entire or toothed, the midrib slender; stem leaves decreasing in size upward, oblanceolate to linear, often subtending a cluster of heads.

HABITAT On dry ledges or rocky banks, on shores, in gravels and sands, or on dunes along the Great Lakes.

FLOWERING June to October.

Hairy Goldenrod

Solidago hispida Muhl. ex Willd.

Downy perennial up to 1 m tall, unbranched or with a few ascending branches. **Heads** yellow, borne in a simple or branched, slender, cylindric, wand-like inflorescence and in small clusters in the axils of the leaves. **Basal leaves** rather thick, oblanceolate to narrowly obovate, crenate to serrate, somewhat downy, at least beneath, the **stem leaves** becoming much smaller upward, velvety to the touch.

HABITAT In dry to moist rocky places, often in calcareous soil (one variety in peaty soil).

FLOWERING July to October.

Early Goldenrod

Solidago juncea Ait.

Mostly smooth, stiffly erect perennial up to 13 dm tall. **Heads** yellow, borne in a dense panicle as broad as long or broader, the racemes strongly 1-sided. **Basal leaves** tufted and persistent, narrowly elliptic, tapering to long, winged petioles, sharply toothed, 15–40 cm long; upward on the stem the leaves become sessile, smaller, and less toothed; margins of leaves and petioles hairy. This is one of the earliest goldenrods to bloom.

HABITAT In dry open places and in open woods.

FLOWERING June to October.

Gray Goldenrod

Solidago nemoralis Ait.

Grayish-green, finely downy perennial with solitary or tufted, usually somewhat arching stems 1.5–6 dm tall. **Heads** yellow, crowded, mostly on one side of the spreading or recurved branches of the terminal panicle; rays 5–9, disk flowers 3–6. Rosette and basal leaves mostly tufted, oblanceolate to obovate, petioled, round-toothed, **stem leaves** thick, rough, decreasing in size upward, at least the upper ones entire, mostly subtending small tufts of reduced leaves.

HABITAT In dry, sterile, often sandy or clayey soil in open places or in thin woods.

FLOWERING August and September.

Field Sow-Thistle

Sonchus arvensis L.

Coarse, glandular to smooth perennial up to 12 dm tall, creeping extensively by underground rootstocks; the stems leafy below, almost naked above; juice milky. **Heads** bright yellow, 3–5 cm wide, resembling dandelions, several on long peduncles in an open corymb. **Flowers** all ligulate; involucre 2.5 cm high, of uniformly

HAIRY GOLDENROD

EARLY GOLDENROD

GRAY GOLDENROD

colored, overlapping bracts of 3 lengths; achenes oblong, the pappus copious, of very soft, white, simple bristles which usually fall off together. **Leaves** lanceolate, the upper leaves deeply pinnately cut, with backward-pointing segments spiny-toothed, up to 3 dm petioled; upper leaves small, entire, clasping.

HABITAT In fields and waste places, along roads, and on gravelly shores. Introduced from Europe; now a widespread weed.

FLOWERING July to October.

Common Tansy

Tanacetum vulgare L.

Smooth, erect, strongly aromatic, bitter and acrid perennial up to 1.5 m tall, branching only at the inflorescence. **Heads** numerous (20–200), yellow, mostly without rays or with a few small obscure ones, hemispheric, 5–10 mm broad, borne in flat-topped terminal corymbs. Involucre of dry overlapping scales. **Leaves** very numerous, alternate, deeply and finely 1–3 times pinnately divided into toothed linear-oblong segments, the lower divisions of the leaves often smaller than the upper; basal leaves up to 3 dm long and half as wide.

HABITAT In fields, along roads, shady shores, and waste places. Introduced from Europe.

FLOWERING July to September.

NOTES This species was cultivated in western Europe during the Middle Ages for its supposed medicinal properties. It is poisonous, and there are records of deaths due to drinking too much tea made from the leaves. Animals may be poisoned by this species, but because of the bitter taste they seldom eat it.

Lake Huron Tansy

Tanacetum bipinnatum (L.) Schultz-Bip.

SYNONYM *Tanacetum huronense* Nutt.

Very similar to **Common Tansy**, but somewhat shorter; long-hairy to woolly throughout; the heads fewer (1–30) and larger (the disk 1.3–2 cm broad).

HABITAT Uncommon on sands and gravels of lakeshores and on river banks.

FLOWERING June to August.

FIELD SOW-THISTLE

COMMON TANSY

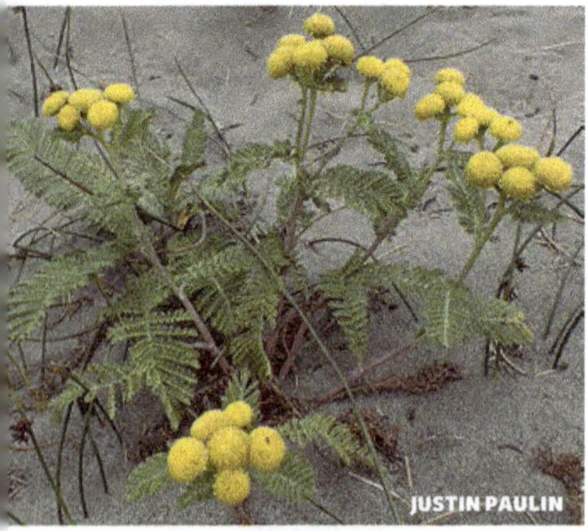

LAKE HURON TANSY

Common Dandelion

Taraxacum officinale G.H. Weber ex Wiggers

Ever-present, rosette-forming perennial 5–50 cm tall from a deep tap root; juice copious, milky. **Heads** numerous, opening only in sunshine, bright golden-yellow, 2–5 cm wide, solitary at the ends of the naked, hollow, flowering stalks. **Flowers** 150–200, all ligulate and perfect, the outer flowers the longest; involucre double, the inner bracts the longer, becoming erect after flowering, then reflexed and exposing the globular head of brown to straw-colored achenes; each achene crowned with a fine, pale, parachute-like pappus. **Leaves** basal, horizontal or ascending, narrowed to the slender, usually slightly winged, petiole-like base, oblanceolate, usually coarsely and irregularly wavy-toothed or pinnately cut, sometimes nearly entire, 6–40 cm long.

HABITAT Abundant in lawns, grasslands, open ground, and disturbed places. Introduced from Europe.

FLOWERING April to September.

USES The leaves of dandelions are a well-known early spring green. They can be served either boiled or raw. Dandelion wine (made from the blossoms) is a favorite with some people.

COMMON DANDELION

RED-SEEDED DANDELION

Red-Seeded Dandelion

Taraxacum erythrospermum Andrz. ex Bess.

Quite similar to the preceding species in general aspect, but the leaves more dissected, usually cut almost or quite to the midrib into narrow, widely spaced lobes; the terminal lobe typically quite slender and not much larger than the lateral lobes; several small lobes usually present between the larger ones; the achenes red to purplish at maturity.

HABITAT In thin, dry soil of fields, pastures, and lawns, along roadsides and in other disturbed areas. Native of Eurasia now established in North America, but much less common than the preceding species.

FLOWERING April to July.

Yellow Salsify

Tragopogon dubius Scop.

SYNONYM *Tragopogon major* Jacq.

Stout, glabrous perennial up to 1 m tall; juice milky, soon brown. **Heads** yellow, large, solitary at the ends of branches, the peduncles enlarged upward. **Flowers** all ligulate, the rays of the central flowers about half as long as those of the outer flowers, the rays 5-lobed at the end; achenes linear, 2.5–4 cm long, borne in large globose heads; pappus of numerous long, plumose bris-

YELLOW SALSIFY

MEADOW SALSIFY

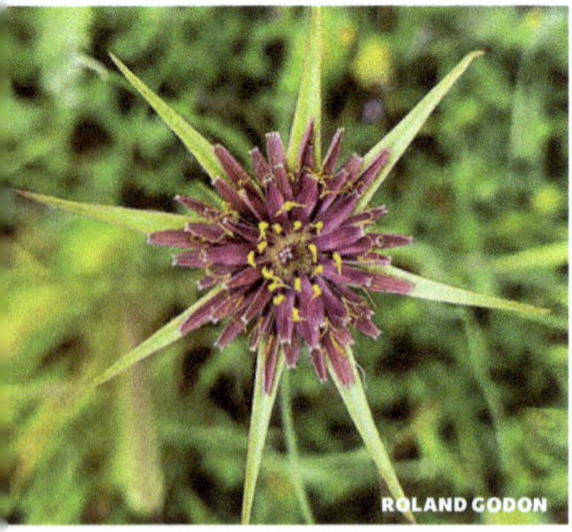

PURPLE SALSIFY, OYSTER-PLANT

tles attached to the long beak of the achene; involucre simple, usually of 8 bracts, the bracts longer than the flowers and becoming 4–7 cm, long in fruit. **Leaves** grasslike, alternate, linear, entire, clasping, parallel-veined.

HABITAT Along roadsides and in waste places. Like the other species of *Tragopogon* found here, this one is native to Europe but is now widely naturalized in North America.

FLOWERING May to July.

NOTES The heads close up shortly after being picked and also when the sun sets and during cloudy weather.

USES The roots are edible and are frequently used in soups or as a vegetable. The tops are sometimes eaten as greens. American Indians used the coagulated juice for chewing-gum.

Yellow Goat's-Beard

Tragopogon pratensis L.

Quite similar to the preceding species, but generally larger and more robust, the peduncle slenderly cylindric throughout, scarcely enlarged even in fruit, the involucre about as long as or shorter than the flowers, elongating to 1.8–3.8 cm in fruit, the achenes 1.5–2.5 cm long.

HABITAT Along roadsides, in fields and waste places.

FLOWERING May to August.

Purple Salsify, Oyster-Plant

Tragopogon porrifolius L.

Similar in general aspect to the 2 preceding species, but readily distinguished by the purple flowers. The peduncle is thickened below the head.

HABITAT In waste places and fields, along roads and railways.

FLOWERING May to July.

Tall Ironweed

Vernonia gigantea (Walter) Trel.

SYNONYM *Vernonia altissima* Nutt.

Hairy, leafy-stemmed perennial up to 2 m tall. **Heads** 13–30-flowered, usually purple, rarely white, numerous, borne in loose cymes 1–5 dm wide. **Flowers**

perfect, longer than the involucre; achenes cylindrical, ribbed, the ribs bearing minute hairs; pappus purple, double, with an outer series of minute scalelike bristles and inner series of copious hairlike bristles. **Leaves** alternate, spreading or loosely ascending, lanceolate to narrowly ovate or lance-oblong, sharply and finely toothed, 3–8 cm broad, hairy beneath.

HABITAT In rich damp soil.

FLOWERING August to October.

SIMILAR SPECIES Missouri Ironweed, *Vernonia missurica* Raf., is very similar to Tall Ironweed, but the heads are 34–55-flowered; the lower surface of the leaves is finely and densely woolly; and the pappus is typically tawny. It is found in rich low ground and on prairies; it flowers July to September. These two species of Ironweed frequently hybridize, and intermediate forms are common.

Rough Cocklebur

Xanthium strumarium L.

SYNONYM *Xanthium italicum* Moretti

Coarse, weedy, usually many-branched annual up to 1.5 m tall. **Heads** unisexual; staminate heads uppermost, many-flowered, the **flowers** tubular, greenish; pistillate heads with 2 flowers consisting of a pistil and a slender, threadlike corolla, enclosed by an ovoid, prickly involucre; fruit a brown spiny bur with 2 incurved beaks, up to 3 cm long. **Leaves** alternate, thick and large, cordate-ovate, coarsely dentate, very rough, long-petioled.

HABITAT In low wet ground, along lakes and streams, on flood plains, and in cultivated ground and waste places.

FLOWERING September to November.

NOTES The seedlings and young leaves of this species are said to be highly poisonous to stock. Common cocklebur is considered native to America, as is Michigan's other species, ***Xanthium spinosum*** L., reported from only several locations. As the name implies, plants are spine-covered.

TALL IRONWEED

MISSOURI IRONWEED

ROUGH COCKLEBUR

BALSAMINACEAE *Touch-Me-Not Family*

This widely distributed family of two genera and over 1000 species is most abundant in the tropics; one genus (*Impatiens*) occurs in Michigan. It includes a few ornamentals, such as Balsam and Touch-me-not.

The Michigan species are characterized by their simple leaves; the showy, horizontal flowers with one of the sepals petal-like, sac-like, and spurred; and by the fruit, which opens explosively when ripe. The actual structure of the flowers and the number of floral parts are differently interpreted by different authorities.

SPOTTED TOUCH-ME-NOT

PALE TOUCH-ME-NOT

Spotted Touch-Me-Not

Impatiens capensis Meerb.

Succulent, quickly wilting, many-branched, smooth and sometimes glaucous annual up to 2 m tall. **Flowers** few, in axillary racemes, hanging horizontally on slender pedicels, orange, spotted with red (but variable and sometimes unspotted, sometimes pale or nearly cream-color), up to 2.5 cm long; perianth much modified, irregular, sepals apparently 3, 2 being small, ovate, and green, and one (the lower one) large, forming an inflated sack which is longer than broad, spurred at the base, the spur bent back along the sack; petals seemingly 3, the upper one broader than long, the others 2-lobed and presumably each representing a united pair; stamens 5, the filaments short and flat, each bearing a scale on the inner side, the scales united over the stigma, the anther also united; ovary 5-celled; capsule oblong, opening violently by 5 coiling. valves and shooting out the ridged seeds when mature; self-pollinating, non-opening flowers are also frequently produced. **Leaves** alternate, thin, ovate to elliptic, 3–10 cm long, bright green above, whitish beneath, the margin with coarse, rounded teeth.

HABITAT In low wet ground, often in shade, along roads, in thickets, along streams, in springy places.

FLOWERING July to September.

USES American Indians used the fresh juice to wash nettle stings, for skin rash, and for poison ivy dermatitis. The seeds are eaten by several kinds of birds, and hummingbirds seek the nectar.

SIMILAR SPECIES Pale Touch-Me-Not, *Impatiens pallida* Nutt., is very similar, but the flowers are pale yellow and sparingly spotted; the spur is bent at right angles and is less than one-fourth as long as the sack; the sack is broader than long; and the plants tend to be stouter, taller, and more glaucous. **Balsam**, *Impatiens balsamina* L., with solitary, purple, rose, or white flowers, is a rather common garden flower that appeals greatly to children because of the explosive capsules.

BERBERIDACEAE *Barberry Family*

This family of perennial herbs and shrubs includes 13 genera and about 700 species, mostly of north temperate regions; 5 genera occur in Michigan. Several members of the family are used as ornamentals, Barberry and Oregon Grape, for example, being well-known shrubs.

It is difficult to characterize the family, since the genera composing it are not closely related. In our genera the stamens (the same number as the petals or twice as many) are borne in two rows, those of the outer row being opposite the petals; there is a single pistil with superior ovary; the petioles are usually dilated at the base. In *Podophyllum*, the anthers open by longitudinal slits; in the other genera, they open by pores.

Blue Cohosh

Caulophyllum thalictroides (L.) Michx.

Glaucous, unbranched perennial up to 1 m tall. **Flowers** inconspicuous, greenish or yellowish, 1–1.5 cm wide, borne in few-flowered, loose, terminal racemes or panicles. Petals 6, greenish, bronze, purplish, or yellowish, thick and gland-like, much smaller than the sepals, having a short claw at the base and kidney-shaped or hooded above, borne at the base of the sepals and opposite them; sepals 6, petal-like, yellowish-green, ovate-oblong, spreading, having small basal bracts; stamens 6; pistil 1, the style short, the stigmatic surface a line along one side; seeds 2, appearing as a pair of blue, globose, pea-like drupes when mature. **Leaves** usually 2, one placed a little below, the other just at the base of the inflorescence; lower leaf large, 2–3 times compound in 3's, with the main divisions petioled; the upper leaf small (becoming larger after flowering) and with the leaflets 2–3-lobed, obovate, and wedge-shaped at base.

HABITAT In rich woods.

FLOWERING late May to mid-June.

Twinleaf

Jeffersonia diphylla (L.) Pers.

Low, unbranched perennial up to 3 dm tall. **Flowers** white, solitary, terminal, about 2–3 cm broad. Petals 8, white, oblong, flat; sepals usually 4 (3–5), petal-like, falling off as the flower opens; stamens 8, prominent; pistil solitary, the ovary ovoid, the stigma 2-lobed; capsule pear-shaped, opening near the top by a peaked

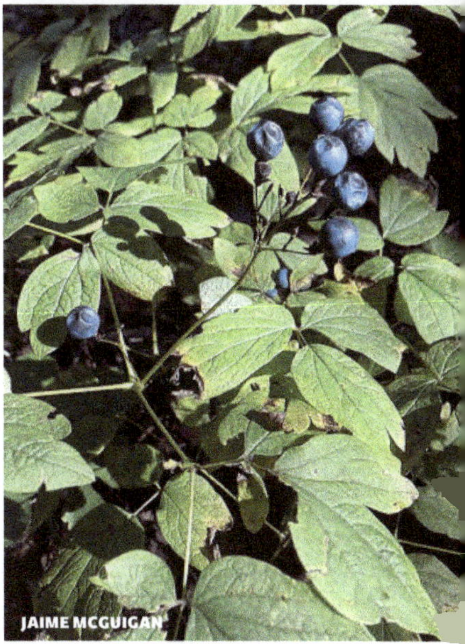

BLUE COHOSH

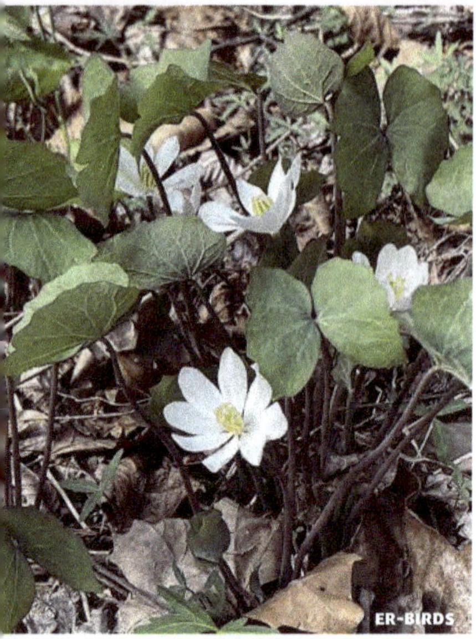

TWINLEAF

cap-like lid; seeds numerous, with a fleshy lateral appendage (aril). **Leaves** basal, becoming taller than the flower after flowering time, divided into two half-ovate leaflets and somewhat like an outspread butterfly in shape, up to 1 dm wide when mature.

HABITAT In rich damp woods. Rare. In Michigan found only in the southern part of the Lower Peninsula.

FLOWERING April and May.

NOTES The genus name honors President Thomas Jefferson. The flowers resemble those of **Bloodroot** (*Sanguinaria canadensis* L.), but they are much shorter-lived and not quite so pretty; the two plants are easily distinguished by their leaves and juice. Ants feed upon the seeds of Twinleaf and often carry them from one place to another, thus helping to spread the species.

May-Apple, Mandrake

Podophyllum peltatum L.

Low, erect perennial 3–5 dm tall with 1 or 2 large, umbrella-like leaves from rootstocks that creep just below the surface and often form large colonies. **Flowers** solitary, drooping on stout peduncles from the fork between the leaves, creamy to white, waxy, ill-scented, often with sickly sweet odor. Petals 6 or 9, 2–3 cm long; sepals 6, soon falling; stamens twice many as the petals; pistil one, large, the ovary superior; berry large, 2–5 cm long, lemon-shaped, becoming yellow, the pulp slightly acid, the seeds numerous. **Leaves** of the flowering stalk 2, umbrellalike, the stem attached to the middle, roundish in outline, deeply 3–5-parted; leaf of nonflowering plant only one, deeply 5- to 9-lobed.

HABITAT In open woods and thickets in meadows, and along partly shaded roadsides.

FLOWERING April to early June.

MAY-APPLE, MANDRAKE

USES The rootstock and, to a lesser degree, other parts of this plant are bitter and contain a poisonous substance that is used in medicine. Some of the Michigan Indians dried, roasted, and boiled the roots and used the liquid, sweetened with maple sugar or honey, as a cure for rheumatism and liver ailments. The green fruit is harmful, but when ripe it is edible in moderate quantities—if one likes the flavor. American Indians gathered the fruits before they were fully ripe and buried them in the ground to ripen, believing this improved the flavor. The fruits are occasionally used for preserves and in making beverages.

BORAGINACEAE *Borage Family*

This widely distributed family includes 94 genera and about 2,000 species; 13 genera occur in Michigan. Virginia Bluebells, Heliotrope, Lungwort, Forget-me-not, and a few others are grown as ornamentals.

Members of this family are easily recognized: plants are usually rough-hairy, the leaves simple and alternate; flowers borne in 1-sided coiled racemes which straighten as the flowers open; the flower parts are mostly in 5's, the petals and sepals both united, the stamens borne on the corolla tube, the ovary deeply 4-lobed and with a single style. The flowers often change from pink to blue between bud and maturity.

KEY TO BORAGINACEAE (BORAGE FAMILY) SPECIES

1 Corolla slightly irregular .... ***Echium vulgare***
1 Corolla regular 2
  2 Flowers yellow to orange .... ***Lithospermum caroliniense*** or ***L. canescens***
  2 Flowers not yellow or orange 3
3 Flowers reddish purple .... ***Cynoglossum officinale***
3 Flowers usually blue, sometimes pink or white 4
  4 Flowers trumpet-shaped, the tube much longer than the calyx .... ***Mertensia virginica***
  4 Flowers flat and circular in outline, the tube scarcely if at all longer than the calyx .... ***Myosotis scorpioides*** or ***M. laxa***

Common Hound's-Tongue

Cynoglossum officinale L.

Softly hairy biennial, branching only at the inflorescence. **Flowers** reddish-purple, about 1 cm wide, borne on 1-sided racemes in a panicle. Corolla with 5 rounded lobes, the tube short, about the same length as the calyx; stamens not extending beyond the corolla; nutlets 4, becoming large and conspicuous, covered with short hooked bristles, over-topped by the style and partially enclosed by the enlarged, flattened calyx. **Leaves** alternate, softly hairy, close to the stem, lanceolate, the upper ones sessile, the lower petioled.

HABITAT In pastures, along roads, and in waste places; often abundant. Introduced from Europe; now a troublesome weed.

FLOWERING May through July.

NOTES The fruits stick firmly to the fleece of sheep and to clothing.

SIMILAR SPECIES Another, more delicate, species, **Northern Wild Comfrey**, *Andersonglossum boreale* (Fernald) Jim. Mejías, J.I. Cohen & Naczi (synonym *Cynoglossum boreale* Fern.), has smaller, attractive blue flowers, in which the delicate style is hidden by the nutlets; the leaves are all below the middle of the stem.

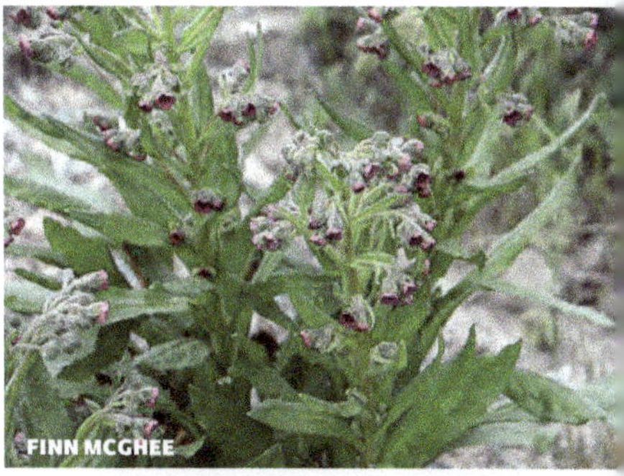

COMMON HOUND'S-TONGUE

NORTHERN WILD COMFREY

COMMON VIPER'S-BUGLOSS

HAIRY PUCCOON

HOARY PUCCOON

Common Viper's-Bugloss

Echium vulgare L.

Bristly-hairy, erect, stiff biennial becoming freely branched, up to 9 dm tall. **Flower buds** pink, the flowers bright blue to violet (sometimes whitish or pink), becoming reddish-purple to purplish-rose in age; flowers numerous, clustered in short, densely hairy, coiled spikes which straighten as the flowers open. Corolla funnel-shaped, unequally 5-lobed, 12–20 mm long; calyx teeth 5, covered with stiff bristles; stamens 5, borne on the corolla, unequal, the 2 longest extending beyond the corolla tube; pistil 1, the ovary superior, deeply 4-cleft, the style long, slender, extending beyond the corolla tube; nutlets 4, rough, dark brown. **Leaves** bristly, entire, oblong to linear-lanceolate, those on the stem sessile, the basal leaves long-petioled, up to 15 cm long.

HABITAT In poor soil in pastures and old fields, common in waste places in calcareous soil. Introduced from Europe; now a troublesome weed.

FLOWERING June to September.

Hairy Puccoon

Lithospermum caroliniense (J.F. Gmel.) MacMill.

SYNONYM *Lithospermum croceum* Fern.

Hairy, erect perennial up to 6 dm tall from deep, purple-staining roots. **Flowers** bright orange-yellow, 1.5–2 cm long, borne on short pedicels in compact, leafy 1-sided coiled racemes. Corolla tubular below with a flat spreading limb above, 1.5–2.5 cm wide, the tube bearded within at base; calyx lobes 9–11 mm long, strongly keeled. **Leaves** 33–45 below the inflorescence, alternate, sessile, crowded, linear-lanceolate, up to 5 cm long and usually 5–6 mm wide, entire, having long, conspicuous, rather stiff hairs.

HABITAT On sand dunes, hillsides, and open lakeshores, in dry, open oak woods, and in jack-pine plains.

FLOWERING late May to early August.

Hoary Puccoon

Lithospermum canescens (Michx.) Lehm.

Quite similar to the **Hairy Puccoon**, but smaller (up to 4 dm tall), hoary with dense soft down; flowers bright yellow, the limb of the corolla 1–1.5 cm wide; calyx lobes 3–6 mm long; leaves 8–30 below the inflorescence, the roots thick, red.

HABITAT On dry sandy or clayey hillsides, in dry open woods, along roadsides, and on prairies.

FLOWERING May and June.

USES American Indians used the small, hard nutlets for beads.

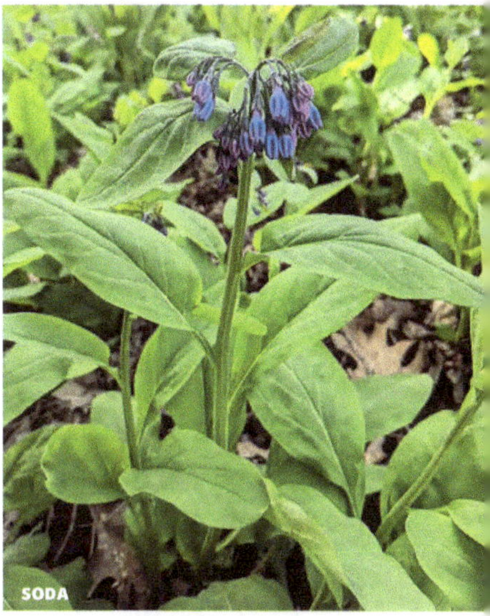

VIRGINIA BLUEBELLS

Virginia Bluebells

Mertensia virginica (L.) Pers. ex Link.

Smooth, often glaucous, fleshy, erect, simple or branched perennial up to 7 dm tall. **Flowers** showy, blue (becoming purple or pinkish-purple), or rarely white, borne in short racemes in a terminal panicle. Corolla trumpet-shaped (the cylindric tube enlarged and bell-shaped at the end), 18–25 mm long; calyx short, the 5 lobes pointed; stamens 5, borne on the corolla tube; pistil solitary, the ovary superior, 4-cleft; nutlets 4, ovoid, roughened. **Leaves** alternate, the upper ones oblong or lanceolate and sessile or clasping at the base, the lower ones oblong to obovate, narrowed to margined petioles.

HABITAT In rich woods and thickets, in low meadows and along streams. Sometimes seen outside of cultivation in Michigan and naturalized in woods.

FLOWERING April to June.

STATUS Threatened in Michigan.

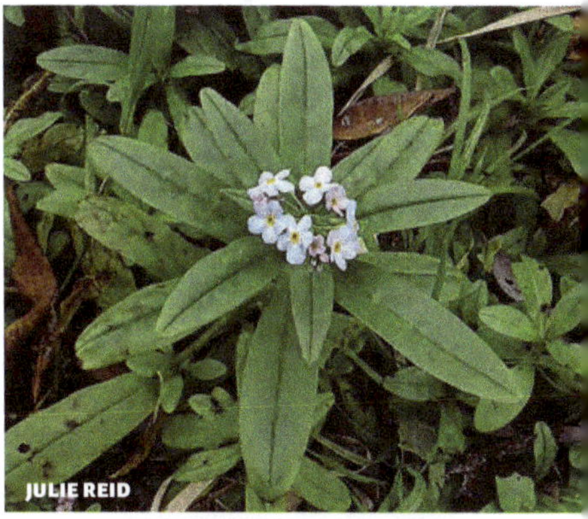

TRUE FORGET-ME-NOT

True Forget-Me-Not

Myosotis scorpioides L.

Slender, weak, appressed hairy perennial with angled stems which are erect at first, partially reclining later and rooting freely at the base. **Flowers** sky blue with yellow center, borne in coiled terminal racemes, the main branches lacking basal leaves. Corolla 6–9 mm wide, saucer-shaped with a short tube which is longer than the calyx, the spreading lobes broad, slightly notched at the tip, becoming white toward the center and each bearing a conspicuous yellow crest; calyx tubular, the 5 lobes hairy, shorter than the tube; stamens 5, attached to the corolla tube, not extending beyond it; pistil 1, the ovary deeply 4-lobed, the style over-topping the 4 nutlets. **Leaves** alternate, sessile, oblong to oblanceolate, narrowed to the base, 2.5–7 cm long.

HABITAT Along margins of brooks, in quiet water in marshes, or in very moist woods. Introduced from Europe.

FLOWERING May to September.

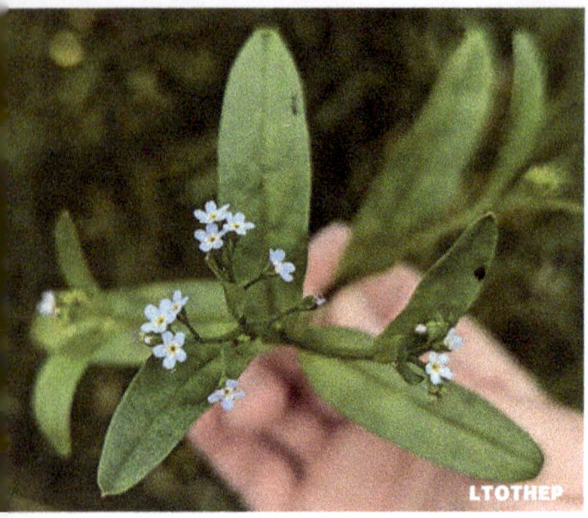

BAY FORGET-ME-NOT

Bay Forget-Me-Not

Myosotis laxa Lehm.

Similar to the preceding species, but the flowers smaller (3–6 mm broad) and often paler, the corolla tube about as long as the calyx, the style shorter than the nutlets and the calyx tube; the main branches of the inflorescence usually with 1 or 2 leaves near the base.

HABITAT In wet, open or partially shaded ground, tamarack swamps, shallow water.

FLOWERING May to September.

BRASSICACEAE *Mustard Family*

A large family of pungent or acrid, nonpoisonous plants comprising 348 genera and 4,060 species, distributed chiefly in the cooler regions of the Northern Hemisphere; 45 genera and a large number of species occur in Michigan. The family is of considerable importance for food crops and for ornamentals. Cabbage, cauliflower, broccoli, rutabaga, turnips, Brussels sprouts, radish, water cress, and kohlrabi, as well as mustard and horseradish, belong here. About 50 genera are cultivated for ornamentals, including Sweet Alyssum, Rock Cress, Stock, Rocket, Candytuft, Wallflower, and Honesty.

It is believed that many of the species have been weeds since prehistoric times, and that they spread over Europe and America as agriculture developed.

The members of this family are easily recognized by the flowers borne in bractless racemes, the 4 petals typically narrowed below into a claw, the 4 sepals, 6 stamens (2 shorter than the others), and the single pistil, usually with a 2-celled superior ovary. However, the classification of the species is difficult. Ripe capsules, as well as flowers, are necessary for the identification of most species.

KEY TO BRASSICACEAE(MUSTARD FAMILY) SPECIES

1 Flowers yellow 2

1 Flowers not yellow 5

  2 Pods slender and cylindric 3

  2 Pods not slender and cylindric 4

3 Stem leaves clasping or having small ears at base .... ***Barbarea vulgaris***

3 Stem leaves not as in alternate choice .... ***Sisymbrium altissimum*** or ***S. officinale***

  4 Pods plump, ellipsoid to nearly globose .... ***Rorippa palustris***

  4 Pods flattened, circular in outline .... ***Alyssum alyssoides***

5 Plants fleshy, growing on shores of the Great Lakes; pods plump, in 2 sections .... ***Cakile edentula***

5 Not as in alternate choice 6

  6 Plants aquatic, rooting readily at nodes; flowers white .... ***Nasturtium officinale***

  6 Not as in alternate choice 7

7 Leaves palmately parted into 3 or more leaflets or segments .... ***Cardamine diphylla*** or ***C. concatenata***

7 Leaves not as in alternate choice 8
  8 Flowering stem rising from a definite basal rosette 11
  8 Flowering stem not rising from a definite rosette 9
9 Lower and upper leaves different; marginal teeth, if present, blunt . . . . *Cardamine bulbosa* or *C. pensylvanica*
9 Lower and upper leaves similar 10
  10 Flowers small, white, petals deeply cut in two at apex . . . . *Berteroa incana*
  10 Flowers large, usually purple or lavender; petals entire at apex . . *Hesperis matronalis*
11 Pedicels and pods erect, close to stem at maturity . . . . *Turritis glabra*
11 Pedicels and pods not as in alternate choice 12
  12 Pods 7–11 cm long, hanging down . . . . *Borodinia canadensis*
  12 Pods 2–4.5 cm long, usually outspread at maturity . . . . *Arabidopsis lyrata*

Pale Alyssum

Alyssum alyssoides (L.) L.

Low annual, up to 2.5 dm, simple or branched at the base, the stem, leaves, and fruits with star-shaped hairs. **Flowers** pale yellow to nearly white, about 2 mm wide, the petals narrowly oblong, gradually narrowed to the base; **pods** borne on spreading pedicels, circular, flattened, each cell with 2 seeds. **Leaves** oblanceolate, 6–15 mm long, entire.

HABITAT In waste land. A common weed; introduced from Europe.

FLOWERING May and June.

SIMILAR SPECIES Golden-Tuft, *Aurinia saxatilis* (L.) Desv. (synonym *Alyssum saxatile* L.), a related but attractive plant, is a low matted perennial with bright golden-yellow flowers which sometimes spreads from cultivation. The common garden **Sweet Alyssum**, *Lobularia maritima* (L.) Desv., may persist for some time after escaping from cultivation. It has honey-scented white or purple flowers.

PALE ALYSSUM

Lyre-Leaved Rock Cress

Arabidopsis lyrata (L.) O'Kane & Al-Shehbaz
SYNONYM *Arabis lyrata* L.

Tufted, freely branching biennial or perennial up to 3.5 dm tall from a basal rosette, hairy below. **Flowers** white, about 5 mm, broad, borne on slender, erect to spreading pedicels, the raceme elongating in maturity; **pods** spreading or somewhat ascending, linear, slightly flattened, 1.3–3 cm long, the seeds in a single row. **Rosette leaves** spatulate to oblanceolate, 2.3–5 cm long,

LYRE-LEAVED ROCK CRESS

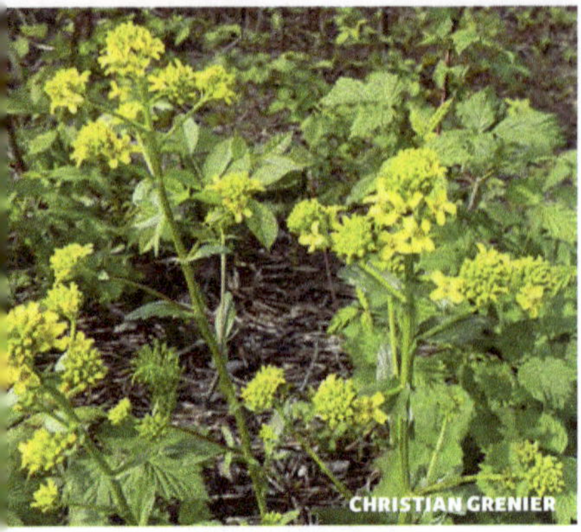

YELLOW ROCKET

HOARY ALYSSUM

SICKLEPOD

usually pinnately lobed with the terminal lobe the longest; **stem leaves** scattered, spatulate to linear, tapering at the base.

HABITAT In rocky places, gravels, and sands, on ledges or cliffs.

FLOWERING April to September.

Yellow Rocket

Barbarea vulgaris Ait. f.

Glabrous, many-stemmed biennial or perennial up to 8 dm tall from a basal rosette of dark-green leaves, unbranched below, freely branching above. **Flowers** in terminal racemes, bright yellow, 5–7 mm wide, the petals spreading above the sepals; pistil long and slender; pod 1.8–2.5 cm long, on a slender pedicel, beaked, with seeds in a single row in each cell. **Basal rosette leaves** pinnately cut, the terminal lobe rounded and much the largest; **stem leaves** alternate, clasping, the upper ones nearly round or ovate and coarsely toothed or lobed.

HABITAT In fields, cultivated ground, along roads and brooks, in woods and waste ground. Introduced from Europe.

FLOWERING April to August.

Hoary Alyssum

Berteroa incana (L.) DC.

Grayish-green, very hairy, stiffly erect annual up to 1 m tall, sparsely branched above. **Flowers** small, white, numerous, borne on ascending pedicels in slender, terminal racemes. Blade of the petals deeply cut; **pod** flattened, oblong. **Leaves** alternate, lanceolate, with entire or slightly wavy margin.

HABITAT In fields, along roads, and in waste places. Introduced from Europe; now a common weed.

FLOWERING June to September.

Sicklepod

Borodinia canadensis (L.) P.J. Alexander & Windham

SYNONYM *Arabis canadensis* L.

Simple or sparingly branched biennial up to 9 dm tall from a basal rosette, glabrous except for the sparsely hairy base.

Flowers cream-white, borne in long, lax racemes, the pedicels soon drooping. Petals slightly longer than the sepals; **pods** flat, curved, 7–10 cm long, hanging down, the seeds in 1 row, winged. **Rosette leaves** obovate to lanceolate, toothed or cut into backward-pointing segments, usually hairy on the midrib, soon wilting; **stem leaves** oblong-lanceolate to elliptic, tapering to the base, often finely toothed.

HABITAT In rich woods, thickets, or on rocky banks.

FLOWERING April to June.

SIMILAR SPECIES **Shepherd's-Purse**, *Capsella bursa-pastoris* (L.) Medic., is another white-flowered, common weed. It has triangular to inverted heart-shaped, flat pods. The upper leaves are sessile and arrow-shaped; those of the basal rosette are usually deeply pinnately cut. This widespread weed is reported to have been in Michigan as long ago as 1839. It is extremely variable; one author has described 63 forms.

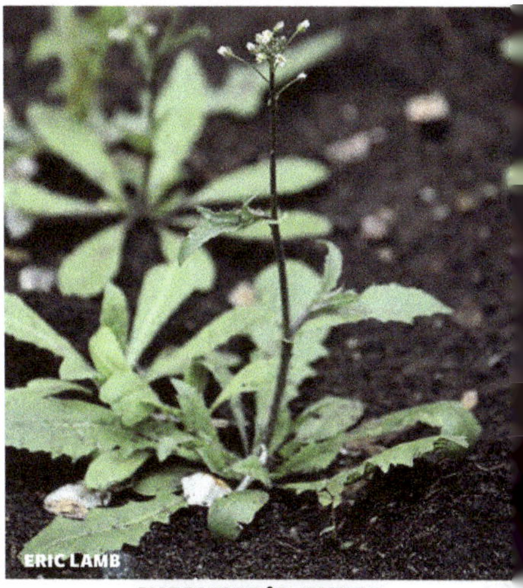
ERIC LAMB

SHEPHERD'S-PURSE

American Sea-Rocket

Cakile edentula (Bigelow) Hook.

Fleshy, spreading, somewhat reclining, glabrous, bushy perennial up to 3 dm tall. **Flowers** lavender, 5–7 mm wide, borne in compact terminal racemes; **pod** short, ovoid, 2-jointed, the upper joint the longer, flattened, and beaked; each joint 1-seeded or the lower seedless. **Leaves** alternate, sessile, fleshy, obscurely veined except for the midrib, oblanceolate to obovate, the margin wavy, toothed, or sometimes lobed, the lower leaves 7–12 cm long.

HABITAT On sandy or gravelly beaches, particularly on the shores of the Great Lakes.

FLOWERING July to September.

USES The leaves and young stems have the pungent flavor of horseradish. American Indians mixed the powdered roots with flour as an extender in times of food shortage.

JOHN D. REYNOLDS

AMERICAN SEA-ROCKET

Bulbous Cress

Cardamine bulbosa (Schreb. ex Muhl.) B.S.P.

Glabrous, simple or branched perennial from a short thick tuber, up to 5 dm tall at flowering time. **Flowers** white, on spreading pedicels in a loose raceme. Sepals greenish with

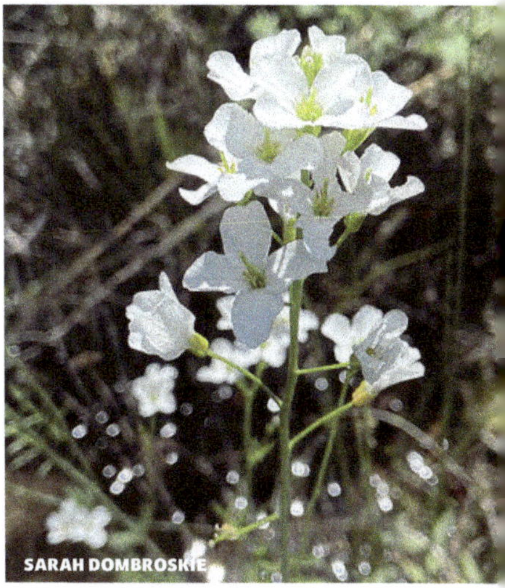
SARAH DOMBROSKIE

BULBOUS CRESS

white margins; **pods** slenderly cylindric, tapering to a pointed apex. **Stem leaves** 4–14, scattered, ovate, rounded, or lanceolate, entire or sparsely and coarsely toothed, the upper leaves sessile; **basal leaves** few, oblong or kidney-shaped, or nearly round, long-petioled.

HABITAT In low woods, moist ravines, around springs, and in wet meadows and shallow water.

FLOWERING April to May.

SIMILAR SPECIES *Cardamine douglassii* Britt. is quite similar to *C. bulbosa,* but the stems are generally hairy and the flowers are a pinkish-purple.

CUT-LEAF TOOTHWORT

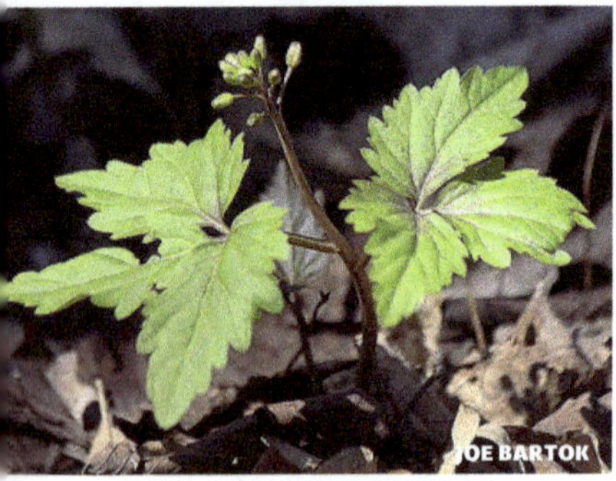

TWO-LEAF TOOTHWORT

Cut-Leaf Toothwort

Cardamine concatenata (Michx.) O. Schwarz

SYNONYM *Dentaria laciniata* Muhl.

Low perennial from deep-seated yellowish-brown, readily separable, scarcely toothed tubers. **Flowers** white or purplish. Petals 1–2 cm long, about twice as long as the sepals; **pods** lanceolate, long-pointed, 2.5–5 cm long, including the beak. **Leaves** 3 (rarely 2), in a whorl or slightly separated on the stem, 3-foliolate, the leaflets simple to deeply 5–9-cleft.

HABITAT In rich moist woods and on calcareous banks.

FLOWERING March to May.

Two-Leaf Toothwort

Cardamine diphylla (Michx.) Alph. Wood

SYNONYM *Dentaria diphylla* Michx.

Glabrous, stout, unbranched perennial up to 3 dm tall at flowering time, the rootstock long, continuous, prominently toothed, often branched, the annual segments not (or only slightly) tapering at the ends. **Flowers** white, borne in a loose few-flowered raceme. Sepals green, linear, pointed, about half as long as the petals; pistil long and thin; **pod** flat, lanceolate, the seeds in a single row, rarely maturing. **Stem leaves** 2, nearly opposite, above the middle of stem, 3-foliolate, 10–15 cm broad, the leaflets sessile, coarsely and bluntly toothed or lobed; **basal leaves** similar to stem leaves but the leaflets shorter and broader.

HABITAT In rich damp woods.

FLOWERING April to early June.

NOTES This species tends to grow in large masses and spreads rapidly from the rootstocks. The rootstocks are crisp and edible, with a mustard flavor.

Pennsylvania Bittercress

Cardamine pennsylvanica Muhl. ex Willd.

Variable, fibrous-rooted perennial up to 5 dm tall. **Flowers** white (rarely rose), 6–17 mm wide, the petals 3 times as long as the sepals. **Leaves** pinnately compound, the leaflets numerous, the margins entire or slightly toothed;

lower leaves with long petioles and rounded leaflets; upper leaves with short petioles and oblong or linear leaflets, the terminal leaflet is the largest.

HABITAT In wet meadows, lawns, shallow water, bogs, springs.

FLOWERING May to July.

USES This species is said to be edible and to be an excellent substitute for water cress.

PENNSYLVANIA BITTERCRESS

Dame's Rocket

Hesperis matronalis L.

Biennial or perennial with one to several hairy, erect stems 6 dm or more tall. **Flowers** lavender, rose-purple, or white (the white-flowered plants often the most numerous in a group), becoming fragrant in the evening, borne in loose terminal or axillary racemes. Claws of the petals longer than the sepals, the blade broad; sepals pale yellowish-green or purplish; **pod** cylindrical, slender, 5–14 cm long. **Leaves** numerous, opposite or nearly so (no rosette), lanceolate, sharply toothed, having a short petiole.

HABITAT In open woods and thickets and along roads and railroads. Introduced from Europe.

FLOWERING May to August.

NOTES This species was formerly very popular in the garden; it has escaped from cultivation and become naturalized in many places.

DAME'S ROCKET

Watercress

Nasturtium officinale Ait. f.

Aquatic, creeping or floating perennial, freely rooting at the nodes, often forming large floating mats. **Flowers** small, white, in racemes which are at first compact, soon elongating. Petals twice as long as the sepals, obovate; pistil cylindric, the style thick, the stigma slightly 2-lobed; **pods** 1–2.5 cm long, 2-celled, with 2 rows of seeds in each cell. **Leaves** somewhat fleshy, alternate, pinnately compound, the leaflets 3–11, roundish, entire or the margin wavy, the terminal leaflet the largest.

HABITAT In clear cold water and in swamps. Introduced from Europe; now widespread.

FLOWERING April to October.

WATERCRESS

USES This is the species that is grown as the water cress of commerce. It is used in salads and sandwiches.

BOG YELLOWCRESS

TALL TUMBLEMUSTARD

■ Bog Yellowcress

Rorippa palustris (L.) Bess.

SYNONYM *Rorippa islandica* auct. non (Oeder) Borbás

Extremely variable simple or branched annual or biennial up to 1.3 m tall. **Flowers** small (3–5 mm wide), yellow, borne on elongate threadlike pedicels, and having small nectar glands; sepals loosely spreading, longer than the petals; pedicels at maturity widely spreading; **pods** slenderly ellipsoid to nearly globose, 2–10 mm long, 1–4 mm thick, straight or curving; seeds minute. **Leaves** variable, lanceolate to oblong-ovate in general outline, pinnately divided or cleft, or merely toothed.

HABITAT On wet shores, in damp openings and waste places.

FLOWERING May to September.

■ Tall Tumblemustard

Sisymbrium altissimum L.

Coarse, erect, freely and loosely branching annual or biennial up to 1.7 m tall. **Flowers** pale yellow, 7–12 mm wide, borne in loose racemes that become greatly elongated after flowering begins; sepals straw-colored; **pod** slender, about the same thickness as the pedicel, 4.5–10.5 cm long. **Leaves** petioled, deeply pinnately cut; segments of the upper leaves threadlike, those of the lower leaves broader, toothed.

HABITAT In light soil in cultivated and waste land. Introduced from Europe.

FLOWERING May to August.

NOTES The stems frequently break off near the ground at the end of the growing season, and the plants are blown considerable distances, scattering seeds as they go.

■ Hedge-Mustard

Sisymbrium officinale (L.) Scop.

Stiffly branching annual up to 8 dm tall, hairy at least below. **Flowers** pale yellow, about 3 mm long, borne in spikelike racemes, the pedicels short, closely appressed to the stem, 2–3 mm long at maturity, thickened (as wide as the pod at the summit); **pod** 10–15 mm long, closely appressed to the stem. **Lower leaves** pinnately cut, the segments oblong to ovate; **upper leaves** sessile, arrow-shaped to lanceolate and entire.

HABITAT In waste places. A common weed; introduced from Europe.

FLOWERING May to October.

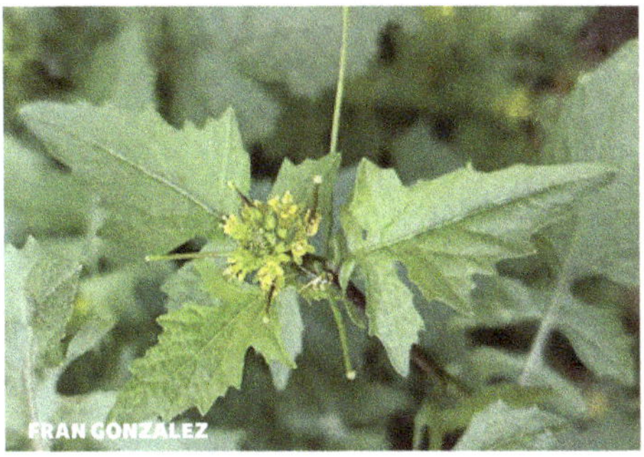

HEDGE-MUSTARD

Tower-Mustard

Turritis glabra L.

SYNONYM *Arabis glabra* (L.) Bernh.

Stoutish, erect biennial up to 1.2 m tall, hairy at the base, smooth above. **Flowers** creamy or yellowish, borne in a slender straight raceme; pods slender, erect or nearly so, 4–9.5 cm long, up to 1 mm thick. **Basal leaves** spatulate or oblanceolate, entire or irregularly toothed; **lower stem leaves** overlapping, the **upper leaves** less crowded, lanceolate or oblong, sessile or clasping.

HABITAT In dry soil.

FLOWERING May to June.

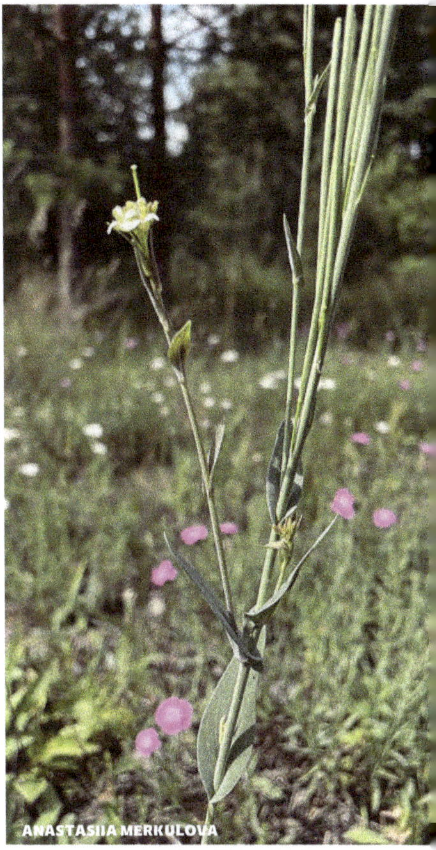

TOWER-MUSTARD

CACTACEAE *Cactus Family*

This primarily American family includes 150 genera and about 1,750 species; one genus occurs in Michigan. The members are important as ornamentals, many species being grown in houses and greenhouses and, in the Southwest, cultivated in the garden. Over 1,200 kinds are cultivated.

Eastern Prickly-Pear

Opuntia cespitosa Raf.

SYNONYM *Opuntia humifusa* Raf.

Prostrate perennial often forming mats, the stems composed of pale to deep green fleshy segments up to 2.5 dm long, the lower segments nearly round and bearing 4–9-mm. long brown leaves (which soon fall) and numerous persistent clusters of barbed bristles, many of the clusters including stiff spines up to 5 cm long. **Flowers** showy, yellow, sometimes with

EASTERN PRICKLY-PEAR

a red star-shaped eye-spot, opening in the sun for 2 days or more, borne along the margins of the newer segments. Petals 8–12; sepals numerous; stamens very numerous, inserted on the tube formed by the union of sepals and petals; ovary inferior; **fruit** green to dull purple, 2–5 cm long, pulpy, edible but bearing some bristles around the base.

HABITAT On sand dunes and open, sandy plains. Infrequent in Michigan.

FLOWERING June and July.

CAMPANULACEAE *Bluebell Family*

This family of 84 genera and nearly 2,400 species, largely temperate and subtropical, is important for its large number of ornamentals; 6 genera occur in Michigan. Over 100 species of bellflower and about 20 of *Lobelia* are cultivated, as are smaller numbers of several other genera.

The family is characterized by the milky juice (often scanty), alternate simple leaves, corolla of united petals, 5 stamens, the single pistil with an inferior ovary, and the fruit a capsule. Two distinct subfamilies are recognized; some authorities consider them two separate families.

KEY TO CAMPANULACEAE (BELLFLOWER FAMILY) SPECIES

1 Corolla regular; anthers separate (subfamily CAMPANULOIDEAE) 2

1 Corolla irregular; anthers united into a tube around the style (subfamily LOBELIODEAE) .... *Lobelia*, 4

2 Stem leaves narrowly ovate to heart-shaped ovate; corolla wheel-shaped .... *Campanulastrum americanum*

2 Stem leaves linear or nearly so; corolla bell-shaped 3

3 Erect, smooth plants; corolla about 1–2 cm long .... *Campanula rotundifolia*

3 Weak, usually reclining, rough-stemmed plants; corolla usually smaller .... *Palustricodon aparinoides*

4 Flowers bright red .... *Lobelia cardinalis*

4 Flowers blue, violet, or whitish.................... 5

5 Flowers 1.5–3.5 cm long, bright blue; corolla tube having openings along the side .... *Lobelia siphilitica*

5 Flowers to 1.6 cm long, light blue to whitish; corolla tube lacking openings on the side .6

6 Leaves linear or narrowly lanceolate; bracts on the pedicels at about the middle; lower lip of corolla glabrous within .... *Lobelia kalmii*

6 Some or all the leaves more than 1 cm wide; bracts near the base of the pedicels; lower lip of corolla hairy within .... *Lobelia spicata*

Harebell, Bluebell

Campanula rotundifolia L.

Extremely variable, erect, usually many-stemmed, smooth perennial up to 5 dm tall with scant milky juice. **Flowers** 1–15, blue, pale lavender, or whitish, somewhat nodding on slender pedicels from the axils of the upper leaves. Corolla bell-shaped with broad, spreading lobes; calyx tube united to the ovary, the lobes very slender; stamens 5, united only at the base; pistil 1, the ovary inferior, the style not extending beyond the corolla; **capsule** nodding, many-seeded, opening by pores at the base. **Leaves** numerous, alternate, sessile, linear to narrowly lanceolate, becoming much smaller upward; basal leaves (usually lacking at flowering time), broadly ovate to somewhat heart-shaped, long-petioled, usually toothed.

HABITAT In a variety of habitats: grassy fields, marshy flats, mossy banks, open woods, rock crevices, sandy shores. Common and widespread.

FLOWERING early June through September.

NOTES This is the well-known **Bluebell** of Scotland and northern England. It is extremely variable in almost all respects, but the variations are said to be readily produced by changes in environment and are of no particular taxonomic significance. In meadows the plants are tall and many-flowered, with thin, elongate stem leaves. In drier, more exposed places they are shorter, the foliage is firmer and thicker, and the upper leaves are smaller. In mountains or the far north the plant seldom exceeds 2 dm in height and may have only a single flower.

HAREBELL, BLUEBELL

Tall Bellflower

Campanulastrum americanum (L.) Small
SYNONYM *Campanula americana* L.

Erect, usually unbranched annual, up to 2 m tall. **Flowers** light blue, solitary or clustered in the axils of the upper leaves. Corolla wheel-shaped, 2–2.5 cm broad, the 5 lobes long and spreading; calyx 5-lobed; style long, declined, then curved upward, extending beyond the corolla; **capsule** opening by terminal pores. **Leaves** alternate, petioled, narrowly ovate to elliptic, coarsely toothed.

HABITAT In moist rich soil, usually in woods.

FLOWERING mid-July to mid-September.

TALL BELLFLOWER

Cardinal-Flower

Lobelia cardinalis L.

Coarse, stiffly erect, usually unbranched perennial up to 1–8 m tall, the juice acrid and milky, usually growing in patches from numerous short basal offshoots. **Flowers** very showy, deep rich red, 3–4 cm long, borne in a somewhat 1-sided terminal raceme 1–5 dm long, on pedicels which are shorter than the subtending bract. Corolla tube straight, split down the middle on the upper side and having openings along the sides, 2-lipped, the upper lip of 2 usually erect lobes, the lower lip spreading and 3-lobed; stamens 5, united into a tube which encloses the slender style and protrudes through the cleft corolla; ovary inferior, **capsule** 2-celled, many-seeded, opening at the top. **Leaves** alternate, the upper ones sessile, the lower ones petioled, oblong-lanceolate to lanceolate.

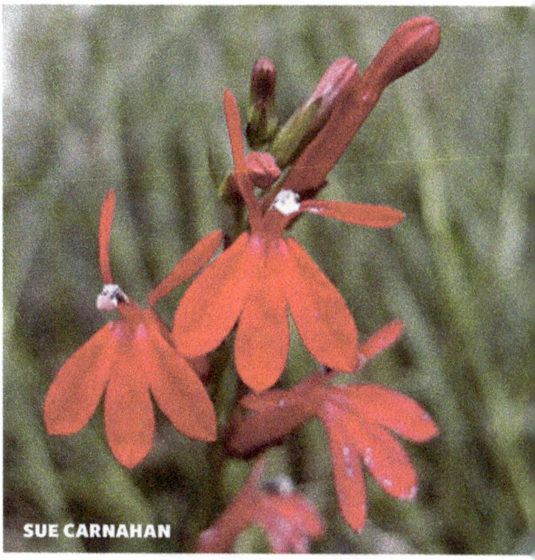

CARDINAL-FLOWER

BROOK LOBELIA, KALM'S LOBELIA

GREAT BLUE LOBELIA

PALE SPIKE LOBELIA

HABITAT In low wet places, in meadows, along streams, swamp borders, and shores.

FLOWERING July and August.

NOTES This beautiful species can be grown on wet soil in the garden, but it is said that it tends to die-out after flowering.

USES American Indians made a cough medicine and a love charm from it.

Brook Lobelia, Kalm's Lobelia

Lobelia kalmii L.

Slender, erect or reclining, simple to diffusely branched perennial up to 5 dm tall, with scant, acrid juice. **Flowers** violet to blue, with a conspicuous white eye-spot, borne on threadlike pedicels which have small bracts at about the middle. Flowers quite similar to those of the two preceding species but smaller (7–16 mm long) and lacking the openings along the sides of the corolla tube. **Leaves** sessile or the basal ones tapering to a petiole, linear to narrowly oblanceolate, entire or sparingly toothed, 1–4 cm long, less than 1 cm wide.

HABITAT In damp soil, usually in calcareous regions, in meadows, sedge mats, bogs, ditches, along shores.

FLOWERING July to September.

Great Blue Lobelia

Lobelia siphilitica L.

Smooth or sparsely hairy, coarse, erect, unbranched perennial up to 1.5 m tall, with milky juice, the stems ridged, leafy to the top, often with short offshoots at the base. **Flowers** purplish-blue, with some white on the lower lip or throat (rarely all white), 2.3–3.3 cm long, borne in dense terminal leafy racemes. Flower structure similar to that of the preceding species. **Leaves** lanceolate, sharply and irregularly toothed, becoming progressively smaller above.

HABITAT Along streams and shores, in swamps and low rich woods.

FLOWERING August to October.

Pale Spike Lobelia

Lobelia spicata Lam.

Quite similar to the preceding species, but often tall, usually unbranched, the flowers 9–12 mm

long, the pedicels with a pair of small bracts at or near the base; leaves ascending, obovate to oblanceolate, 5–10 cm long, some or all more than 1 cm wide.

HABITAT In rich moist or dry soil, in meadows, fields, and thickets, and around lakes; often a weed.

FLOWERING June to August.

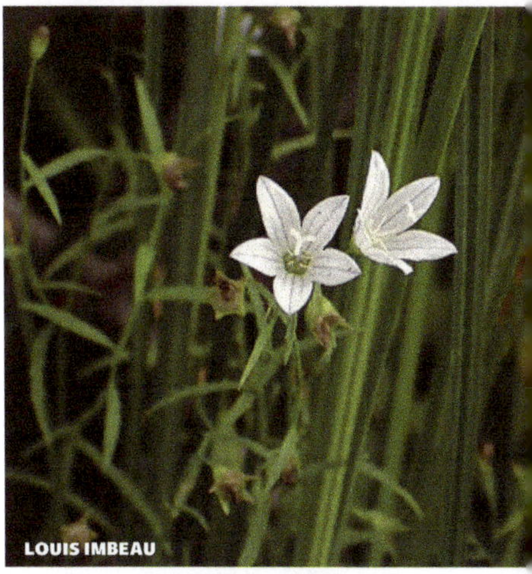

LOUIS IMBEAU

MARSH BLUEBELL

Marsh Bluebell

Palustricodon aparinoides (Pursh) Morin

SYNONYM *Campanula aparinoides* Pursh

Slender, weak, branching perennial, usually reclining on other plants, the stems rough, 3-angled, somewhat zigzag, 3–10 dm long, with scant milky juice. **Flowers** pale blue with darker lines, or pale lavender, or whitish and fading to lavender or bluish, solitary on slender peduncles. Corolla bell-shaped, 4–5 mm long; **capsule** subglobose, erect, opening near the base. **Leaves** alternate, linear, entire or slightly toothed, hairy on margin and midrib, the principal ones 2–5 cm long and not over 6 mm wide.

HABITAT In swales, wet meadows, boggy lake margins, open grassy marshes, marly flats on lakeshores.

FLOWERING late June into September.

CAPRIFOLIACEAE *Honeysuckle Family*

This family of 33 genera and about 860 species occurs mostly in the northern hemisphere; 11 genera occur in Michigan. The family now includes former members of Dipsacaceae and Valerianaceae. It includes many of our well-known ornamentals. These include Honeysuckle, Snowberry and Coralberry, Beauty-bush, Twinflower, and Weigelias. Former members Viburnum and Elderberry are now placed in the Viburnaceae (Viburnum Family).

The members of this family have opposite leaves without stipules. The corolla is 4–5-lobed, the calyx tube is attached to the inferior ovary, the stamens are borne on the corolla tube and do not exceed the lobes in number.

Wild Teasel

Dipsacus fullonum L.

SYNONYM *Dipsacus sylvestris* Huds.

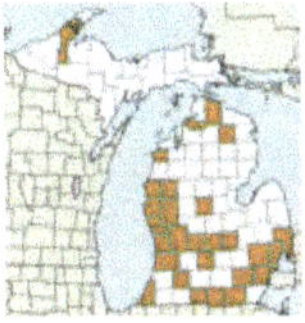

Stout, coarse biennial with ridged prickly stems up to 2 m tall. **Flowers** small, white, shorter than the long-pointed bracts, borne in large, dense, ovoid-ellipsoid heads surrounded by an involucre of linear prickly leaves. Corolla nearly regular, 4-lobed; calyx small; sta-

CHRISTINE YOUNG

WILD TEASEL

mens 4, distinct; pistil 1, the ovary inferior. **Leaves** opposite, sessile, and partly clasping, lance-oblong, often prickly on margin and on veins beneath.

HABITAT Along roadsides, in pastures, old fields, and waste places. Introduced from Europe; now a common weed.

FLOWERING July to October.

USES The stalks and heads are persistent and can be found for a considerable time after flowering. They are sometimes used in winter bouquets, either in their natural state or painted. Formerly grown commercially because the ripe inflorescences were used by textile mills for raising the nap on cloth.

TWINFLOWER

Twinflower

Linnaea borealis L.

Slender, creeping, somewhat woody evergreen perennial with long, prostrate stems which give rise to numerous, erect, leafy branches up to 10 cm tall. **Flowers** delicate, fragrant, nodding in pairs at the top of the erect, threadlike peduncles, white tinged with pink, rose, or rose-purple (sometimes almost entirely rose). Corolla bell-shaped, spreading, 5-lobed, 10–15 mm long, attached to the top of the ovary; calyx tube adherent to the densely hairy, glandular ovary; stamens 4, in 2 unequal pairs; ovary 1, inferior, the style long, slender, extending beyond the corolla tube; **capsule** 3-celled but one-seeded. **Leaves** opposite, hairy, short-petioled, nearly round, oval, or broadly elliptic, often somewhat 3-lobed or crenate at apex, sometimes entire.

HABITAT In cold, moist or dry woods and peaty bogs.

FLOWERING June to August, occasionally again late in the fall.

NOTES This circumpolar species is one of our most attractive plants and often occurs in large patches. It is quite common in some areas in the northern part of the state. It was named for the great Swedish botanist, Linnaeus, the father of our system of classifying plants.

TINKER'S-WEED

Tinker's-Weed

Triosteum perfoliatum L.

Coarse, hairy perennial up to 1.2 m tall, glandular above and leafy to the top. **Flowers** yellowish or

greenish to dull purple, sessile, erect, 3 or 4 in the leaf axils. Corolla tubular to bell-shaped, 5-lobed, scarcely longer than the calyx; calyx 5-lobed, the lobes linear-lanceolate; stamens 5, borne on the corolla tube and nearly equaling it; pistil 1; **fruit** a dry, dull orange-yellow, globose drupe superficially resembling a small, hard tomato, 3-seeded, the calyx tube persistent. **Leaves** opposite, sessile or united at the base, downy, soft to the touch beneath, oblong-ovate to ovate, the middle ones violin-shaped.

HABITAT In thin or rocky soil in or at the edges of woods and thickets.

FLOWERING mid-May to mid-June.

Swamp Valerian

Valeriana uliginosa (Torr. & Gray) Rydb.

Smooth, coarse, many-stemmed perennial 4–12 dm tall from a basal rosette. **Flowers** white, small (about 5 mm wide) in clustered headlike terminal cymes which are very compact at first. Corolla tubular, with a small swelling on one side at the base, the limb regular and 5-lobed; calyx tube adherent to the ovary, the limb composed of about 10 elongate plumose bristles which are inrolled at flowering time but which unroll and spread as the fruit matures; stamens 3, borne on the corolla tube; pistil 1, the ovary inferior; **fruit** an achene. **Basal leaves** simple or cleft (if pinnately cut, the terminal leaflet the largest); **stem leaves** opposite, smooth, pinnately cut, with 3–11 divisions or leaflets.

HABITAT In wet woods and meadows, in bogs and calcareous swamps; often associated with tamarack and arbor vitae.

FLOWERING late May to July.

SIMILAR SPECIES The familiar **Garden Heliotrope**, *Valeriana officinalis* L., which often persists as an escape, is quite similar, but the flowers are pinkish, all of the leaves are pinnate, the divisions being nearly equal, and at least the lower leaves are hairy beneath.

USES The thick, strong-scented roots of some species are used as antispasmodics.

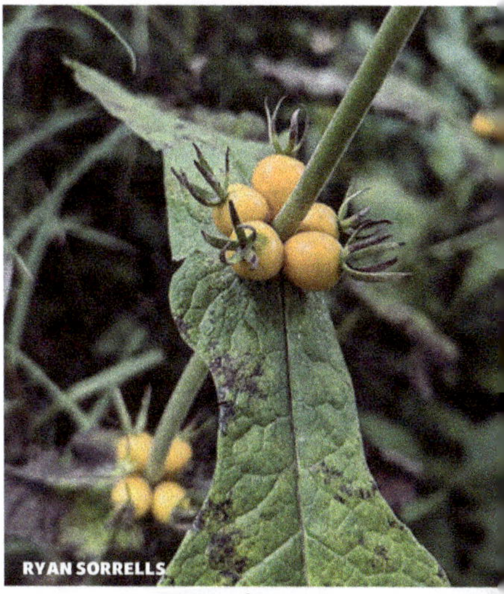

TINKER'S-WEED

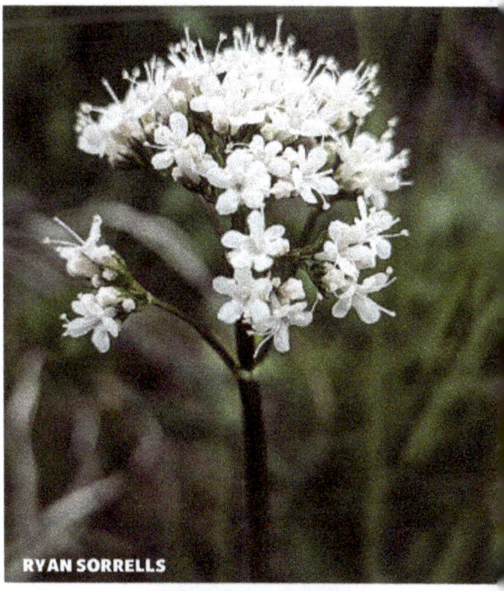

SWAMP VALERIAN

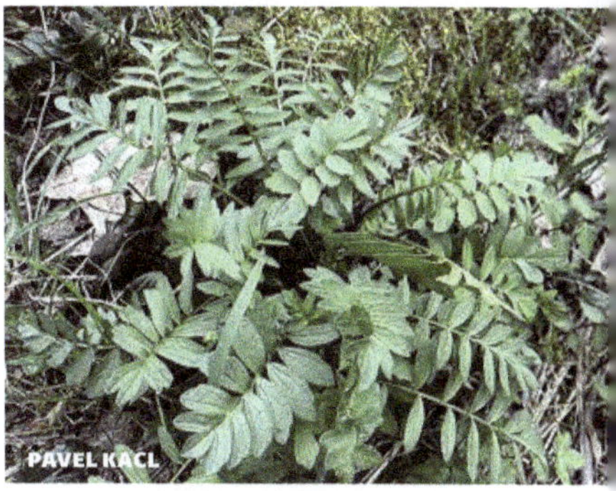

GARDEN HELIOTROPE

CARYOPHYLLACEAE *Pink Family*

This primarily north-temperate family includes 105 genera and 2,625 species; 20 genera occur in Michigan. Many ornamentals belong to the family, the florist's carnation (*Dianthus caryophyllus*) being perhaps the best-known. Baby's-breath (*Gypsophila*), Catchfly (*Silene*), Maltese Cross (*Lychnis*), Sweet William, Bouncing-Bet, and pinks are also familiar flowers. Several well-known weeds—Chickweed, Corn Cockle, Sleepy Catchfly, Bladder Campion—likewise belong to this family.

The family is characterized by opposite, entire, simple, mostly narrow leaves, often united at the base; usually swollen nodes; symmetrical flowers with 4–5 sepals and 4–5 usually notched petals; stamens the same as or twice the petals in number; and a single pistil with superior ovary. Although the leaves of many species of this family have parallel veins, the 4–5 sepals and petals readily distinguish the plants from the monocots.

CARYOPHYLLACEAE

1 Sepals nearly or entirely separate to base; petals not greatly narrowed at base 2
1 Sepals united into a tube or cuplike structure; petals having a long, narrow base 4
  2 Petals entire or nearly so, not notched at apex; leaves needlelike .. *Sabulina michauxii*
  2 Petals notched or deeply cut at apex 3
3 Styles usually 3; capsule ovoid to oblong .... *Stellaria media*
3 Styles usually 5; capsule cylindric .... *Cerastium arvense*
  4 Styles 5 5
  4 Styles 2 or 3 6
5 Lobes of calyx much longer than tube; styles opposite petals ...... *Agrostemma githago*
5 Lobes of calyx shorter than tube; styles alternate with petals........ *Silene latifolia* or *S. coronaria*
  6 Flowers deep pink to red; calyx having 1–3 pairs of bracts at base ... *Dianthus armeria*
  6 Flowers pale or white; calyx lacking basal bracts........ 7
7 Flowers numerous, large, in large clusters compact at first; plants stout; styles usually 2 .... *Saponaria officinalis*
7 Flowers in loose, spreading clusters; styles usually 3 .... *Silene*, 8
  8 Leaves at middle of stem opposite 9
  8 Leaves at middle of stem in 4's .... *Silene stellata*
9 Petals conspicuous; calyx inflated .... *Silene vulgaris*
9 Petals inconspicuous or lacking; calyx tight over capsule in fruit .... *Silene antirrhina*

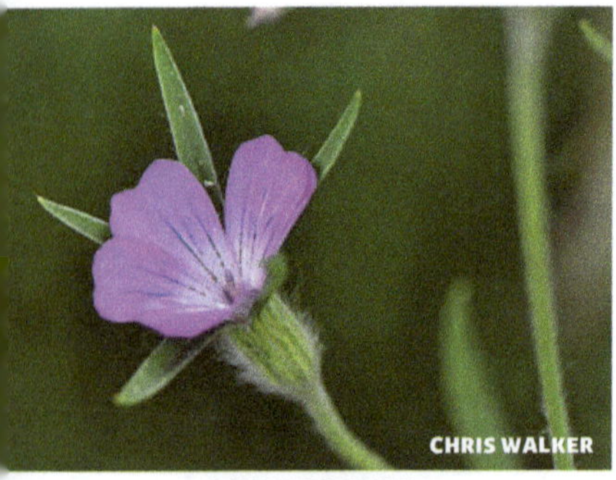

CORN COCKLE

Corn Cockle

Agrostemma githago L.

Tall, silky annual or biennial often 1 m or more tall. **Flowers** purplish-red, spotted with black, 2.5–4 cm wide. Petals 5, narrowed to a claw at the base, borne on the stem of the ovary; calyx ovoid and lobed at the apex, the lobes leaflike and longer than the petals; stamens 10, attached to the stem of the ovary; pistil 1, the styles 5, opposite the petals; **capsule** 1-celled. **Leaves** opposite, sessile, linear, silky.

HABITAT In grain fields, along roadsides, and in waste places. Introduced from Europe; now a common weed.

FLOWERING June to September.

Field Chickweed

Cerastium arvense L.

Matted or tufted, grayish-green, often glandular perennial up to 4 dm tall, the tough basal branches bearing withering but persistent leaves and many conspicuous axillary leafy tufts. **Flowers** white, few to many, borne on simple or freely branching ascending stems. Petals 5, 2-lobed, 2–3 times as long as the sepals; sepals 5, separate; stamens 10; styles usually 5, opposite the petals; **fruit** a cylindric capsule equaling or exceeding the sepals in length, opening by 10 teeth. **Leaves** mostly on the lower 2/3 of the stem, sessile or nearly so, linear or narrowly ovate, 0.5–6.5 cm long.

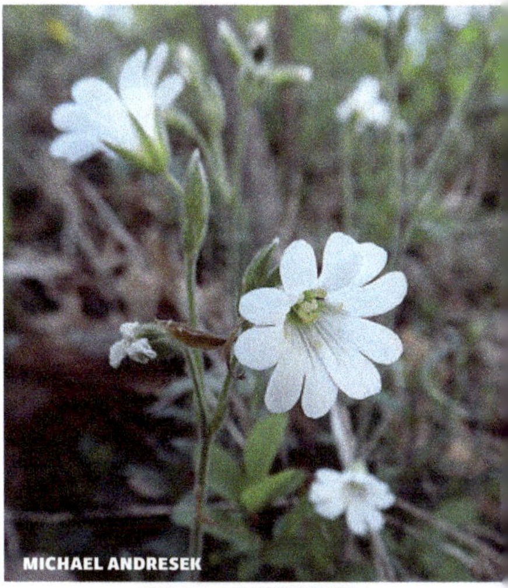
MICHAEL ANDRESEK

FIELD CHICKWEED

HABITAT In gravelly, rocky, or sandy areas, abandoned fields, meadows, and grasslands.

FLOWERING April to August.

SIMILAR SPECIES Another common Chickweed, ***Cerastium glomeratum*** Thuill., is a short-lived, introduced perennial which has leafy basal offshoots and few or no axillary tufts; the hairy leaves are oblong; the petals are narrow, deeply cleft, and about the same length as the sepals, or slightly shorter.

Deptford Pink

Dianthus armeria L.

Stiff, simple or sparsely branched, somewhat hairy annual or biennial up to 8 dm tall. **Flowers** few, showy, red or roseate with white dots, borne in flattish terminal clusters which are subtended by strongly ribbed, narrow bracts. Petals 5, toothed on margin, attached to the stem of the ovary; calyx long, cylindrical, 5-toothed, with narrow basal bracts which are about as long as the tube; stamens 10, pistil 1; styles 2; **capsule** opening at the apex by 4 teeth, the seeds flattish. **Stem leaves** 5–10 pairs, opposite, linear to linear-lanceolate, entire, 3–8 cm long, 2–8 mm, wide; **basal leaves** numerous.

HABITAT In dry fields, along roads, and on dry sandy shores. Introduced from Europe.

FLOWERING May to July.

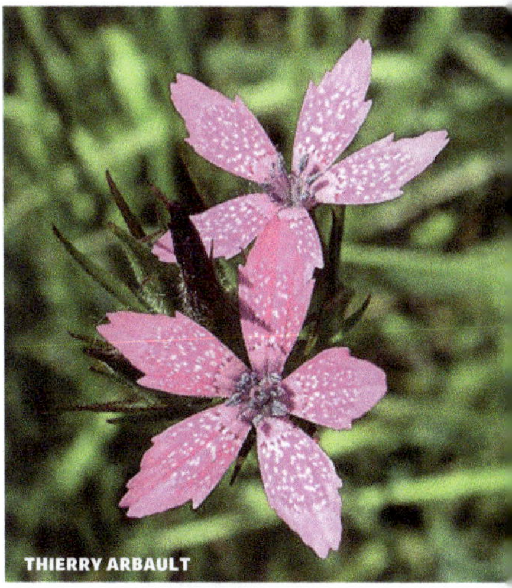
THIERRY ARBAULT

DEPTFORD PINK

ELAINE WOLSHOCK

ROCK SANDWORT

THIERRY ARBAULT

BOUNCING BET, SOAPWORT

RACHEL STRINGHAM

SLEEPY CATCHFLY

Rock Sandwort

Sabulina michauxii (Fenzl) Dillenb. & Kadereit
SYNONYM *Arenaria stricta* Michx.

Low, spreading, loosely tufted perennial 15–25 cm tall, the stems numerous, very leafy in the lower half, nearly naked above. **Flowers** white, about 8 mm wide, borne on slender pedicels in loose, terminal or axillary, 3–30-flowered cymes. Stems, pedicels, flowers, and fruit drying and persisting on the plant, giving it a somewhat ragged appearance. Petals 5, obovate, attached with the stamens to a basal disk; sepals 5, 3-ribbed, about half as long as the petals; stamens 10, white; styles 3; **capsule** splitting into 3 segments, the seeds kidney-shaped, nearly black. **Leaves** opposite, rigid, linear or needlelike, at least the lower ones usually with a cluster of smaller leaves in the axil.

HABITAT In wet sand, along bog margins, on sand dunes; frequent in the jackpine plains.

FLOWERING late June through July.

Bouncing Bet, Soapwort

Saponaria officinalis L.

Nearly glabrous, sparingly branched perennial 3–6 dm tall, spreading rapidly by branching, underground runners to form dense colonies. **Flowers** white or pinkish to pinkish lavender, frequently double, about 2.5 cm wide, borne in dense terminal corymbs. Petals 5, obcordate, shallowly notched at the apex and extending into a long narrow claw; calyx long, tubular, 5-toothed at the apex; stamens 20; pistil 1, the styles 2, long and threadlike; **pod** shorter than the calyx. **Leaves** sessile, opposite, narrowly lanceolate to ovate, strongly 3-veined.

HABITAT Along roadsides and railroads, on sandy open ground, and in waste places. Introduced from Europe; now a common weed.

FLOWERING late July and August (sometimes a few flowers as late as October).

USES This species has long been known for its therapeutic properties. The ancient Greeks and Romans cultivated it for medicine and as a soap. The crushed leaves and stems make a soapy lather.

Sleepy Catchfly

Silene antirrhina L.

Erect or ascending annual or biennial up to 8 dm tall, usually with dark glutinous spots on the stem. **Flowers** pink or purplish (sometimes partly white), borne on slender pedicels in loose spreading panicles. Petals small, soon falling, sometimes lacking; calyx spindle-shaped, not inflated, 5-toothed, the teeth often purple; stamens 10; styles 3; capsule nearly sessile, closely covered by the calyx. **Leaves** opposite, sessile, lanceolate, oblanceolate, or linear.

HABITAT In dry or rocky open woods, fields, waste places, in sandy soil, and jackpine plains.

FLOWERING May to September.

Mullein-Pink, Rose Campion

Silene coronaria (L.) Clairv.

SYNONYM *Lychnis coronaria* (L.) Desr.

Erect, densely white-woolly perennial up to 1 m tall, the stems conspicuously swollen at the nodes. **Flowers** few, showy, purplish-red, about 3.5 cm wide, borne on stiffly erect peduncles. Petals 5, purplish-red, becoming white at the base; calyx with prominent ribs and twisted, linear, pointed teeth, becoming inflated. **Leaves** densely hairy, almost velvety in appearance, thick, the veins obscure except for the midrib, the stem leaves opposite, oblong to ovate, sessile; rosette leaves spatulate, narrowed to winged petioles.

CALEB CATTO

MULLEIN-PINK, ROSE CAMPION

HABITAT Along roads, in fields, rocky woods, clearings. Introduced from Europe and a garden escape, now thoroughly naturalized throughout the United States.

FLOWERING June to August.

White Campion

Silene latifolia Poir.

SYNONYM *Lychnis alba* Mill.

Erect or ascending, glandular and sticky, unisexual biennial or short-lived perennial with 4–8 forking stems up to 12 dm tall from a basal rosette. **Flowers** white, 2–2.5 cm wide, borne in few-flowered cymes, opening in the evening. Petals 5, heart-shaped above and deeply cut; calyx tubular, inflated, ellipsoid in the staminate flowers, ovoid in the pistillate, 5-toothed; stamens 10, in 2 whorls, the outer whorl attached to the petals, the inner stamens more or less united at the base; **capsules** narrowed at the top; styles 5, alternate with the petals. **Stem leaves** opposite, sessile, lance-oblong, entire, hairy; the **basal leaves** petioled.

FRANCK CABOT

WHITE CAMPION

HABITAT In waste places, along roads, and in fields. Introduced from Europe.

FLOWERING late May to September.

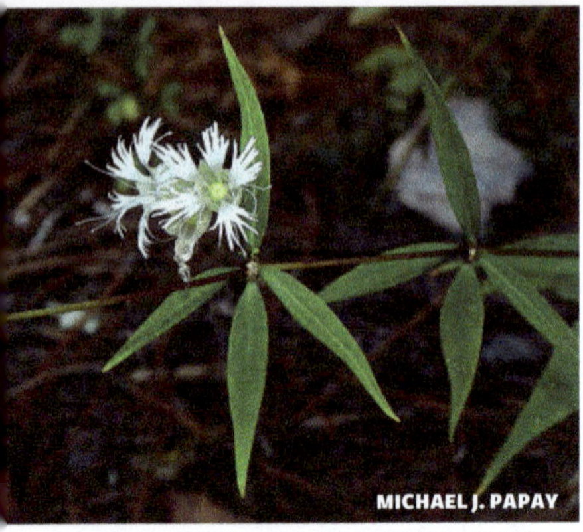

STARRY CAMPION, WIDOW'S-FRILL

BLADDER CAMPION

COMMON CHICKWEED

NOTES This persistent weed of field crops is a frequent contaminant of clover and forage-grass seed. It is easily confused with **Night-flowering Catchfly**, *Silene noctiflora* L., which has smaller perfect flowers with narrower corolla lobes and only 3 styles.

Starry Campion, Widow's-Frill

Silene stellata (L.) Ait. f.

Perennial with several stiff unbranched stems up to 8 dm tall. **Flowers** white, borne in a loose panicle, the petals fringed; calyx bell-shaped, somewhat inflated. **Leaves** mostly in whorls of 4, linear-lanceolate to ovate-lanceolate.

HABITAT Uncommon in southern Michigan in woods and clearings.

FLOWERING July to September.

STATUS Threatened in Michigan.

Bladder Campion

Silene vulgaris (Moench) Garcke

SYNONYM *Silene cucubalus* Wibel

Erect or partly decumbent, whitish, glabrous perennial up to 4.5 dm tall, the stems branching at the base. **Flowers** white or occasionally pinkish, about 2 cm long, borne on long peduncles in loose, spreading panicles. Petals 5, attached with the stamens to a disk at the base of the ovary, narrowing to a slender claw at the base, the blade broad and cleft at the apex; calyx greatly inflated, 5-toothed, whitish with 20 greenish or light brown veins connected by a network of veinlets; stamens 10, borne on the disk; pistil 1, the ovary stalked, the styles 3; **capsule** 1-celled, opening at the apex, stalked, covered by the greatly inflated calyx. **Leaves** opposite, sessile, lance-ovate to narrowly oblong.

HABITAT Along roads, on gravelly shores, borders of fields, in meadows, and waste places. Introduced from Europe; now a common weed,

FLOWERING April through August.

Common Chickweed

Stellaria media (L.) Vill.

Small, densely matted annual or perennial with weak stems and threadlike branches. **Flowers** small, white, star-shaped. Petals

4–5, deeply cleft, shorter than the calyx, sometimes lacking; sepals 4–5, distinct; stamens 3–10; styles usually 3. **Leaves** ovate with rounded base; upper leaves sessile; lower leaves petioled, often fringed with hairs toward the base or on the petioles.

HABITAT In lawns, along roadsides, and in moist waste ground, often forming continuous mats under shrubbery. Introduced from Europe.

FLOWERING over a long period.

NOTES This widespread weed is difficult to eradicate. It is said that its seeds, if buried deeply, will retain viability for 30 years. Birds eat the capsules, seeds, and leaves. American Indians used the plant in making medicine for sore eyes.

CELASTRACEAE *Bittersweet Family*

This mostly tropical family includes 98 genera and 1,350 species; 3 genera, 2 of which are shrubs, occur in Michigan. The herb *Parnassia* ranges into alpine and arctic regions.

Fen Grass-of-Parnassus

Parnassia glauca Raf.

Smooth perennial with flowering stems (usually leafless but sometimes with a single sessile leaf) up to 6 dm tall. **Flowers** white or whitish, solitary, upright, terminal, 2.2–3.2 cm wide. Petals 5, separate, spreading out nearly flat, white, strongly veined with greenish or yellowish, about 3 times as long as the sepals, each petal with a staminode at base; staminode deeply cleft into 3 gland-tipped, threadlike segments shorter than (or about as long as) the stamens; sepals 5, blunt, united very slightly at the base, becoming reflexed; stamens 5, alternate with the petals; pistil 1, the ovary green, ovoid, 1-celled; stigmas 4, sessile; **capsule** many-seeded. **Basal leaves** parallel-veined, ovate, to nearly round, rounded or heart-shaped at base, occasionally winged along the petiole, rather thick, 2.5–5 cm long, long-petioled.

HABITAT In low wet places, swamps, meadows, along roadsides and in ditches; in calcareous soils.

FLOWERING July to October.

FEN GRASS-OF-PARNASSUS

SMALL-FLOWERED GRASS-OF-PARNASSUS

Small-Flowered Grass-of-Parnassus

Parnassia parviflora DC.

Generally smaller plants than the above, the 1–25 flowering stems up to 3.5 dm tall. **Flowers** white, 10–15 mm broad, solitary at the top of the slender stems; sometimes one small, sessile clasping leaf at or below the middle. The staminode shorter than the stamens, cleft into 5–9 (or even more) threadlike segments, each tipped with a globular, clear-yellow gland; sepals very slightly united at base, spreading in flower, ascending and persisting in fruit. **Basal leaves** oval or oblong, narrowed at base, the blades 5 mm–3.5 cm long, petioled.

HABITAT In wet calcareous soil. Frequent in roadside ditches along with *Lobelia kalmii*.

FLOWERING July and August.

CISTACEAE *Rockrose Family*

This small family of 8 genera and 170–200 species is especially abundant in the Mediterranean region; 3 genera occur in Michigan.

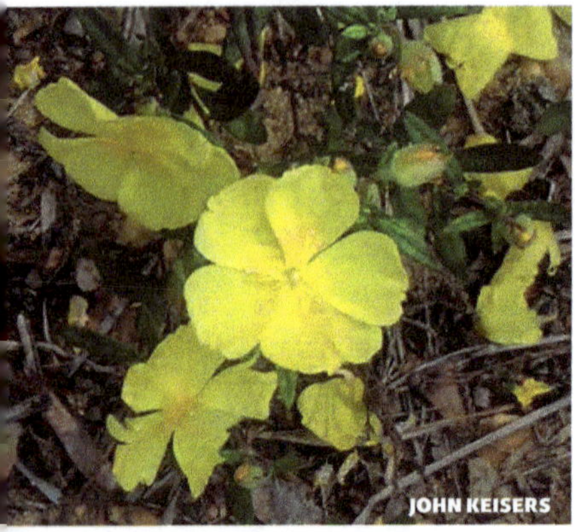

CANADA FROSTWEED

Canada Frostweed

Crocanthemum canadense (L.) Britton

SYNONYM *Helianthemum canadense* (L.) Michx.

Erect perennial 2–5 dm tall, unbranched at first but branching after the first flowers bloom. **Flowers** of 2 kinds; petaled flowers showy, yellow, 2.5–3 cm wide, solitary at the top of the stem, opening in sunshine; petals 5, broadly obovate, soon withering and falling; sepals 5, the 2 outer ones much the smaller; stamens numerous; pistil solitary, the ovary superior, ovoid, the style short; capsule 1-celled, many-seeded; self-pollinating, non-opening flowers small, numerous, lacking petals, producing larger capsules than the petaled flowers. **Leaves** scattered, elliptic, tapering evenly to both ends, 1–3.5 cm long, green above, whitish and hairy beneath.

HABITAT In dry, open, sandy woods, clearings, and barrens.

FLOWERING mid-May to mid-July.

NOTES It is said that the name Frostweed was applied because crystals of ice shoot from the cracked bark at the base of the plants in late autumn.

CONVOLVULACEAE *Morning-Glory Family*

This primarily tropical family includes 57 genera and about 1,650 species; Michigan has 4 genera. Its importance comes principally from the sweet potato and the common morning-glory. The troublesome parasitic dodder (*Cuscuta*) also belongs here. The Michigan leafy, green members are characterized by their twining or trailing habit and the milky juice. The large funnel-shaped flowers are borne in the axils of the leaves. (Dodder is brownish or yellowish, the leaves are reduced to scales, and the flowers are small and clustered.)

CALYSTEGIA, CONVOLVULUS

1 Plants twining or trailing, stems forking freely 2

1 Plants partially erect, twining only at the tip; stem unbranched or sparsely branched .... *Calystegia spithamaea*

2 Corolla 4–8 cm long; calyx covered by 2 large, leaflike bracts .... *Calystegia sepium*

2 Corolla up to 3 cm long; bracts small, well below the calyx ...... *Convolvulus arvensis*

Hedge Bindweed

Calystegia sepium (L.) R. Br.

SYNONYM *Convolvulus sepium* L.

Freely branching, smooth to hairy, creeping or twining vine up to 3 m long, often climbing on other plants, fences, etc. **Flowers** showy, white to roseate, borne on peduncles in the axils of the leaves. Corolla 4–8 cm long and nearly as broad; calyx subtended by and covered by 2 broad, ovate to heart-shaped, leaflike bracts, 1.5–3.5 cm long, the sepals unequal. **Leaves** heart-shaped to triangular and arrow-shaped at base, 5–10 cm long.

HABITAT In low or damp ground in the open or in partial shade; often along roads or railways.

FLOWERING mid-May to August.

NOTES The plant spreads rapidly and is not desirable in the garden.

HEDGE BINDWEED

Low False Bindweed

Calystegia spithamaea (L.) Purssh

SYNONYM *Convolvulus spithamaeus* L.

Hairy, grayish-green, partially erect or sometimes prostrate, simple or sparsely branched perennial with stems up to 5 dm long, the tip twining or not, often elongating and eventually bending downward. **Flowers** showy, white, sometimes pink, 1–4 on peduncles from the axils of the lower leaves. Corolla funnel-shaped, 4–7 cm long; calyx completely covered by the 2 large leaflike bracts; stamens 5, borne on the base of the corolla tube; pistil 1, the ovary superior, 2-celled, the style long, 2-cleft at the apex; **fruit** a globose capsule. **Leaves** oblong-oval, usually somewhat heart-shaped at base, hairy, prominently veined; lower leaves small.

HABITAT In poor, sandy or rocky open soil and in thin woods, often under aspens and in old burns.

FLOWERING late May through July

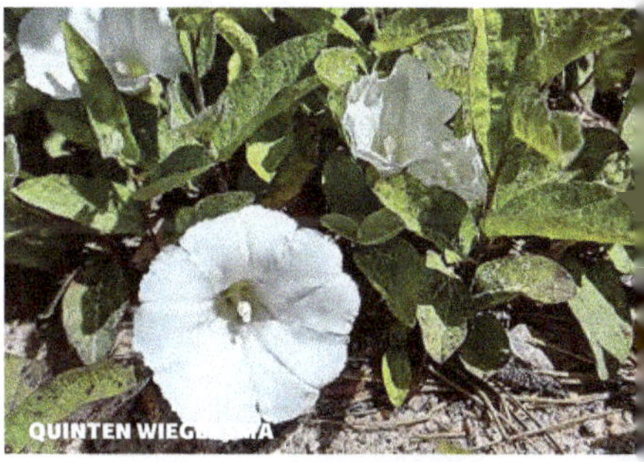

LOW FALSE BINDWEED

Field Bindweed

Convolvulus arvensis L.

Freely branching, trailing or climbing perennial, the stems up to 1 m long, often forming tangled masses. **Flowers** showy, white, often tinged with red outside. Corolla 15–20 mm long; bracts of the peduncle small, pointed, borne below the calyx. **Leaves** typically ovate with arrow-shaped or hastate base but quite variable and sometimes linear, oblong, or ovate with a heart-shaped base.

FIELD BINDWEED

HABITAT In fields, lawns, along roadsides, and in waste places. Introduced from Europe; now a common weed.

FLOWERING June to September.

NOTES This species is persistent and hard to eradicate. Its deep, spreading roots may penetrate to a depth of more than 20 feet. If proper control measures are not taken this weed may so overrun a farm that productivity is seriously impaired.

CORNACEAE *Dogwood Family*

This widely distributed family includes 2 genera and 85 species of tropical and temperate plants. A single genus, *Cornus*, is present in North America.

Flowering Dogwood, a tree, and several shrubby species of *Cornus* are widely grown as ornamentals. Most members of this family are woody, but a few species are herbaceous.

BUNCHBERRY, DWARF CORNEL

Bunchberry, Dwarf Cornel

Cornus canadensis L.

Low, usually colonial perennial, the short, erect flowering stems up to 2.5 dm tall from horizontal, woody, forking rootstocks. **Flowers** small, numerous, creamy, greenish, or yellow, in a single terminal cluster subtended by 4 large, white, petal-like bracts, the whole structure resembling a single 4-petaled flower. Petals 4, borne on the margin of the disk, spreading; calyx tubular, with 4 minute teeth; pistil 1, the ovary inferior, 2-celled; **fruit** a bright, shining, red drupe. **Leaves** 2–3 opposite pairs so closely crowded near the top of the stem that they appear whorled, broadly ovate, strongly pinnately veined, entire, 4–7 cm long; 1–2 opposite pairs of small bracts may be present below on stem.

HABITAT In moist woods, thickets, and clearings. Mostly in northern part of Michigan.

FLOWERING May to July; fruits ripe late July to October.

NOTES The fruits are said to be edible and are eaten by birds. This plant is as pretty in fruit as in flower, but is, unfortunately, difficult to grow in the garden.

CRASSULACEAE *Stonecrop Family*

A family of succulent herbs with 38 genera and approximately 1,400 species distributed over most of the world; 5 genera are known from Michigan. A number of species, such as Hen-and-chickens (*Sempervivum*), Kalanchoe, Life-plant (*Bryophyllum*), and various sedums, are grown as ornamentals.

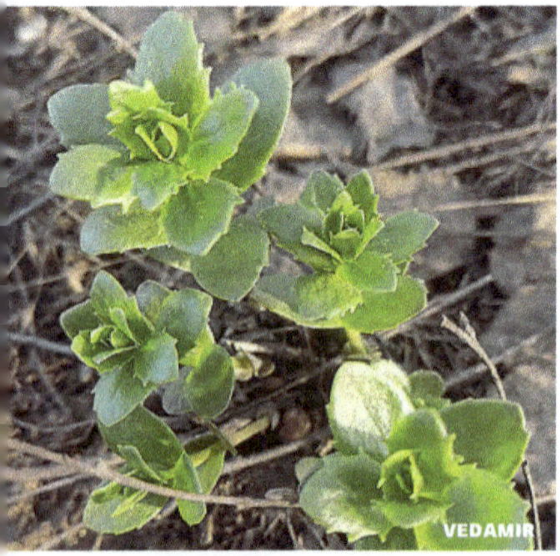

LIVE-FOREVER, ORPINE

This family is characterized by fleshiness; symmetrical flowers borne in cymes; petals, sepals, and pistils the same in number (3–30); stamens the same number or twice as many; all parts separate.

Live-Forever, Orpine

Hylotelephium telephium (L.) H.Ohba

SYNONYM *Sedum purpureum* (L.) Link

Coarse, erect, succulent perennial, the usually many-flowering stems 2–8 dm tall from fleshy, carrot-like tubers. **Flowers** in compact corymbs, purple-red to deep roseate; petals widespreading, 3 times as long as the sepals; nectar-bearing scales present, longer than broad; stamens nearly equaling the petals; follicles nearly erect. **Leaves** succulent, alternate or in whorls of 3, broadly oblong or elliptic, coarsely toothed, the larger ones 4–10 cm long.

HABITAT Along roadsides, on banks, or in open woods. Spread from cultivation.

FLOWERING July to September.

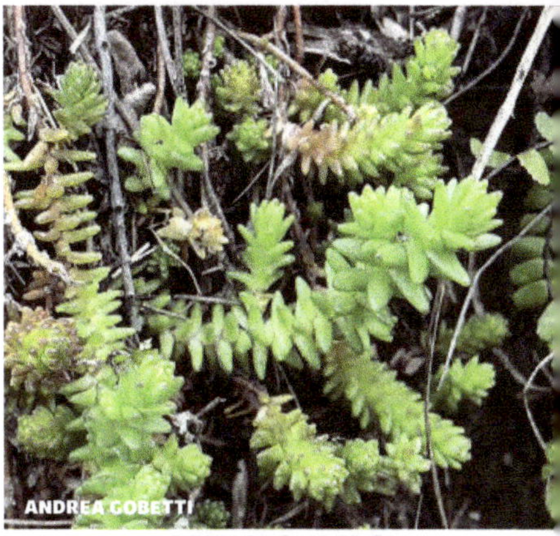

MOSSY STONECROP

Mossy Stonecrop

Sedum acre L.

Low, matted, creeping and freely rooting, fleshy evergreen perennial forming moss-like carpets, the erect flowering stems 2–8 dm tall. **Flowers** yellow, 1 cm or more wide, borne on very short pedicels in small, forked, leafy racemes. Petals 4 or 5, separate, narrow, somewhat fleshy; sepals much shorter than the petals; stamens twice as many as the petals; follicles spreading, long-beaked, the seeds numerous. **Leaves** sessile, alternate, fleshy, thick and ovoid, usually closely overlapping, having a small spur at the base.

HABITAT In sand, on rocks, and in dry, open places and in lawns. Introduced from Europe.

FLOWERING in June and July.

CUCURBITACEAE *Gourd Family*

This mainly tropical and subtropical family, comprising about 100 genera and 965 species, is important both for food and for ornament; 6 genera are reported for Michigan. Pumpkins and squash, cucumber, muskmelons, watermelons, and citron, as well as gourds, belong here.

The members of this family are easily recognized by the succulent, usually weak, tendril-bearing stems, and the unisexual flower with inferior ovary which develops into a special type of fleshy or membranous fruit.

Wild Cucumber, Balsam-Apple

Echinocystis lobata (Michx.) Torr. & Gray

High-climbing, herbaceous annual vine with angular, grooved, nearly smooth stems often several meters long, bearing single leaves and 3-forked tendrils at the nodes. **Flowers** greenish-white, glandular-hairy, unisexual, the staminate flowers numerous, in many-flowered compound racemes borne singly in the leaf axils, the pistillate flowers short-stalked, one or a few from the same axils as the staminate racemes. Corolla wheel-shaped, 6-parted nearly to the base, the lobes linear, spreading, usually twisted; calyx small, 6-lobed; stamens united; pistil 1, the ovary inferior, prickly, 2-celled; **fruit** ovoid, 3–5 cm long,

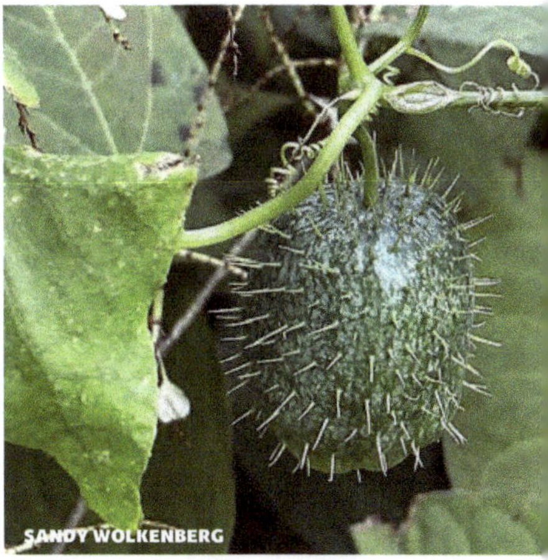

WILD CUCUMBER, BALSAM-APPLE

fleshy, becoming dry, covered with weak prickles. **Leaves** thin, alternate, bright green, rough on both sides, roundish in general outline (but deeply palmately 5-lobed), sharply toothed.

HABITAT In rich moist soil, in thickets or woods or along streams.

FLOWERING July to September.

DROSERACEAE *Sundew Family*

An insectivorous family of 3 genera and about 180 species, most of which grow in bogs. One species is almost cosmopolitan; the others occur in the Mediterranean region, Australia, and North America. Michigan is home to a single genus, with 4 species.

The family is characterized by rosettes of leaves, which are generally covered with sticky, insect-catching glands.

KEY TO DROSERA (SUNDEW) SPECIES

1 Leaf blades nearly round, or broader than long .... *D. rotundifolia*

1 Leaf blades longer than broad, linear or spatulate 2

  2 Leaf blades linear, stipules adnate to base of petiole .... *D. linearis*

  2 Leaf blades obovate to spatulate, stipules nearly free from base of petiole *D. intermedia*

ROUND-LEAF SUNDEW

Round-Leaf Sundew

Drosera rotundifolia L.

Low-growing perennial from rosettes of sticky-haired leaves. **Flowers** white (rarely pink), borne in a 1–25 flowered, 1-sided raceme on a slender, leafless flowering stem which is usually simple but may be forked once, the undeveloped apex of the raceme nodding, so that the full-blown flower is always the highest; flowers opening only in sunlight. Petals 5, spatulate; calyx tube short, free from the ovary, usually 5 (4–8)-lobed; stamens 5; pistil 1, the ovary 1-celled; styles usually 3 or 5, deeply 2-parted so that there appear to be 6 or 10; **capsule** stalked, the seeds numerous, spindle-shaped. **Leaves** all basal, the blades nearly round, broader than long, coiled from the apex to the base of the petiole in bud, pale yellowish-green, the upper surface thickly covered with long gland-tipped reddish hairs which exude a sticky fluid that glitters in the sun; petioles long, hairy.

HABITAT In bogs, peaty or moist acid soil, or wet sand.

FLOWERING June to August.

Linear-Leaf Sundew

Drosera linearis Goldie

In general aspect somewhat like the preceding species, but the **flowers** 1–8, the **leaves** linear (up to 8 cm long and 3 mm wide), the petioles short, not hairy, and the stipules united to the base of the petiole.

HABITAT In bogs, on limy shores, and in beach pools.

FLOWERING June to August.

LINEAR-LEAF SUNDEW

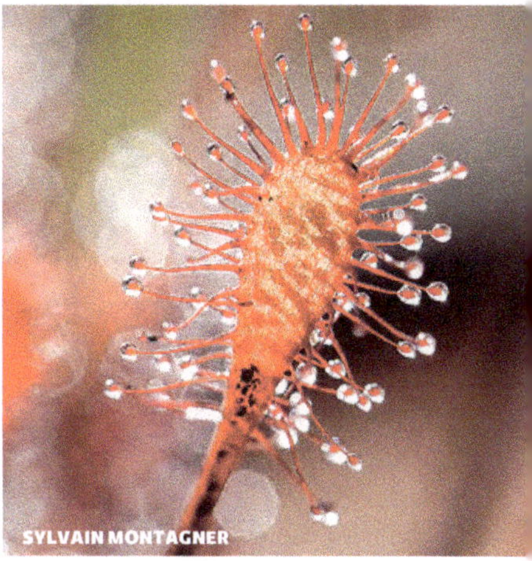

SPOON-LEAF SUNDEW

Spoon-Leaf Sundew

Drosera intermedia Hayne

Quite similar to the two preceding species but distinguished by the spatulate or obovate leaves with the stipules nearly free from the base of the petiole.

HABITAT In bogs and on wet shores.

FLOWERING June to August.

Another, very similar species, ***Drosera anglica*** Huds., occurs in the Upper Peninsula. It has narrow leaf blades 1.5–5 cm, long, and the stipules are united, except at the tip, to the base of the petiole.

ERICACEAE *Heath Family*

This is a family of 121 genera with about 4,250 species , which grow in acid soil throughout the temperate regions, in the subarctic, and in the mountains of the tropics; 19 genera are present in Michigan. Its chief importance is for ornamentals, which include such evergreen shrubs as azaleas, rhododendron, heather, heath, mountain laurel, and leather-leaf. Leaves of salal from the northwest are used a great deal by midwestern florists under the name of lemon-leaf. Cranberries and blueberries, also included in this family, are important fruits. The family also includes plants which lack chlorophyll and grow saprophytically or parasitically.

Most authorities consider the **Wintergreen Family** (Pyrolaceae) to be a part of this family, and is treated as such here. The chief points of difference are in their herbaceous growth form, and in having the petals usually separate; the heaths are woody and have the petals united, and the anthers are upright and open by terminal pores.

KEY TO ERICACEAE (HEATH FAMILY) SPECIES

1 Plants herbacous (may be woody at base) 2
1 Plants woody shrubs, upright or trailing 10
  2 Plants not green, lacking true leaves 3
  2 Plants having normal, green leaves 4
3 Corolla urn-shaped, petals united . . . . ***Pterospora andromedea***
3 Corolla of separate petals . . . . ***Monotropa uniflora*** or ***M. hypopithys***

4 Stem leafy; flowers borne in a flattish-topped corymb . . . . *Chimaphila umbellata* or *C. maculata*

4 Leaves basal; flowers solitary or in a raceme 5

5 Flower solitary . . . . *Moneses uniflora*

5 Flowers in a raceme 6

6 Flowers borne principally on one side of the raceme; corolla longer than broad . . . . *Orthilia secunda*

6 Flowers borne on all sides of the racemes; corolla broader than long7

7 Leaf blade nearly as long as or longer than the petiole, usually 2.5–6 cm long 8

7 Leaf blade usually shorter than the petiole, usually 1–3 cm long . . . . . . *Pyrola chlorantha*

8 Leaf blade thin; sepals as broad as long, or broader . . . . *Pyrola elliptica*

8 Leaf blade thick, usually lustrous; sepals distinctly longer than broad 9

9 Petals thick, white or creamy . . . . *Pyrola americana*

9 Petals thin, pink to pale purple . . . . *Pyrola asarifolia*

10 Erect or nearly erect shrubs 11

10 Low, prostrate or creeping shrubs 12

11 Leaves having a dense rusty wool beneath; flowers white . *Rhododendron groenlandicum*

11 Leaves whitened beneath; flowers rose or pink . . . . *Kalmia polifolia*

12 Corolla having a slender tube enlarged at apex; leaves round or heart-shaped at base; flowering in very early spring . . . . *Epigaea repens*

12 Not as in alternate choice 13

13 Flowers pale rose; corolla deeply lobed . . . . *Vaccinium macrocarpon* or *V. oxycoccos*

13 Not as in alternate choice 14

14 Leaves and berries with flavor of wintergreen . . *Gaultheria hispidula* or *G. procumbens*

14 Not as in alternate choice . . . . *Arctostaphylos uva-ursi*

BEARBERRY, KINNIKINICK

Bearberry, Kinnikinick

Arctostaphylos uva-ursi (L.) Spreng.

Trailing evergreen shrub with long grayish to reddish branches. **Flowers** 5–12 in dense terminal racemes; corolla white, sometimes pink-tipped, or pale pink, 5-cleft into short rounded lobes; stamens 10; anthers with 2 awns as long as the filaments; **fruit** a dry, red, mealy drupe. **Leaves** alternate, leathery, shiny on upper surface, obovate, entire, short-petioled.

HABITAT In rocky and sandy ground.

FLOWERING May to July.

Prince's Pine, Pipsissewa

Chimaphila umbellata (L.) W. Bart.

Low evergreen perennial with erect, leafy stems up to 2.5 dm tall from creeping underground rootstocks. **Flowers** 2–8, roseate, pink or white, nodding in loose, terminal flat-topped corymbs. Petals 5, concave, nearly round, inserted on the calyx; calyx 5-lobed, spreading; stamens 10, rose to rose-violet, inserted with the petals, opening by pores; ovary globose, large, the style nearly lacking, the stigma flattened, sticky; **capsule** erect, 5-lobed, splitting from the apex downward, many-seeded. **Leaves** whorled or scattered, green and shining on upper surface, paler beneath, thick, oblanceolate, wedge-shaped at the base, 2–9 cm long, toothed or nearly entire.

HABITAT In dry woods, on dune slopes, in upland coniferous forests, and under aspen. Mostly in northern Michigan.

FLOWERING July and August.

PRINCE'S PINE, PIPSISSEWA

Striped Wintergreen

Chimaphila maculata (L.) Pursh

Similar to the preceding species, but up to 2 cm broad, very fragrant; less common; the flowers 1–5, white, **leaves** lanceolate to ovate-lanceolate, rounded at the base, the upper surface variegated with white.

HABITAT In dry woods, on sand dunes, and in shore woods.

FLOWERING June to early August.

STRIPED WINTERGREEN

Trailing Arbutus

Epigaea repens L.

Prostrate or trailing, somewhat shrubby, freely rooting evergreen perennial with bristly rusty hairs at least on the stem. **Flowers** pink or white, having a rich spicy fragrance, borne in terminal or axillary clusters. Corolla salverform, the tube hairy inside, 5-lobed; sepals 5; stamens 10; **capsule** 5-lobed and 5-celled, many-seeded. **Leaves** alternate, green on both sides, thick, prominently veined, more less rusty-hairy when young, becoming smooth in age, oval, oblong or ovate, rounded to somewhat heart-shaped at base; petioles slender, about half as long as the blade, rusty-hairy.

TRAILING ARBUTUS

HABITAT In sandy or rocky woods, frequently under pine or aspen, in upland second growth, and on wooded dunes. Occurring sparingly in southern Michigan, more commonly in the northern part of the Lower Peninsula, and quite commonly in the Upper Peninsula.

FLOWERING mid-April to early June.

NOTES Picking the flowers usually loosens the roots, causing the plants to die out. The flowers often bloom before the snow is gone and can be readily located under its cover by their fragrance. Contrary to popular belief, this is not always the first wildflower to bloom. In the tip

CREEPING SNOWBERRY

CREEPING WINTERGREEN

BOG LAUREL

of the Lower Peninsula, I have found it still in bud in late April when Dutchman's-breeches, Squirrel-corn, Hepatica, Spring-beauty, and *Erythroniums* were in full bloom and Trillium was coming into bloom.

Creeping Snowberry

Gaultheria hispidula (L.) Muhl. ex Bigelow

Delicate, creeping, matted evergreen perennial with stiffly hairy, barely woody stems 1–3 dm long. **Flowers** white, solitary and nodding on short peduncles in the axils of the leaves. Corolla bell-shaped, deeply 4-cleft; calyx 4-toothed; stamens 8; pistil 1, the ovary 4-celled; **berry** white, fleshy, juicy, delicately acid and aromatic. **Leaves** abundant, 2-ranked, firm, dark green on upper surface, bristly-hairy beneath, ovate, 5–10 mm long, short-petioled.

HABITAT In mossy, mostly coniferous, woods and in bogs and swamps.

FLOWERING April and May.

Creeping Wintergreen

Gaultheria procumbens L.

Low-growing evergreen shrub with a wintergreen odor when crushed, the stems somewhat woody, more or less erect, leafy above, 1–2 dm tall from slender creeping rootstocks. **Flowers** white, 7–10 mm long, solitary from the axils of the leaves, drooping on often reddish pedicels which have 2 small fleshy bracts at the base of the flower. Corolla urn-shaped, roughly 5-sided, 5-toothed; stamens 10, attached at the base of the corolla tube; pistil 1, the ovary 5-angled, borne above the glandular disk; fruits red, berrylike, mealy, formed by the fleshy calyx growing out and enclosing the many-seeded **capsule**, often persistent and in good edible condition when the ensuing year's flowers are in bloom. **Leaves** alternate, at first light green and tender, soon hard and leathery, sometimes quite brittle, elliptic to narrowly obovate, narrowed to the base or sometimes nearly round, 2–5 cm long, the margin slightly rolled under and with a few low bristle-tipped teeth; petiole short (2–5 mm long).

HABITAT In sterile woods and clearings; frequent under conifers.

FLOWERING July and August.

Bog Laurel

Kalmia polifolia Wangenh.

Slender, straggling shrub up to 1 m tall, the branches 2-winged. **Flowers** showy, deep pink to crimson, 1–2 cm wide, borne on threadlike pedicels (1–4 cm long) in terminal corymbs. Corolla shallowly saucer-shaped, with 10 small pouches; calyx 5-parted, free from the ovary; stamens 10, the anthers at first in the pouches of the corolla but soon free as the filaments become straight and erect; ovary superior; **capsule** globose, many-seeded, persistent. **Leaves** opposite, green on upper surface, whitened beneath, lanceolate, the margin usually rolled under; veins obscure except for the midrib, which is prominent on lower surface.

HABITAT In open swamps and bogs, in marshy places, around lakes, and in sphagnum mats.

FLOWERING May and June.

One-Flowered Wintergreen

Moneses uniflora (L.) Gray

Small, low, evergreen perennial, with underground stems giving rise to ascending leafy tips, the flowering stalk up to 1.3 dm tall. **Flowers** solitary, nodding, fragrant, white or sometimes pinkish, waxy. Petals 5, spreading, concave; calyx whitish, 5-lobed; stamens 10, the filaments green or white, the anthers opening by pores; pistil 1, the ovary superior, globose, the style slender, the stigma 5-lobed; **capsule** 5-celled. **Leaves** short-petioled, nearly round, 1–2 cm long, the margin with small rounded teeth or nearly entire.

HABITAT In cool mossy woods.

FLOWERING June to August.

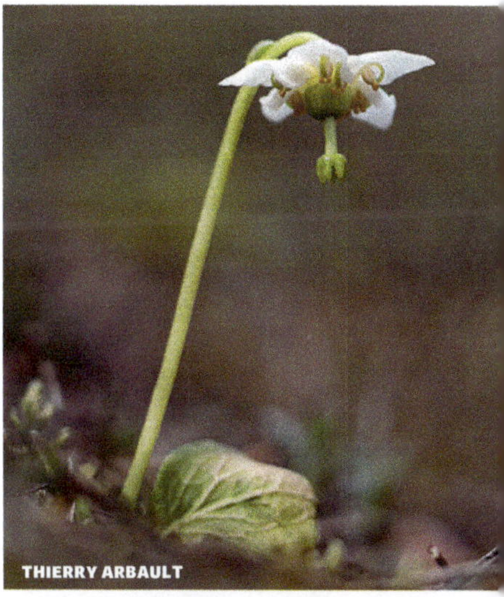

ONE-FLOWERED WINTERGREEN

Indian-Pipe, Ghost-Pipe

Monotropa uniflora L.

Stiffly erect, succulent, usually clustered saprophyte or parasite up to 3 dm tall, waxy white, rarely pink or reddish, darkening in age or when bruised, the roots matted, fibrous, brittle, growing on roots of other plants or on decayed vegetable matter. **Flowers** solitary, white, waxy, nodding, 10–17 mm long, oblong bell-shaped. Petals 4–5, separate, firm, somewhat translucent; sepals 2–5, bractlike and soon falling, often lacking; stamens 8 or 10, shorter than the petals, the anthers opening by 2 chinks; pistil 1, the ovary superior, the style short and thick, straight, the stigma broad, funnel-shaped, sticky; **capsule** ovoid, many-seeded, borne upright owing to the straightening of the stem after flowering. True leaves lacking, the stem clothed with alternate scales or bracts.

INDIAN-PIPE, GHOST-PIPE

HABITAT In woodland humus.

FLOWERING June to August.

USES It is said that American Indians made a medicinal tea from this species.

PINESAP

ONE-SIDED WINTERGREEN

Pinesap

Monotropa hypopitys L.

SYNONYM *Hypopitys americana* (DC.) Small

Resembles **Indian-pipe**, but with a cluster of flowers, the plants tawny, yellowish or reddish, becoming brown in age or when bruised, the stem up to 4 dm tall, the flowers drooping in a compact raceme, the terminal flower with parts in 5's, the others with parts in 4's or 3's, the stamens twice as many as the petals.

HABITAT In woodland humus.

FLOWERING June to September.

One-Sided Wintergreen

Orthilia secunda (L.) House

SYNONYM Pyrola secunda L.

Small evergreen up to 2 dm tall. **Flowers** greenish or greenish-yellow, bell-shaped, longer than broad, borne in a 1-sided terminal raceme on a scape bearing 2–5 small green bracts. Filaments and style straight. **Leaves** in a basal rosette or somewhat scattered on the short stem, somewhat leathery, elliptic to ovate, narrowed to the tip, somewhat toothed, 1.5-4 cm long, the blade longer than the petiole.

HABITAT In dry, sometimes in moist, woods.

FLOWERING June to August.

Pinedrops

Pterospora andromedea Nutt.

Slender, leafless, glandular-hairy, reddish, purplish, or brownish perennial up to 15 dm tall, parasitic on the roots of other seed-bearing plants. **Flowers** white to reddish, bell-shaped, 6–7 mm long, drooping in a long, many-flowered raceme. Corolla urn-shaped, constricted at the apex and shallowly 5-lobed; calyx deeply 5-parted; stamens 10, the anthers opening by longitudinal slits and having 2 long slender awns; pistil 1, the ovary superior, 5-celled, the stigma 5-lobed. True leaves lacking, but the stem covered with brownish to reddish lanceolate bracts, particularly toward the base.

HABITAT In dry woods, chiefly under pine.

FLOWERING June to August.

STATUS Threatened in Michigan.

AMERICAN WINTERGREEN

American Wintergreen

Pyrola americana Sweet

Stemless evergreen perennial, the flowering stalk up to 3 dm tall, 1–7-bracted, more or less 4-angled. **Flowers** white, creamy white, or tinged with rose, fragrant, nodding in a 5–21-flowered raceme. Petals 5, waxy, concave, separate, thick, and firm, obscurely veined; sepals oblong or ovate-oblong, nearly twice as long as wide, not overlapping at the base; stamens 10, the filaments white, flattened, curved upward, the anthers purple, opening by pores; ovary superior, 5-lobed, the style long, bent abruptly down at the base, then arching upward near the apex, the stigma 5-lobed; **capsule** globose, flattened, 5-lobed, the seeds minute, innumerable. **Leaves** all basal, leathery, lustrous on upper surface, broadly oblong to nearly round, not heart-shaped, the blade decurrent on the petiole, about the same length as the petiole (2.5–7 cm), the margin slightly inrolled, with slightly rounded teeth.

HABITAT In dry woods and clearings, often in sandy soil.

FLOWERING June to August.

Pink Wintergreen

Pyrola asarifolia Michx.

Extensively creeping evergreen perennial quite similar to the preceding species, but the **flowers** pink to pale purple, 4–22 in a raceme, the scape with 1–3 ovate, dry bracts, the petals thin, conspicuously veiny at least when dry, the sepals triangular, distinctly longer than broad, slightly overlapping at the base, the **leaves** kidney-shaped, nearly round, or broadly elliptic, truncate or heart-shaped at base.

HABITAT In rich woods and thickets, and in bogs, chiefly in calcareous soil.

FLOWERING in July and August.

PINK WINTERGREEN

Shinleaf

Pyrola elliptica Nutt. (not illustrated)

Quite similar to the two preceding species, the **flowers** white, the leaf blades thin, usually longer than the petiole, the sepals about as broad as long and the bracts on the scape linear and pointed.

HABITAT In dry or moist woods.

FLOWERING June to August.

GREEN-FLOWER WINTERGREEN

LABRADOR-TEA

Green-Flower Wintergreen

Pyrola chlorantha Sw.

SYNONYM *Pyrola virens* Schweigger

Low perennial, the erect scapes up to 3 dm tall from an evergreen rosette. **Flowers** pale green or greenish-white, nodding in a simple 2–13-flowered raceme. Flower structure as as in the preceding species. **Leaves** basal, leathery, dark green on upper surface, lighter beneath, nearly round or broadly ovate, usually shorter than the petiole, the teeth slightly rounded.

HABITAT In rather dry coniferous woods and in thickets.

FLOWERING June and July.

Labrador-Tea

Rhododendron groenlandicum (Oeder) Kron & Judd

SYNONYM *Ledum groenlandicum* Oeder

Evergreen shrub up to 1 m tall. **Flowers** white, about 1 cm broad, borne in dense terminal clusters. Petals 5, separate; calyx minute, with rounded lobes; stamens 5–7, spreading 135 longer than the petals, the filaments threadlike; pistil 1, the ovary superior, green, covered with a sticky mucilaginous substance, the style long; **capsule** slender, many-seeded, splitting from the base upward. **Leaves** alternate, green on the upper surface, the midrib depressed, densely woolly beneath with rusty hairs, linear-oblong or oblong, up to 2.5 cm long, the margin entire and rolled under; petioles short, stout.

HABITAT In bogs and damp thickets and along wet shores.

FLOWERING May and June.

NOTES Labrador-tea blooms at the same time as Bog Laurel, and the two together make a most attractive sight. It is said that this species is poisonous to sheep. It was used to make a tea by American Indians, and by the colonists during the Revolutionary War.

Large Cranberry

Vaccinium macrocarpon Ait.

Trailing evergreen shrub with frequently branched stems, rooting freely at the nodes, often forming dense mats. **Flowers** 1–10, pale rose, nodding on slender upright reddish pedicels rising from the end of a stem opposite bracts near the top, the elongate leafy stem continuing to grow and extending beyond the origin of the pedicels. Corolla so deeply 4-cleft that it appears to be composed of long, separate petals, the lobes purplish-red at base, curving abruptly back, inserted on the disk; calyx 4-lobed, adherent to the ovary below; stamens 8; pistil 1, the ovary inferior; **berry** green at first, becoming red, juicy, tart. **Leaves** alternate, oblong-elliptic, 6–17 mm long, green on upper surface, paler beneath, short-petioled.

HABITAT In boggy meadows, bogs, and swamps and on wet shores.

FLOWERING June and July.

NOTES Cranberries are remarkable for their keeping qualities and were especially prized by the early colonists for that reason. In the early days, the entire supply came from wild berries, but they have now been cultivated for over a hundred years. American Indians used the berries as an article of commerce. The plant was originally called "crane berry" because the shape of the flowers suggests the head and neck of a crane.

LARGE CRANBERRY

Small Cranberry

Vaccinium oxycoccos L.

Similar in general aspect to the preceding species, but the plants generally smaller, the **flowers** only 1–4, the pedicels rising from the end of a stem that does not continue to grow, the bracts on the pedicels red, borne at or below the middle, the **leaves** strongly whitened beneath, ovate, oblong-ovate, or triangular.

HABITAT In boggy or peaty soil.

FLOWERING June and July; berries ripening August to October.

SMALL CRANBERRY

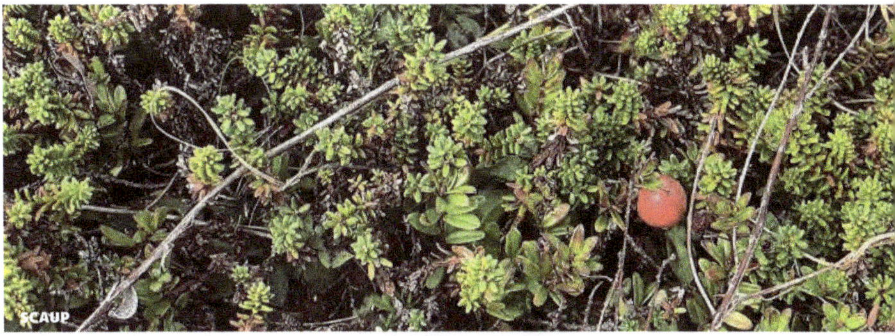

LARGE CRANBERRY

EUPHORBIACEAE *Spurge Family*

This cosmopolitan, but predominantly tropical, family has 228 genera and more than 7,500 species; 4 genera occur in Michigan. Our only economically important members are such ornamentals as Poinsettia, Crown-of-thorns, and Castor Bean. Rubber, tung oil, castor oil, cassava, and tapioca are derived from primarily tropical species.

The members of the family included here are distinguished by the milky juice and the greatly reduced, unisexual flowers. The flowers are borne in a calyx-like, cup-shaped involucre or cyathium. The staminate flowers consist of a single-stalked stamen, the pistillate flowers of a stalked pistil surmounted by 3 curved styles. Several staminate flowers surround a single pistillate flower in the involucre, the whole structure looking like a single flower. In maturity the stalk of the ovary elongates, and the capsule (usually 3-sided) extends beyond the involucre.

KEY TO EUPHORBIA SPECIES

1 Leaves (at least upper ones) having broad, white, petal-like margins .... *E. marginata*
1 Leaves not bordered with white 2
  2 Leaves oblong-oval to linear .... *E. corollata*
  2 Leaves narrowly linear to threadlike .... *E. cyparissias*

FLOWERING SPURGE

CYPRESS SPURGE

Flowering Spurge

Euphorbia corollata L.

Bright green, glabrous or somewhat hairy perennial up to 1 m tall, simple below, many branched above; juice milky. **Flowers** greatly modified and reduced, the involucre resembling a small white flower with 5 rounded corolla lobes, each with a crescent-shaped, green gland at the base. **Leaves** numerous (75 or more below the inflorescence), alternate, sessile, firm, glabrous, oblong, oval, or linear, entire, obscurely veined; opposite, bractlike leaves in the inflorescence.

HABITAT In dry open places, along roads and railways, in fields, clearings, and waste places.

FLOWERING June to October.

Cypress Spurge

Euphorbia cyparissias L.

Erect perennial up to 4 dm tall, usually growing in large patches from extensively creeping and forking rootstocks; juice milky. Involucres in terminal umbels, yellowish-green, with 4 crescent-shaped glands; **capsule** stalked, nearly globose, rarely developing. **Leaves** narrowly linear, numerous above, but few, scattered, and reduced below; leaves in the inflorescence usually in pairs, broadly ovate, yellowish-green, becoming red or purplish.

HABITAT Along roadsides, in old fields and waste places. Introduced from Europe, now widely distributed.

FLOWERING April to August.

Snow-on-the-Mountain

Euphorbia marginata Pursh

Bright green, glabrous or hairy annual up to 1 m tall, topped by a usually 3-rayed terminal umbel. Inflorescence like that of the preceding species but the lobes of the bell-shaped involucre finely fringed, the bracts of the inflorescence and the upper leaves with conspicuous, broad, white, petal-like margins. **Lower leaves** broadly oblong to ovate; stipules lanceolate, soon falling.

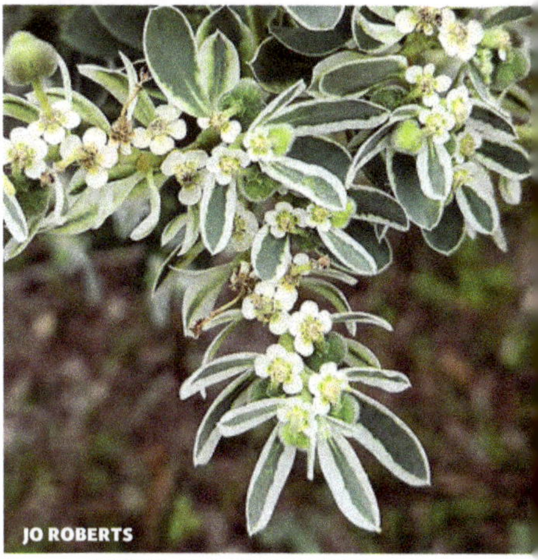

JO ROBERTS

SNOW-ON-THE-MOUNTAIN

HABITAT In dry soil in waste places. Native to the Great Plains, frequently cultivated; spreading readily.

FLOWERING from June to October.

NOTES Several additional species of *Euphorbia,* mostly low or prostrate, occur in waste ground in Michigan.

FABACEAE *Pea Family*

The Pea Family, in the broadest sense, is one of the largest of the flowering-plant families. It has 796 genera and and nearly 20,000 species, placing it third in size (below the Orchid and Aster families); 44 genera occur in Michigan. In economic importance it is second only to the grass family. Garden peas, beans, soybeans, lentils, and peanuts, all widely used table foods, belong to this family. The clovers, alfalfa, soybeans, and vetch are important both as forage and fodder plants. Peas are also important as rotation crops to increase the nitrogen content of the soil. Well over 100 species of the family are grown for ornamentals, among the most common ones being Sweet Pea, Lupine, Red-bud, and Wisteria. Kentucky Coffee-tree, Locust, and Honey-locust are trees which are often grown in Michigan. Many members of the family are poisonous; locoweeds in particular cause a heavy loss of livestock in the western mountains.

The family is characterized by usually irregular flowers, which are butterfly-like ("papilionaceous"), with a large upper petal or "standard," two side petals or "wings," and 2 petals united into a "keel"; the stamens usually 10; the simple pistil ripening into a pod or legume which opens along 2 sides. The 10 stamens may be arranged in one of 3 fashions: the filaments distinct and separate, the filaments all united into a tube around the pistil (*monadelphous*) or the filaments of 9 stamens united into a tube, the tenth filament free (*diadelphous*). The leaves are usually compound and have stipules.

KEY TO FABACEAE (PEA FAMILY) SPECIES

1 Plants trailing, twining, or climbing 2
1 Not as in alternate choice 5
 2 Plants having tendrils 3
 2 Plants lacking tendrils 4
3 Wings of flower attached to keel; style threadlike, having a tuft of hairs at tip ...... *Vicia*
3 Wings free, or nearly free, from keel; style flattened, bearded down inner side .. *Lathyrus*
 4 Leaves having 5–7 leaflets; ovary coiled .... *Apios americana*
 4 Leaves having 3 leaflets; ovary and pod straight .... *Amphicarpaea bracteata*
5 Leaves pinnately compound with 10–18 pairs of leaflets; flowers nearly regular .... *Chamaecrista fasciculata*
5 Not as in alternate choice 6
 6 Leaves wheel-like, the 7–11 leaflets radiating from top of petiole ..... *Lupinus perennis*
 6 Leaves having 3 leaflets 7
7 Leaflets having fine teeth on margin 8
7 Leaflets having entire margin 9
 8 Flowers borne in dense heads; terminal leaflets sessile or nearly so .... *Trifolium*
 8 Flowers borne in long, slender racemes; terminal leaflet stalked .... *Melilotus alba* or *M. officinalis*
9 Pods covered with hooked hairs; breaking readily into sections .... *Desmodium canadense* or *Hylodesmum nudiflorum*
9 Pods lacking hooked hairs; not readily breaking into sections 10
 10 Flowers yellow .... *Baptisia tinctoria*
 10 Flowers white to purple .... *Lespedeza*

HOG-PEANUT

Hog-Peanut

Amphicarpaea bracteata (L.) Fern.

Low, twining or trailing perennial with brown, threadlike stems to 2.6 m long. **Flowers** of 2 kinds, the petaled ones pale lilac or purple to white, 9–13 mm long, borne in nodding 2–13-flowered racemes from the upper axils. Standard narrowed to the base, longer than the wings and keel; calyx slightly irregular, the tube short-cylindric, the teeth apparently 4; 10th stamen free; pods from upper flowers stalked, 1.5–3 cm long, flat, oblong, pointed at both ends, 3-seeded; unpetaled flowers small, borne on threadlike creeping branches, self-fertile, the stamens few, free; **pods** of unpetaled flowers often borne underground, fleshy, obovate or pear-shaped, usually with a single large seed. **Leaves** 3-foliolate, the leaflets broadly ovate, sharply pointed, 2–6 cm long.

HABITAT In moist woodlands.

FLOWERING August and September.

USES The underground seeds are said to be good to eat and to have a flavor somewhat like that of raw peanuts. They are often abundant and generally appear just under the surface of the ground, where they are easily found by hogs—hence the name. Hog-peanuts were highly valued by American Indians. Some tribes obtained considerable quantities by robbing rodents' nests of their stored supplies.

Groundnut

Apios americana Medik.

Perennial twining herb from slender rootstock with numerous tuberous enlargements resembling a string of beads; juice milky. **Flowers** reddish-brown, purplish-brown, or mauve, about 1 cm long, borne in short, dense, often branching racemes from the axils of the leaves, fragrant, the odor somewhat like that of violets, the peduncle shorter than the leaves. Standard round, bent backward; wings oblique, adherent to the keel, elongated, incurved, and horseshoe-shaped, broadest at the base; calyx 2-lipped, with 4 teeth very small, the other about as long as the tube; 10th stamen free; ovary coiled; **pod** slightly curved or straight, 5–10 cm long. **Leaves** pinnately compound, the leaflets usually 5–7, ovate-lanceolate, pointed at apex, rounded at base, 2.5–7 cm long.

HABITAT On lakeshores, in moist rich thickets, and along streams.

FLOWERING July to September.

USES Groundnuts are considered one of our best wild foods by some authorities and are said to be good raw, boiled, or roasted. They reportedly were eaten by the Pilgrims and constituted an important food for American Indians.

AGUJACERATOPS

GROUNDNUT

Wild Indigo, Rattleweed

Baptisia tinctoria (L.) R. Br.

Slender, smooth, bushy-branched perennial up to 1 m tall, somewhat glaucous when young. **Flowers** yellow, 1–1.3 cm long, borne in numerous, loose, terminal racemes. Standard and wings about the same length, the sides of the standard turned back, the petals forming the keel nearly separate, straight; stamens 10, separate; ovary and pod stalked, the **pod** papery to woody, inflated, 8–15 mm long. **Leaves** 3-foliolate, the leaflets usually 1–2 cm long, obovate, wedge-shaped at base, blackening on drying, minute stipules.

HABITAT In dry, open woods and clearings.

FLOWERING late May to September.

USES Indian children used the dried stalks with the inflated seed pods as rattles. The young

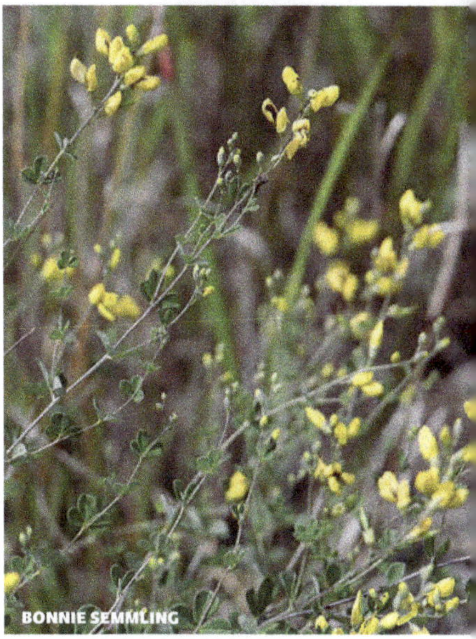
BONNIE SEMMLING

WILD INDIGO, RATTLEWEED

shoots were eaten like asparagus. This species makes a showy garden plant, but is somewhat difficult to transplant; it does best in acidic soil.

SIMILAR SPECIES **White or Prairie False Indigo**, ***Baptisia lactea*** (Raf.) Thieret (synonym *Baptisia leucantha* Torr. & Gray), has white flowers, sometimes tinged with purple, borne in stout, erect racemes; the stipules are slender, 5–10 mm long and soon falling. This rare species of southern Michigan is listed as **Threatened**. **Cream False Indigo**, ***Baptisia leucophaea*** Nutt., has cream-colored flowers borne in somewhat one-sided drooping racemes, and the stipules are up to 4 cm long, conspicuous, and leaflike. Rare in southern Michigan, and listed as **Threatened**.

PARTRIDGE-PEA

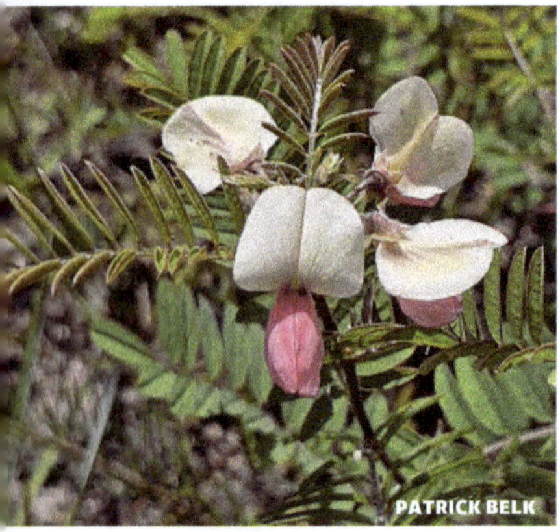

GOAT'S-RUE

Partridge-Pea

Chamaecrista fasciculata (Michx.) Greene

SYNONYM *Cassia fasciculata* Michx.

Nearly erect, rather slender annual up to 9 dm tall. **Flowers** yellow or sometimes white, nearly regular, 1–6 in short, axillary, bracted racemes. Petals 5, often with a purple spot at the base, slightly unequal in size, 1–2 cm long, spreading; sepals scarcely united at the base; stamens 10, distinct, very unequal, the anthers much longer than the filaments; **pod** linear, oblong, 4–13-seeded. **Leaves** pinnately compound, sensitive to touch, the leaflets 10–18 pairs, oblong, 1–2 cm long, stipules persistent, streaked, folding upward when touched.

HABITAT In moist or dry, usually sandy soil, on prairies and in open woods, along roadsides, and in old fields. Occurs only in the southern part of the state.

FLOWERING July to September.

SIMILAR SPECIES **Wild Senna** (*Senna hebecarpa* (Fernald) H.S. Irwin & Barneby), with 4–10 pairs of larger leaflets (2–5 cm long) is more common in southern Michigan. It grows up to 2 m tall, and the flowers are borne in racemes from the axils of the leaves, the flowers are 10–15 mm long, the 3 upper stamens lack normal anthers, and the segments of the legumes are much shorter than broad.

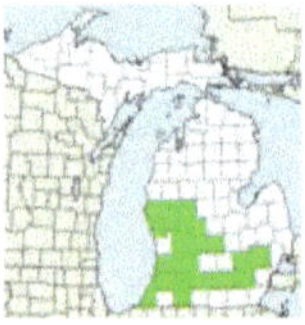

Goat's-Rue, ***Tephrosia virginiana*** (L.) Pers., is another erect pea with pinnately compound leaves, but in this there is an odd terminal leaflet in addition to the pairs of leaflets, and the leaves are not sensitive to touch. This species grows to be up to 6 dm, tall and is covered with long silky hairs. The numerous large flowers have yellow or yellowish standards and pink to pale purple wings. They are borne in compact terminal racemes. This is an attractive native species which grows in dry, sandy woods and openings and is conspicuous on sandy plains in southwestern Michigan.

Showy Tick-Trefoil

Desmodium canadense (L.) DC.

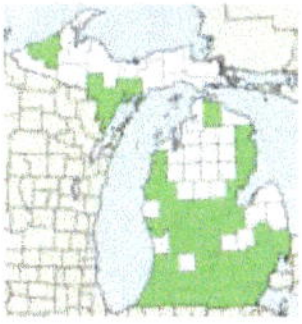

Erect, hairy perennial up to 1.3 m tall. **Flowers** rose-purple changing to blue, 8.5–14 mm, long, borne on densely hairy, often sticky pedicels in racemes, the lanceolate to lance-ovate bracts conspicuous before flowering but then soon falling. Standard rounded to ovate; wings very slightly adherent to the curved keel; calyx 2-lipped; the 10th stamen free; **pod** flattened, composed of jointed, 1-seeded divisions which are densely covered with small hooked hairs. **Leaves** 3-foliolate, alternate, borne on the stem up to the inflorescence, with stipules; leaflets lance-ovate, thick, hairy on both surfaces, paler beneath, the larger leaflets 4.8–10.5 cm long.

HABITAT In open woods and waste places.

FLOWERING July and August.

NOTES Everyone who has hiked in the woods and fields of southern Michigan is familiar with the fruits of this and related species, which cling so tenaciously to clothing, especially to wool socks.

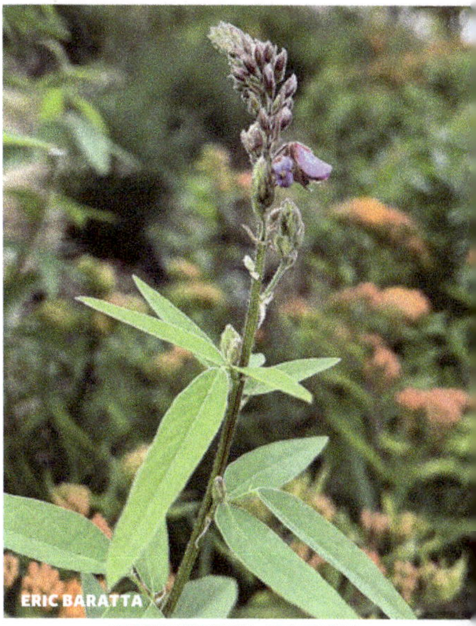

SHOWY TICK-TREFOIL

Naked-Flower Tick-Trefoil

Hylodesmum nudiflorum (L.) H. Ohashi & R.R. Mill

SYNONYM *Desmodium nudiflorum* (L.) DC.

Erect or ascending perennial up to 2 m tall, usually forked at the base, giving rise to a leafless flowering stem and a leafy stem. **Flowers** rose to purple, rarely white, borne on slender, finely hairy pedicels in racemes. Calyx obscurely 2-lipped; stamens 10, all united into a tube; **pod** with 1–4 sections, densely covered with hooked hairs, the stipe longer than the pedicel. **Leaves** 3-foliolate, whorled at the apex of the leaf stem; terminal leaflets obovate to rhombic, 4.5–12 cm long, finely downy above, glaucous and with some long hairs below; stipules soon falling.

HABITAT In woods.

FLOWERING July and August.

NOTES The several other species of *Desmodium* or *Hylodesmum* occurring in Michigan are distinguished mainly by their stipules and fruits. ***Hylodesmum glutinosum*** (Muhl. ex Willd.) H. Ohashi & R.R. Mill, with the leaves crowded at the middle of the stem, is especially noticeable in the woods of the Lower Peninsula.

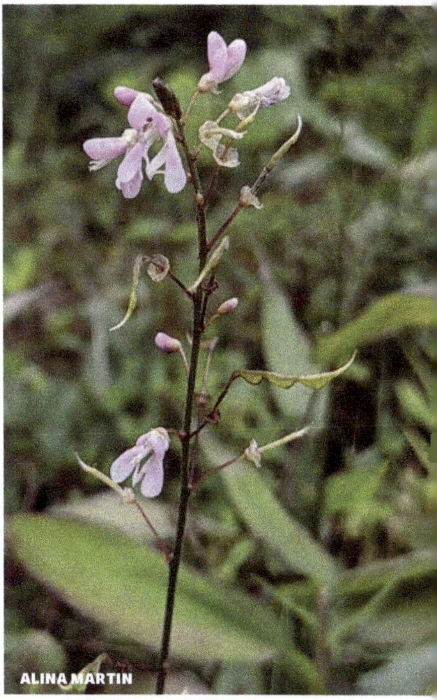

NAKED-FLOWER TICK-TREFOIL

KEY TO LATHYRUS SPECIES

1 Flowers blue, purplish, or rose-purple, rarely white 2
1 Flowers cream to yellowish-white .... *L. ochroleucus*
2 Stipules nearly as large as leaflets, having 2 basal lobes .... *L. japonicus*
2 Stipules not as large as leaflets, having only 1 basal lobe .... *L. palustris*

BEACH PEA

PALE VETCHLING

Beach Pea

Lathyrus japonicus Willd.

Somewhat fleshy, stiffly branching, climbing or partially erect, glaucous perennial with stout, sharply angled stems up to 1.5 m long. **Flowers** butterfly-like, deep rose-purple, blue-violet, or crimson, 1.2–3 cm long, in stout 4–10-flowered racemes which are usually shorter than the subtending leaves, the pedicels short, stout and arching. Standard broad; wings only slightly coherent with the upwardly curved keel; calyx irregular; stamen tube not oblique at the apex, the 10th stamen free; style bearded along the inner side (next to the free stamen); **pods** brown, 3–7 cm long. **Leaves** pinnately compound with a forking tendril at the end; leaflets thin, very firm, sessile, elliptic to obovate, blunt and with an abrupt point at the apex, smooth and glaucous, 1–7 cm long, entire; stipules broadly ovate with 2 basal lobes, nearly as large as the leaflets.

HABITAT On sandy or gravelly shores and beaches; common on the shores of the Great Lakes.

FLOWERING in July and August.

USES The fresh stalks and young sprouts were used as food by American Indians.

Pale Vetchling

Lathyrus ochroleucus Hook.

Smooth, more or less trailing or climbing perennial with wingless, sometimes angled, green or purplish-red stems up to 1 m long. **Flowers** cream or yellowish-white, 12–18 mm long, in 5–10-flowered racemes which are much shorter than the subtending leaves. Calyx irregular, the lower teeth considerably longer than the upper. **Leaflets** 2–3 (sometimes 4) pairs, thin, ovate, somewhat glaucous below; stipules half heart-shaped, about half as long as the lower leaflets.

HABITAT In dry or moist woods, on rocky banks or slopes,

FLOWERING in May and June.

SIMILAR SPECIES The introduced **Yellow Vetchling**, *Lathyrus pratensis* L., has bright yellow flowers and only 2 leaflets.

Marsh Pea

Lathyrus palustris L.

Slender, climbing perennial with weak stems up to 1 m long. **Flowers** purple, rose-purple, violet, or whitish, 1–2.5 cm long, borne in racemes which are nearly as long as or slightly longer than the subtending leaf. **Leaflets** 2–5 pairs, firm, linear to ovate; stipules with only one lobe at base, pointed at both ends.

On lakeshores, in meadows and damp thickets.

FLOWERING June to September.

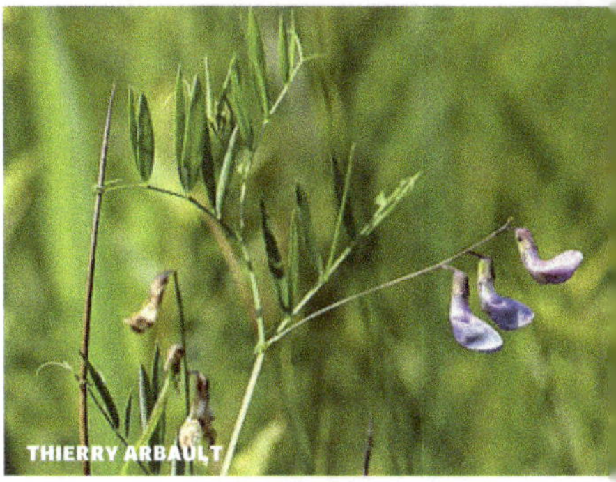

MARSH PEA

SIMILAR SPECIES Two rather common everlasting or perennial peas are quite striking and frequently grow wild. ***Lathyrus latifolius*** L. may be high-climbing. It has broadly winged stems; 2 lanceolate, oblong, or oval leaflets 4–9 cm long; and conspicuous purple, pink, or white flowers about 2.5 cm long, borne in 4–10-flowered racemes. ***Lathyrus sylvestris*** L. is quite similar, but the 2 leaflets are narrowly lanceolate and 1–1.5 dm long; the flowers are about 1.5 cm long.

USES The shelled peas were cooked and eaten by American Indians. However, recent studies have shown that at least some species of *Lathyrus* are poisonous, and it is therefore inadvisable to eat wild peas. The familiar garden Sweet Pea belongs to this genus.

KEY TO LESPEDEZA (BUSH-CLOVER) SPECIES

1 Flowers white to cream or yellowish, borne in dense heads or racemes 2

1 Flowers purple, borne in loose, few-flowered racemes .... ***L. violacea***

  2 Flowers in very dense, subglobose heads, peduncles shorter than subtending leaf, or lacking .... ***L. capitata***

  2 Flowers borne in spikelike racemes on peduncles longer than subtending leaf . ***L. hirta***

Round-Head Bush-Clover

Lespedeza capitata Michx.

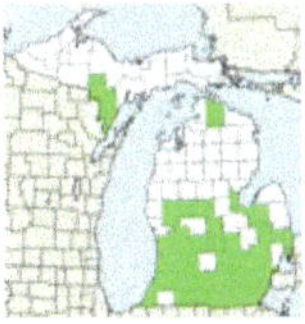

Erect, stiffish perennial with simple or sparsely branched stems up to 1.2 m tall. **Flowers** creamy-white, borne in many dense, subglobose heads or spikes which form a compact inflorescence, the heads sessile or on short peduncles which are usually shorter than the subtending leaves. Flowers of 2 kinds, those lacking petals rare and hidden among the petaled ones; petaled flowers with a purple spot at the base of the standard; calyx very hairy, 7–13 mm long, 5-cleft, the slender lobes nearly equal; the 10th stamen free; **pod** flat, 1-seeded, shorter than the calyx. **Leaves** 3-foliolate, petioles 2–5 mm long; leaflets oblong to lance-linear, thickish, downy, without stipules.

HABITAT In dry open places.

FLOWERING late July to September.

ROUND-HEAD BUSH-CLOVER

HAIRY BUSH-CLOVER

PURPLE BUSH-CLOVER

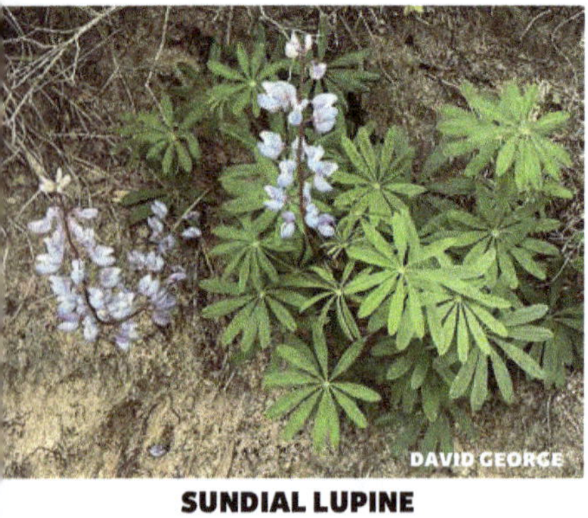

SUNDIAL LUPINE

SIMILAR SPECIES The seeds of some species of bush-clover, particularly the introduced **Korean Lespedeza**, *Kummerowia stipulacea* (Maxim.) Makino, provide food for the bobwhite, but are not attractive to other birds.

Hairy Bush-Clover

Lespedeza hirta (L.) Hornem.

Stout, usually erect, simple or branched perennial up to 1.5 tall, covered with spreading hairs. **Flowers** whitish to yellowish, purple at the base, the petaled flowers in cylindric, spikelike racemes which form a rather spreading inflorescence, the peduncles longer than the subtending leaves, the unpetaled flowers often in separate clusters. **Leaves** definitely petioled, 3-foliolate, the leaflets rounded obovate to lanceolate.

HABITAT In dry soils.

FLOWERING July to October.

Purple Bush-Clover

Lespedeza violacea (L.) Pers.

Freely branching, upright or spreading, glabrous to somewhat hairy perennial 4–8 dm tall. **Flowers** purple, 7–10 mm long, the keel often the longest part; borne in few, loose, few-flowered racemes which are much longer than the subtending leaves; unpetaled flowers numerous, in sessile or nearly sessile clusters in axils of leaves. **Leaves** few, 3-foliolate, the petioles slender, often nearly as long as the leaflets; leaflets elliptic, 1–4 cm long and about half as wide.

HABITAT In dry woods, thickets, and openings.

FLOWERING August to October.

Sundial Lupine

Lupinus perennis L.

Erect, bushy, hairy, many-stemmed, freely branching perennial up to 1 m tall. **Flowers** usually blue or violet (rarely pinkish or white), about 1 cm long, borne short pedicels in loose terminal racemes. Standard nearly round, the sides turned back, wings completely covering the keel; keel curved; calyx strongly 2-lipped and very hairy; stamens 10, the filaments united into a closed tube for half their length; pistil solitary, the ovary superior, covered with long, soft, silky hairs, the style long, tapering, curved, the stigma appearing as a terminal fringe; **pod** 3–5 cm long, hairy, the 2 halves coiling after opening. **Leaves** alternate, compound, circular, the 7–11 leaflets radiating from the top of the petiole like the spokes of a wheel; leaflets narrow, oblanceolate to obovate, abruptly tipped, entire.

HABITAT In open woods and clearings, usually in dry sandy soil.

FLOWERING April to July.

NOTES This is the only native Michigan representative of this genus, which is so well known in the West. Some species are known as **Bluebonnet** in Texas. Several species of Lupine are cultivated for ornament, but they tend to die out in Michigan gardens. **Blue-Pod Lupine**, *Lupinus polyphyllus* Lindl., sometimes escapes to roadsides in the Upper Peninsula.

White Sweet Clover

Melilotus albus Medik.

Erect annual or biennial 1–3 m tall, fragrant on drying. **Flowers** small, white, borne in long, narrow, spikelike, often 1-sided racemes from the axils of the upper leaves. Standard nearly round, a little longer than the wings; wings fastened to the keel; calyx tube bell-shaped, the lobes nearly equal; the 10th stamen free, the anthers all alike; **pod** ovoid, straight, leathery, wrinkled, longer than the calyx. **Leaves** 3-foliolate, the terminal leaflet stalked; leaflets narrowly oblong, pinnately veined, the veins ending in the teeth.

HABITAT Along roads, ditches, and fences and in waste places. This common species was introduced from Europe as a forage and honey plant and is now widespread.

FLOWERING May to October.

WHITE SWEET CLOVER

Yellow Sweet Clover (not illustrated)

Melilotus officinalis (L.) Lam.

Very similar to the preceding species but usually not as tall; the flowers yellow and the standard nearly the same length as the wings. These two species sometimes both treated as *M. officinalis.*

HABITAT In waste or cultivated ground.

FLOWERING May to October.

SIMILAR SPECIES Two other common introduced clovers with toothed 3-foliolate leaves and the terminal leaflet stalked are naturalized in Michigan: **Black Medick**, *Medicago lupulina* L., is a creeping, prostrate annual. The small yellow flowers are crowded into a compact head, and the ovaries and pods are kidney-shaped. The pods are black when mature. **Alfalfa**, *Medicago sativa* L., one of our most valuable forage plants, is a weak-stemmed perennial with blue, blue-violet, or whitish flowers in short cylindric heads. The pod is coiled into a loose spiral of 1–3 turns.

KEY TO TRIFOLIUM (CLOVER) SPECIES

1 Flowers white, pink, or rose to red 2
1 Flowers yellow .... *T. aureum*
 2 Stems creeping, flowers white .... *T. repens*
 2 Stems erect or ascending 3
3 Heads pink and white; their stalks leafless .... *T. hybridum*
3 Heads red, having a pair of leaves at base .... *T. pratense*

LARGE HOP CLOVER

RED CLOVER

WHITE CLOVER

Large Hop Clover

Trifolium aureum Pollich

SYNONYM *Trifolium agrarium* L.

Glabrous or hairy, trailing or ascending, many-branched annual up to 5 dm tall. **Flowers** yellow, becoming brown, borne on short pedicels in short-cylindric, 1–2-cm. long heads. Corolla brown and distinctly streaked in age, calyx 2-lipped; **pod** straight. **Leaves** petioled, 3-foliolate; leaflets sessile or nearly so; stipules linear.

HABITAT Along roadsides, in waste places and dry fields.

FLOWERING May to September.

SIMILAR SPECIES Two other, less common but quite similar, yellow clovers have stalked terminal leaflets. The yellow clovers can be distinguished from the similar **Black Medick**, *Medicago lupulina* L., by their straight pods and 2-lipped calyx.

Red Clover

Trifolium pratense L.

Erect, partially reclining, or ascending biennial or perennial up to 8 dm tall. **Flowers** magenta to nearly white, 13–20 mm long, borne in dense globose to ovoid heads, the heads sessile or on peduncles, which are subtended by a pair of more or less hairy, opposite leaves. **Leaves** alternate, the lower ones long-petioled, the upper ones short-petioled to sessile, palmately 3-foliolate; the leaflets short-stalked, oval to cuneate-obovate, finely toothed; stipules oblong.

HABITAT In clearings and old fields and along roadsides.

FLOWERING May to August.

USES Extensively grown for hay.

White Clover

Trifolium repens L.

Low, smooth perennial with long creeping stems. **Flowers** white, becoming brown, borne on pedicels in dense globose solitary heads 1.52.3 cm in diameter. Corolla butterfly-like, the petals separate, 2–3 times as long as the calyx; standard oblong to ovate, not spreading, claws of the petals more or less united with the stamen tube below; calyx 5-cleft, the teeth bristle-tipped, similar, nearly equal; 10th stamen free; pistil 1, the **legume** usually 4-seeded. **Leaves** palmately 3-foliolate, long-petioled; leaflets broadly obovate, attached at the same point, finely toothed.

HABITAT In fields, waste ground, open pastures, and woods.

FLOWERING May to October.

USES This clover is commonly planted in lawns and as a forage crop. Like all other clovers in Michigan, it is an introduced species.

Alsike Clover (not illustrated)

Trifolium hybridum L.

Quite similar to **White Clover** but differs by having leafy erect or ascending stems up to 5 dm tall; the pink and white heads borne on peduncles which are longer than the subtending leaves.

HABITAT Along roadsides and in clearings statewide; naturalized from Europe.

FLOWERING June to October.

KEY TO VICIA (VETCH) SPECIES

1 Racemes as long as, or longer than, subtending leaves; leaflets 6–12 pairs 2

1 Racemes shorter than subtending leaves; leaflets 4–9 pairs .... *V. americana*

2 Plants sprawling, stems weak, with spreading hairs .... *V. villosa*

2 Plants nearly erect, stems quite stiff, with tiny appressed hairs .... *V. cracca*

American Vetch

Vicia americana Muhl. ex Willd.

Glabrous or nearly glabrous climbing or trailing perennial with stems up to 1 m long. **Flowers** 3–9, bluish-purple to rose-purple, 1.5–2 cm long, in racemes which are shorter than the subtending leaves. Calyx oblique, the lower teeth narrow and tapering, the upper teeth short and broad; **pods** 2.5–3.5 cm long. **Leaflets** 4–9 pairs, nearly opposite, oblong-ovate to elliptic, 1.5–3.5 cm long and less than half as broad.

HABITAT On damp or gravelly shores, in thickets and meadows.

FLOWERING May to July.

AMERICAN VETCH

HAIRY VETCH

Hairy Vetch, Winter Vetch

Vicia villosa Roth

Grayish, long-hairy, weak, climbing or partly reclining perennial, with tough, ridged stems up to 1 m long. **Flowers** 10–30, violet, violet and white, or sometimes all white, 12–20 mm long, strongly overlapping in compact, 1-sided racemes which equal or over-top the subtending leaves. Standard overlapping the wings, the blade less than half as long as the claw; wings attached to the middle of the keel; wings and keel often whitish; calyx irregular, 5-toothed, the upper side enlarged and the pedicel seeming to be attached to the lower surface (instead of at base or ends); stamen tube oblique at apex, the 10th stamen free; style threadlike, with a tuft or ring of hairs at the apex; **pod** beaked, 1–3 cm long. **Leaves** even-pinnate with a forking tendril at the end; leaflets 6–12 pairs, oblong-elliptic, abruptly pointed, obscurely veined; stipules hairy, half arrow-shaped.

HABITAT In fields and along roadsides. Introduced from Europe as a forage plant; now naturalized and widespread.

FLOWERING June to September.

SIMILAR SPECIES **Wood Vetch**, ***Vicia caroliniana*** Walt., is another weak, trailing or climbing vetch. It has 7–20 white to pale-blue flowers borne in a loose raceme which is as long as or longer than the subtending leaf. The calyx is nearly regular, with the teeth nearly equal. This species grows in rich woods, in thickets, and on shores, and flowers from April to June.

TUFTED VETCH, CANADA-PEA

Tufted Vetch, Canada-Pea

Vicia cracca L.

Quite similar to the preceding species, but the plants greener, less hairy, the hairs when present appressed, and the plants generally stronger and more upright in appearance. This species often stands without support, forming large, roundish patches. The racemes are stiffer, the blade of the standard is nearly as long as the claw, and the calyx is merely rounded on the upper side. The leaflets tend to be narrower and upstanding.

HABITAT In fields, waste places, and thickets, on shores, and along roadsides. Naturalized from Europe.

FLOWERING May through August.

GENTIANACEAE *Gentian Family*

This family, worldwide in distribution but most abundant in temperate regions, includes 104 genera and about 1600 species; 8 genera are present in Michigan. Its main importance to us is for ornamentals: gentians, *Centaurium,* and others. The family is characterized by opposite, usually sessile leaves without stipules; the corolla is tubular to bell- or wheel-shaped; the stamens are the same number as the corolla lobes and opposite them; the sepals are united; and the single, superior ovary develops into a many-seeded capsule.

KEY TO GENTIANACEAE (GENTIAN FAMILY) SPECIES

1 Leaves 3-foliolate . . . . (see ***Menyanthes trifoliata*, p. 156**)
1 Leaves simple 2
  2 Corolla (at least of larger flowers) spurred at base . . . . ***Halenia deflexa***
  2 Corolla not spurred at base 3
3 Flowers solitary, corolla 4-lobed, lobes fringed 4
3 Flowers borne in clusters; parts in 5's 5
  4 Upper leaves ovate to broadly lanceolate; ovary and capsule distinctly stalked . . . . ***Gentianopsis crinita***
  4 Upper leaves linear to narrowly lanceolate; ovary and capsule sessile or nearly so . . . . ***Gentianopsis virgata***
5 Corolla without folds, teeth, or secondary lobes in sinuses between lobes . . . . ***Gentianella quinquefolia***
5 Corolla having folds, plaits, or teeth between lobes, these usually different in texture or color 6
  6 Flowers some shade of blue or purplish 7
  6 Flowers yellowish to greenish white . . . . ***Gentiana alba***
7 Corolla nearly closed at top, lobes incurved and shorter than intervening folds . . . . ***Gentiana andrewsii***
7 Corolla open or spreading at top, lobes longer than folds . . . . ***Gentiana puberulenta***

Yellow Gentian

Gentiana alba Muhl. ex J. McNab

SYNONYM *Gentiana flavida* Gray

Stout, smooth, usually unbranched perennial up to 5.5 dm tall. **Flowers** yellowish-white to greenish, 3–4.5 cm long, crowded in terminal and sometimes axillary clusters. Corolla cylindric to bell-shaped, open at the top, the 5 erect primary lobes much longer than the broad, irregularly toothed secondary lobes; calyx 5-lobed; stamens 5, the anthers united; ovary and capsule stalked. **Leaves** opposite, sessile, lanceolate to ovate, rounded or heart-shaped at base, 3–5-nerved, the upper leaves the largest, forming an involucre, the lowest leaves reduced to bracts.

HABITAT In thin woods in wet or dry soil, on rocky shores, and in meadows.

FLOWERING mid-August to mid-October in southern Michigan.

YELLOW GENTIAN

CLOSED BOTTLE GENTIAN

STATUS Endangered in Michigan.

SIMILAR SPECIES **Red-Stemmed Gentian,** ***Gentiana rubricaulis*** Schwein., which looks very similar, is common in northern Michigan. It has pale blue flowers, reddish stems, the corolla lobes are longer than the entire or 1-toothed secondary lobes, and the calyx lobes are narrow and glabrous. Growing in wet meadows; flowering in August and September.

Closed Bottle Gentian

Gentiana andrewsii Griseb.

Smooth, stout, usually many-stemmed, unbranched perennial up to 8 dm tall. **Flowers** blue-violet (rarely pinkish or white), sessile, several in a terminal cluster and in some of the upper leaf axils. Corolla 3–4 cm long, closed or nearly closed at the top, the 5 lobes narrow, truncate, and joined by longer whitish secondary lobes, which are linear, spreading, fringed with hairs; stamens 5, the broad, flat filaments attached to the corolla tube, the anthers coherent; ovary long, stalked. **Leaves** sessile, ovate-lanceolate, 3–7-nerved, the 4–6 upper leaves often larger and forming an involucre, the lowest leaves the smallest.

HABITAT In moist soil in meadows, prairies, fringed at the summit; calyx 5-lobed, and low thickets and on open wooded the sinuses rectangular, the lobes shores.

FLOWERING August to October.

DOWNY GENTIAN

Downy Gentian

Gentiana puberulenta J. Pringle.

SYNONYM *Gentiana puberula* Michx.

Rigid, hairy perennial with 1 to few, often purplish, simple or branched stems up to 5 dm tall. **Flowers** showy, 3.5–5 cm long, blue-purple, crowded at the summit of the stem and in the upper leaf axils. Corolla open-funnel-shaped, 5-lobed, the primary lobes entire, spreading, twice as long as the 2-cleft secondary lobes; stamens 5, anthers free or promptly separating; ovary stalked. **Leaves** 13–19 pairs below the inflorescence, opposite, sessile, lanceolate, the lower ones the smallest; veins parallel.

HABITAT In dry soil, on hills and prairies, often in sand, rarely in low open ground.

FLOWERING late August through October.

Stiff Gentian, Agueweed

Gentianella quinquefolia (L.) Small

SYNONYM *Gentiana quinquefolia* L.

Simple or stiffly branched annual or biennial up to 8 dm tall, the stems strongly angled and winged. **Flowers** pale violet-blue, lilac, or sometimes greenish-white, solitary or clustered in the axils of leaves and at the summit of the stems. Corolla funnel-shaped, 1.6–2 cm long, 5-lobed, the lobes triangular, bristle-tipped; calyx cleft to below the middle; **capsule** on a slender stalk. **Leaves** opposite, the middle and upper ones ovate-lanceolate, clasping by rounded bases, up to 8 cm long; lower leaves oblong-ovate; veins 3–7, parallel.

HABITAT In dry or moist, shaded ground, in thin woods or thickets, on river banks, and in swamps.

FLOWERING August to mid-October.

STATUS Threatened in Michigan.

STIFF GENTIAN, AGUEWEED

Greater Fringed Gentian

Gentianopsis crinita (Froel.) Ma

SYNONYM *Gentiana crinita* Froel.

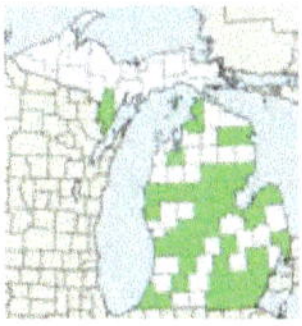

Stiff, erect, simple or branched biennial or annual up to 9 dm tall, the stems somewhat 4-angled. **Flowers** showy, violet, blue or rarely white, 4–5.5 cm long, solitary and upright on stiff, erect peduncles 1–11 cm long at the ends of the branches. Corolla slenderly bell-shaped, 4-lobed, the lobes wide-spreading in sunshine and closing at night or in cloudy weather, rounded and much-fringed at the summit, slightly fringed on sides; calyx 2.5–4 cm long, cleft about halfway to the base, the lobes and tube keeled, the 2 inner lobes the shorter; stamens 4, attached to the corolla tube; pistil 1, the ovary superior, long and slender, stalked; seeds numerous, covered with minute projections. **Leaves** opposite, sessile, the middle and upper ones ovate to ovate-lanceolate, rounded to somewhat heart-shaped at base, parallel-veined.

HABITAT In wet thickets and low woods, along streams, in damp open meadows and fields.

FLOWERING late August to October.

GREATER FRINGED GENTIAN

LESSER FRINGED GENTIAN

SPURRED GENTIAN

■ Lesser Fringed Gentian

Gentianopsis virgata (Raf.) Holub
SYNONYM *Gentiana procera* Holm

Greatly resembling the preceding species, but the plants slightly smaller, 2–6 dm tall, the corolla lobes deeply fringed on the sides but having short, thick teeth at the apex, the capsule and ovary sessile or nearly so, the middle and upper leaves narrowly linear-lanceolate, up to 8 mm wide.

HABITAT On marly lakeshores, in boggy prairies and sandy swamps.

FLOWERING mid-August to October.

■ Spurred Gentian

Halenia deflexa (Sm.) Griseb.

Leafy annual or biennial up to 9 dm tall, simple or branched above. **Flowers** greenish or bronze, borne in terminal and axillary umbels or sometimes solitary. Corolla nearly tubular, 8–15 mm long, 4-lobed to below the middle and having 4 basal spreading spurs; calyx 4-parted; stamens 4, inserted on the corolla tube; **capsule** oblong, flattish. **Leaves** opposite, the upper leaves sessile, lanceolate to ovate, 3–5-nerved; lower leaves petioled, oblong-spatulate, the lowest often forming a rosette.

HABITAT In moist or marshy ground, in wet grassy open places, in woods or cedar bogs.

FLOWERING July to September.

■ GERANIACEAE *Geranium Family*

This is a widely distributed family of 104 genera and about 830 species; 2 genera are in Michigan. It is important primarily for ornamentals, the florist's geranium (*Pelargonium zonale*) being well known. A few species are grown for the aromatic oil in the foliage, and some species of *Geranium* and *Erodium* are cultivated in the garden.

The family is characterized by its regular (or nearly regular) 5-parted flowers, and the beaked fruit which separates elastically to throw out the seeds. The leaves have stipules.

Red-Stem Stork's-Bill

Erodium cicutarium (L.) L'Hér. ex Ait.

Rosette-forming winter annual or biennial up to 2.5 dm tall. **Flowers** roseate or purple, about 1 cm wide, borne in several-flowered, long-peduncled umbels in the axils of the leaves. Petals 5, the upper ones slightly smaller than the lower; sepals 5, bristle-tipped; stamens with anthers 5; ovary deeply 5-lobed; ripe **fruit** long, slender, and pointed, the styles separating below and spirally twisting, bearded on the inner side. **Leaves** pinnately compound, the leaflets linear, finely pinnately cut, sessile; leaves mostly basal at first but stem leaves produced as the stems elongate.

RED-STEM STORK'S-BILL

HABITAT In fields, along roads, and in waste places. Introduced from Europe.

FLOWERING March to November.

NOTES This is a valuable forage plant for small mammals, particularly in the West. Many birds also eat the seeds.

Wild Geranium, Spotted Crane's-Bill

Geranium maculatum L.

Hairy, erect, bushy perennial up to 6 dm tall. **Flowers** rose-purple, rose, bluish, or whitish, usually with darker veins, 2.5–4 cm wide, erect or nodding in loose, few-flowered terminal corymbs, the pedicels and peduncles hairy to long-hairy. Petals 5, separate; sepals 5, separate, lanceolate or narrowly ovate; stamens 10 in 2 circles, the outer stamens maturing first; filaments flattened, woolly at the base, the anthers attached by the middle; ovary deeply 5-lobed, the style compound, the stigma with 5 curved branches; **capsule** elongate, up to 5 cm long, tipped with the persistent style, the 5 divisions separating from the base at maturity, arching up and out to throw out the single seed. **Leaves** deeply 3–5-palmately cut, the segments narrowed at the base, the margin much incised; basal leaves often larger, up to 15 cm wide, long-petioled.

HABITAT In rich or moist woods, thickets, and meadows, along roadsides. Mostly in the southern part of the state.

FLOWERING May and June.

NOTES This is one of the most satisfactory wildflowers for the garden; it flowers freely and is attractive even when not in bloom. It requires little attention and multiplies readily, sometimes too readily. American Indians used the dried and powdered roots medicinally.

WILD GERANIUM

HERB-ROBERT

Herb-Robert

Geranium robertianum L.

Strong-scented, nearly erect, hairy or glandular, many-stemmed and manybranched annual up to 4.5 dm tall. **Flowers** rose to reddish-purple, sometimes white, 1–1.5 cm wide, borne in pairs at the ends of axillary peduncles. Petals 5, veined with a deeper color, the claws very narrow, the blade obovate, indented at the end. **Leaves** compound with 3–5 bright green, pinnately cut leaflets, the lobes rounded and abruptly pointed, the terminal leaflet stalked.

HABITAT In rocky woods, thickets, and ravines and on gravelly shores.

FLOWERING June and July.

SIMILAR SPECIES In addition to some introduced species, we have another native geranium, ***Geranium bicknellii*** Britt. Its leaves are deeply divided but not compound, the petals are roseate, and only slightly longer than the calyx lobes. It is common on rocky outcrops, especially in the northern part of the state.

HYDROPHYLLACEAE *Waterleaf Family*

This is a small but widely distributed family, containing about 300 species, and particularly abundant in western North America. The family is sometimes grouped as a subfamily within the larger Borage family. Three genera occur in Michigan, one of which, *Phacelia,* is rather uncommon.

BROAD-LEAF WATERLEAF

Broad-Leaf Waterleaf

Hydrophyllum canadense L.

Erect, smooth perennial up to 7 dm tall from scaly, toothed rootstocks. **Flowers** whitish to pale blue, borne in coiled, somewhat 1-sided clusters in the axils of the leaves. Corolla bell-shaped, 5-lobed, the tube having nectar-bearing grooves opposite the lobes; calyx lobes 5, linear or awl-like, having minute teeth in the sinuses; stamens 5, extending well beyond the corolla; pistil 1, the ovary superior, the style extending beyond the corolla. **Leaves** mostly basal, palmately 5- to 7-lobed (sometimes compound), heart-shaped at base, unequally toothed.

HABITAT In rich moist woods, borders of woods, and thickets.

FLOWERING June and July.

USES American Indians ate the roots in times of food scarcity.

NOTE Two other species of waterleaf are found in Michigan: **Great Waterleaf**, ***Hydrophyllum appendiculatum*** Michx., is hairy, the flowers are violet, purple, or whitish, and the calyx has a small reflexed lobe in each sinus; **Virginia Waterleaf**, ***Hydrophyllum virginianum*** L., is quite smooth, the flowers are lavender to white, and the leaves are pinnately divided.

VIRGINIA WATERLEAF

HYPERICACEAE *St. John's-Wort Family*

This family is composed of 6 genera and about 700 species, but these numbers vary depending on the authority. The family is of limited economic importance; a few species of St. John's-wort are grown as ornamentals. The members of this family in Michigan (genus *Hypericum*) are characterized by pellucid-dotted or black-dotted, opposite leaves and stamens usually united in 3 or more clusters or fascicles.

KEY TO HYPERICUM (ST. JOHN'S-WORT) SPECIES

1 Plants herbaceous 2
1 Plants shrubby . . . . ***H. kalmianum***
  2 Flowers pinkish . . . . ***H. fraseri***
  2 Flowers yellow 3
3 Styles 3, separate to base; plants of dry places . . . . ***H. perforatum***
3 Styles united into a beak; plants of damp places . . . . ***H. majus***

Marsh St. John's-Wort

Hypericum fraseri (Spach) Steud.

SYNONYM *Hypericum virginicum* var. *fraseri* (Spach) Fern.

Usually simple but sometimes bushy-branched perennial with 1 to several, often purplish stems up to 8 dm tall; usually stoloniferous. **Flowers** relatively few, flesh-color to mauve, large, borne in axillary or terminal clusters. Petals 5; sepals 5, oblong to elliptic, blunt; stamens 9, in 3 clusters which alternate with 3 large, orange glands; **capsule** often rounded. **Leaves** opposite, sessile and heart-shaped at base or clasping; oblong to oblong-ovate, entire, 2–7 cm long, often purplish, dotted with translucent spots; upper leaves smaller.

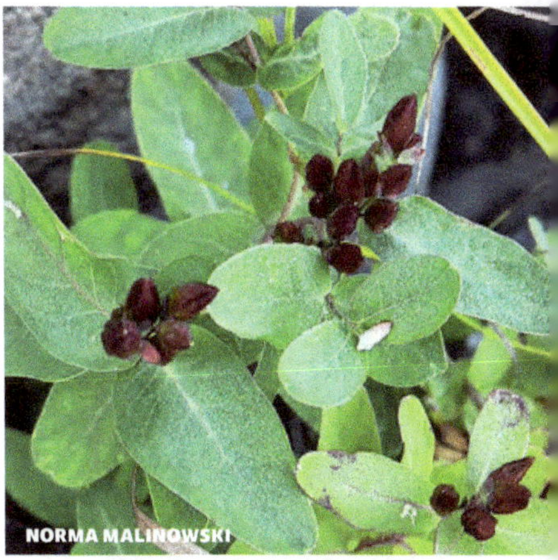

MARSH ST. JOHN'S-WORT

HABITAT In open marshes, swamps, bogs, and in wet sand along lakes and in shallow water.

FLOWERING July and August.

NOTES It is often hard to find these flowers open, even when the weather appears favorable.

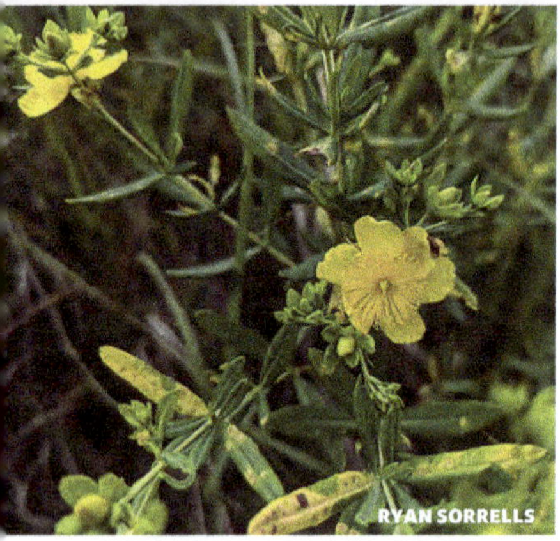

SHRUBBY ST. JOHN'S-WORT

Shrubby St. John's-Wort

Hypericum kalmianum L.

Small, slender, freely branching shrubs up to 6 dm tall. **Flowers** bright yellow, 2–3 cm wide, solitary or in few-flowered, leafy, terminal cymes. Petals 5, separate, spreading, oblique at apex; sepals 5; stamens very numerous, separate, soon falling; pistil 1; styles usually 5, united below; **capsule** about 5 mm long. Branches 4-sided, the branchlets flattened and 2-sided; bark papery, whitish. **Leaves** numerous, opposite, sessile, rather crowded, linear to oblanceolate, entire or somewhat wavy-margined, 3–6 cm long, finely dotted with translucent spots; veins, except for the midrib, obscure.

HABITAT In rocky and sandy soil. In Michigan mostly on the shores of the Great Lakes.

FLOWERING July and August.

GREATER CANADIAN ST. JOHN'S-WORT

Greater Canadian St. John's-Wort

Hypericum majus (Gray) Britt.

Erect perennial with solitary or tufted stems up to 7 dm tall, often reddish below. **Flowers** numerous, small, yellow, borne in axillary or terminal cymes. **Leaves** commonly ascending, opposite, lanceolate, rounded and sessile or clasping at the base, upper ones 3–7-nerved at the base.

HABITAT On margins of dune ponds, in marshes, on wet sandy or gravelly lakeshores, and on river banks.

FLOWERING late June to early September.

SIMILAR SPECIES Several other small herbaceous species, and one very large, tall species, ***Hypericum ascyron*** L., also occur in the state.

Common St. John's-Wort

Hypericum perforatum L.

Smooth, many-branched perennial up to 6 dm tall, having translucent dots and black spots on many parts and frequently producing runners at the base. **Flowers** numerous, bright yellow, about 2.5–3 cm broad, borne in leafy terminal cymes. Petals 5, separate, oblique at apex, unequal in size, frequently with black dots along the margin; sepals 5; stamens numerous, united at the base into 3–5 clusters, often with black dots; pistil 1, the ovary superior, ovoid, 3-celled, the styles 3, spreading; **capsule** 3-celled, many-seeded. **Leaves** opposite, sessile, narrowly oblong to lanceolate, entire, smooth, having small transparent dots (best seen when held up to the light) and black dots. In dryish open ground, often in sandy soil, in pastures, along roads and railways.

HABITAT Introduced from Europe; now a common weed.

FLOWERING July to September.

NOTES This species produces an interesting photosensitization in unpigmented skin of cattle, sheep, and horses. If a white-skinned animal is exposed to strong sunlight after eating these plants, it suffers a severe skin irritation which may result in blistering of the skin and falling hair.

COMMON ST. JOHN'S-WORT

LAMIACEAE *Mint Family*

A large family of some 227 genera and over 7,500 species worldwide; 33 genera occur in Michigan. It is centered in the Mediterranean region, where it often forms a dominant part of the vegetation. Among the volatile aromatic oils derived from this family are spearmint, peppermint, lavender, and rosemary. Hyssop, pennyroyal, thyme, sage, marjoram, savory, and basil, the last sacred to the Hindus, are among the important culinary herbs. Hoarhound is used in medicines. Salvia, Dragonhead, False Dragonhead, Monarda, Skullcap, Coleus, Deadnettle, and many others are grown for ornament.

Most of our representatives of this family can be easily recognized. They have square stems, opposite, simple leaves, and an aromatic odor, especially when crushed. The flowers are borne in pairs or whorls in the axils of the leaves. They are irregular to nearly regular, the petals are united, and the 2 or 4 stamens are borne on the corolla tube. The ovary is usually deeply 4-lobed and the single style is attached at the base of the ovary. The usually 4 nutlets are attached at the base.

The Verbena Family is quite similar to it in many respects, but the foliage is not aromatic; the ovary is merely slightly lobed, with the style borne at the top and the nutlets attached at the side.

KEY TO LAMIACEAE (MINT FAMILY) SPECIES

1 Corolla split on upper side, appearing to have a single 5-lobed lip (the lower), the stamens protruding through the split; ovary only slightly lobed .... ***Teucrium canadense***

1 Corolla not split; ovary deeply 4-parted............ 2

2 Calyx having a small knob-like projection on the upper side near the base; flowers blue .... ***Scutellaria galericulata*** or ***S. lateriflora***

2 Calyx without a projection on upper side; flowers blue or not 3

3 Stamens with anthers 4 4

3 Stamens with anthers 2 12

4 Corolla strongly 2-lipped, the upper lip concave............ 5

4 Corolla regular to weakly 2-lipped, the upper lip flat or nearly so 10

5 Upper pair of stamens longer than the lower pair 6
5 Upper pair of stamens shorter than the lower pair 7
6 Stem erect; flowers whitish, in terminal spikes or racemes .... ***Nepeta cataria***
6 Stem weak, trailing or creeping; flowers blue-purple, in small clusters in axils of leaves .... ***Glechoma hederacea***
7 Calyx decidedly 2-lipped .... ***Prunella vulgaris***
7 Calyx not strongly 2-lipped 8
8 At least basal leaves palmately lobed .... ***Leonurus cardiaca***
8 Leaves not lobed 9
9 Flowers 2–2.5 cm long, borne in elongate, slender, continuous spikes .... ***Phystostegia virginiana***
9 Flowers 1–1.5 cm long, borne in several separate, leafy whorls .... ***Stachys palustris***
10 Flowers mostly in compact, headlike terminal clusters .. ***Pycnanthemum virginianum***
10 Flowers mostly in axils of leaves 11
11 Corolla 2-lipped, throat inflated .... ***Clinopodium arkansana*** or ***C. vulgare***
11 Corolla 4-lobed, upper lobe largest, throat not inflated . ***Mentha arvensis*** or ***M. x piperita***
12 Flowers in headlike clusters 13
12 Flowers a few in axils of ordinary foliage leaves or in panicles 14
13 Corolla greatly elongated .... ***Monarda fistulosa*** or ***M. punctata***
13 Corolla not greatly elongated .... ***Blephilia hirsuta***
14 Calyx distinctly 2-lipped, lobes not all alike .... ***Hedeoma pulegioides***
14 Calyx lobes all essentially and shape same size .... ***Lycopus americanus***

HAIRY WOODMINT

Hairy Woodmint

Blephilia hirsuta (Pursh) Benth.

Hairy perennial up to 1 m tall. **Flowers** white to pale bluish, purple-spotted, borne in dense globose, axillary and terminal clusters. Corolla inflated in the throat, hairy, the 2 lips nearly equal, the upper lip erect, entire, the lower lip spreading, 3-cleft; calyx 2-lipped, the 3 upper teeth tipped with awns; stamens 2, extending beyond the corolla. **Leaves** long-petioled, ovate, round or heart-shaped at the base, strongly toothed.

HABITAT In moist or shady places.

FLOWERING early June to September.

SIMILAR SPECIES *Blephilia ciliata* (L.) Benth., a less hairy species with nearly sessile leaves, also occurs in dry woods in southern Michigan.

Low Calamint

Clinopodium arkansanum (Nutt.) House

SYNONYM *Satureja arkansana* (Nutt.) Briq.

Essentially glabrous, erect, simple or branching perennial up to 4 dm tall, with a Pennyroyal-like odor. **Flowers** rose-purple to lavender, one to a few on slender pedicels in the axils of the leaves. Corolla tubular, somewhat inflated above, 8–15 mm long, 2-lipped; calyx tubular, with 5 sharp-pointed teeth; stamens 4. **Leaves** opposite, often reddish, linear-oblong to linear-oblanceolate, entire, up to 2.5 cm long and less than 6 mm wide.

HABITAT On calcareous shores, banks, and barrens, and on rock.

FLOWERING June to October.

SIMILAR SPECIES Summer Savory, ***Satureja hortensis*** L., long cultivated as a culinary herb, sometimes escapes from cultivation and persists. It resembles Low Calamint, but the internodes are hairy, the flowers nearly sessile, smaller (5–7 mm long), and pale pinkish-purple to white.

Wild Basil

Clinopodium vulgare L.

SYNONYM *Satureja vulgaris* (L.) Fritsch

Erect, hairy, usually simple, slightly aromatic perennial up to 6 dm tall from a creeping base, often growing in large patches. **Flowers** usually rose-purple but at times purple, pink, or white, borne in dense, headlike, axillary and terminal clusters about 2.5 cm in diameter. Corolla tubular, slightly inflated, 2-lipped; calyx tubular, hairy, 2-lipped, the upper lip cleft; stamens 4, ascending under the upper lip of the corolla, of 2 unequal lengths; style cleft into 2 unequal lobes. **Leaves** opposite, elliptic-ovate, bright green on upper surface, pale and hairy beneath, short-petioled, the margin entire, wavy, or crenate.

HABITAT In open woods or thickets, along roads and in rocky or alluvial soil.

FLOWERING June to September.

Ground-Ivy

Glechoma hederacea L.

Smoothish to somewhat hairy, extensively creeping perennial. **Flowers** blue or purplish-blue, 1.6–2.2 cm long, usually 3 in each axil. Corolla 2-lipped, the tube elongate; calyx tubular, slightly oblique. **Leaves** round to kidney-shaped, crenate.

HABITAT In yards, along roads, in moist woods. Introduced from Europe.

FLOWERING April to July.

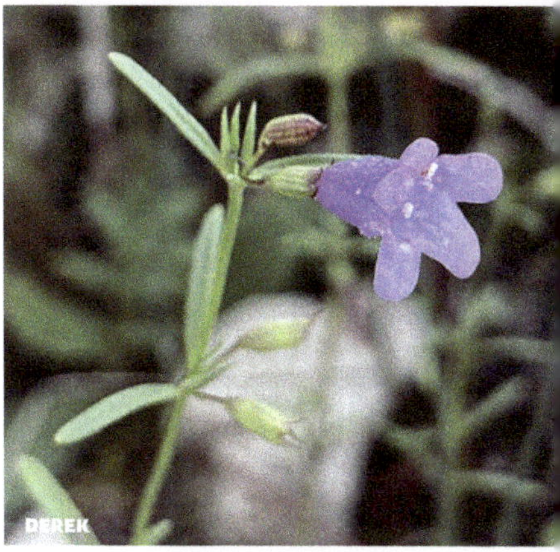

LOW CALAMINT

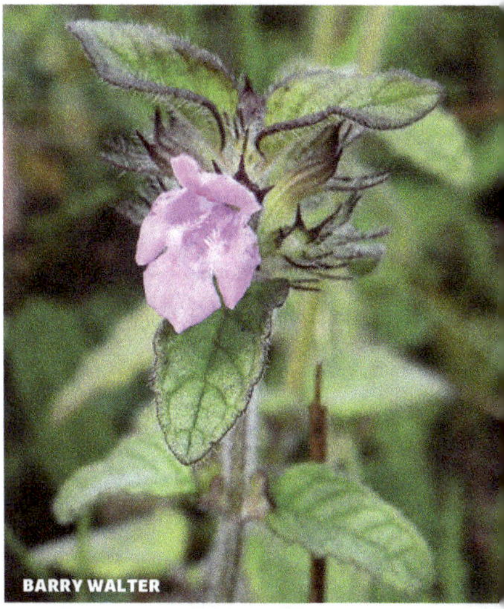

WILD BASIL

GROUND-IVY

AMERICAN PENNYROYAL

COMMON MOTHERWORT

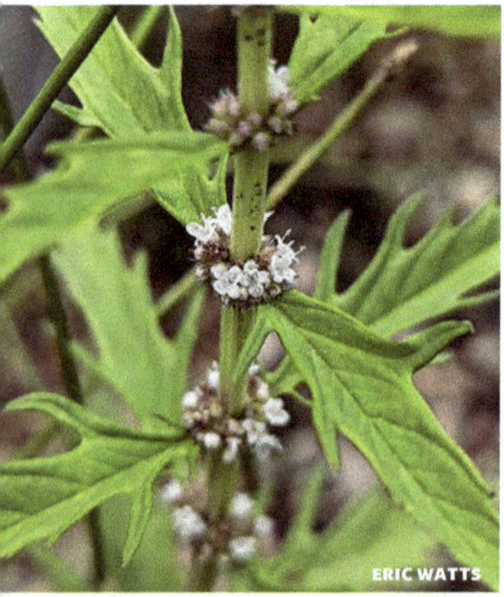

WATER-HOREHOUND, BUGLEWEED

American Pennyroyal

Hedeoma pulegioides (L.) Pers.

Erect, freely branching, hairy, odorous annual 1–4 dm tall. **Flowers** bluish, borne in few-flowered whorls in the axils of the leaves. Corolla scarcely longer than the calyx, weakly 2-lipped, the upper lip flat, erect, notched, the lower lip 3-lobed; calyx tubular, 2-lipped; stamens 2. **Leaves** opposite, small, the larger ones petioled, lanceolate to oblong-ovate, usually toothed.

HABITAT In moist or dry upland woods.

FLOWERING July to September.

SIMILAR SPECIES **Horse-balm**, *Collinsonia canadensis* L., also with 2 stamens, has a spreading panicle of yellowish flowers with a distinct lemon odor. The elongated corolla has 4 nearly equal lobes and a lower, much larger, lobe which is fringed around the margin. The leaves are numerous, large, and ovate. The species grows in rich moist woods.

Common Motherwort

Leonurus cardiaca L.

Coarse, erect, weedy perennial up to 1.5 m tall, usually many branched above, hairy on the 4 angles of the stem and on the nodes. **Flowers** pale purple, borne in whorls in the axils of the upper leaves. Corolla 2-lipped, the upper lip oblong and entire, somewhat arched, densely bearded, the lower lip spreading, 3-lobed; calyx teeth nearly equal, spiny; stamens 4, the upper pair the shortest. **Leaves** opposite, long-petioled; lower leaves broadly rounded at base and palmately lobed, the margin toothed; upper leaves smaller, wedge-shaped at base, more or less 3-lobed, the margin toothed or entire.

HABITAT In farmyards, along railways and roads, in dry open ground, and in moist shaded places. Introduced from Europe; now a common weed.

FLOWERING June to August.

Water-Horehound, Bugleweed

Lycopus americanus Muhl. ex W. Bart.

Simple or branching, erect, glabrous to slightly hairy, non-aromatic, stoloniferous perennial up to 1 m tall. **Flowers** white, small, borne in dense clusters in the axils of the leaves. Corolla nearly regular, bell-shaped, 4-lobed, with one lobe larger than the others; calyx bell-shaped, the teeth tapered to a long sharp point; stamens 2. **Leaves** opposite, ovate-lanceolate, the lower ones pinnately cut and often falling when the fruit is mature, the upper leaves merely toothed.

HABITAT In low wet ground, in swampy areas around lakes, along streams and wet shores.

FLOWERING July to September.

NOTES Medicinal horehound is not derived from this but from a European species. Several other *Lycopus* species also occur in Michigan.

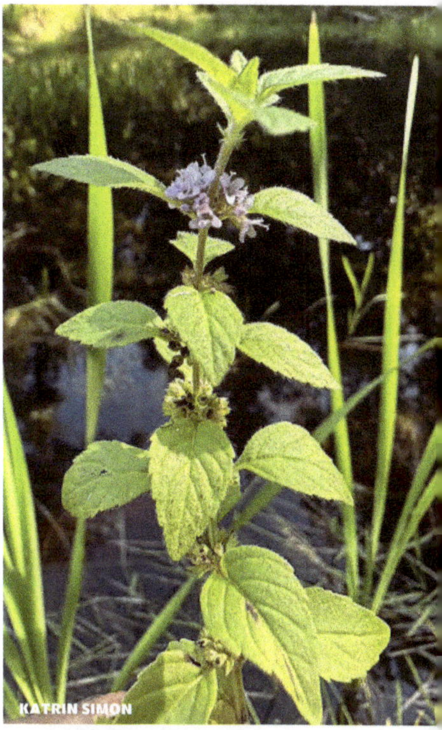

AMERICAN WILD MINT, FIELD MINT

American Wild Mint, Field Mint

Mentha arvensis L.

Erect, usually simple, aromatic perennial 1.5–8 dm tall, spreading by stolons, hairy on the angles of the stem. **Flowers** lilac, pinkish, or purplish, rarely white, small, borne in dense, subglobose clusters in the axils of the leaves. Corolla regular, short-tubular, with 4 spreading lobes, one of which is notched and often larger than the others; calyx 5-toothed, densely glandular; stamens 4, longer than the corolla, **Leaves** opposite, short-petioled, strongly veined, ovate to lanceolate, sharply toothed.

HABITAT In damp, open soil, along shores and streams, in marshes and swamps.

FLOWERING July to September.

NOTES This is our only native true mint. American Indians used the leaves in treating fevers and pleurisy.

PEPPERMINT

Peppermint

Mentha x *piperita* L.

Smooth, usually branched perennial up to 1 m tall, with a strong peppermint odor. **Flowers** purplish, small, borne in crowded, compact, ovoid, terminal spikes which elongate and become loose, the lower flower clusters becoming widely separated. Corolla nearly regular, the tube short, 4-lobed; calyx tubular, the 5 teeth nearly equal; stamens 4, all the same length. **Leaves** opposite, lance-oblong to ovate-oblong, toothed, definitely petioled, the petioles of the principal leaves 4–15 mm long.

HABITAT In damp open ground, in pastures, along roadside ditches and streams. Introduced from Europe, often cultivated and now wild over a large area.

FLOWERING late July to late September.

NOTES Oil of peppermint and peppermint extract are derived from this species. **Spearmint**, *Mentha spicata* L., also a frequent and persistent escape from cultivation, has more slender, elongate spikes, and the leaves are sessile or have petioles only 3 mm long, or less.

WILD BERGAMOT, OSWEGO-TEA

SPOTTED BEEBALM, HORSEMINT

Wild Bergamot, Oswego-Tea

Monarda fistulosa L.

Erect, many-stemmed, gray-green, hairy to smooth, very aromatic perennial up to 1.2 m tall. **Flowers** lavender, borne in solitary, compact, headlike clusters which are terminal or arise from the upper axils, and are subtended by numerous green or lavender-tinged leaflike bracts. Corolla 2–3 cm long, tubular, inflated above, 2-lipped, the upper lip erect, the lower spreading, 3-lobed; calyx tubular, nearly regular, having a ring of hairs at the throat; stamens 2, extending beyond the corolla. **Leaves** gray-green, very strongly scented, strongly pinnately veined, lanceolate, toothed.

HABITAT Along roadsides, in clearings and dry thickets, on open hillsides, and in borders of woods.

FLOWERING June to August.

USES The leaves of this species are sometimes made into tea. American Indians boiled the plant with Sweetflag and bone-marrow to make a perfumed hair dressing.

SIMILAR SPECIES **Scarlet Beebalm**, *Monarda didyma* L., is often cultivated and grows wild in Michigan, both natively (although rare) and as an escape from cultivation. It has bright red flowers 3–4.5 cm long.

Spotted Beebalm, Horsemint

Monarda punctata L.

Simple to branching, grayish-green, long-hairy, odorous perennial up to 1 m tall. **Flowers** cream-color to yellowish, with purple spots, borne in leafy, interrupted spikes, the leaflike bracts conspicuous, recurved, lilac to whitish. Corolla tube elongate, with a slightly inflated throat, the upper lip longer than the stamens. **Leaves** grayish (especially beneath) with fine hairs, lanceolate, coarsely and shallowly toothed.

HABITAT Along roads and railways, in sandy open ground, on low dune ridges, and on sandy plains. Mostly in southern Michigan.

FLOWERING mid-July to mid-September.

SIMILAR SPECIES **Lemon-Mint**, ***Monarda citriodora*** Cerv. ex Lag., somewhat resembles Dotted Monarda but is an annual and has a strong lemon odor. It does not grow wild in Michigan.

Catnip

Nepeta cataria L.

Stout, erect, branched, downy, grayish, strongly aromatic perennial up to 1 m tall. **Flowers** small, white, striped or dotted with rose-purple, crowded in many-flowered whorls in rather dense terminal inflorescences. Corolla tubular, dilated in the throat, strongly 2-lipped, the upper lip erect, large, and spreading, 2-lobed, the lower lip smaller, 3-cleft; calyx tubular, with 5 long narrow lobes, downy; stamens 4, the anthers lying on or close to the lower lip; ovary deeply 4-cleft; nutlets 4. **Leaves** opposite, soft-hairy, ovate to oblong, coarsely toothed.

HABITAT In yards, roadsides, and waste places. Introduced from Europe; now a common weed.

FLOWERING June to September.

USES The leaves are sometimes used for making tea. This plant is well-known as a favorite of cats and is often used in stuffing toys for them.

CATNIP

Obedient-Plant

Physostegia virginiana (L.) Benth.

Smooth perennial, simple or branched at the summit, up to 1.5 m tall. **Flowers** purple to roseate, rarely white, 2.5–3 cm long, borne singly in the axils of small bracts and forming elongate, terminal, wand-like spikes. Corolla funnel-shaped with an inflated throat, 2-lipped, the upper lip erect, nearly entire, the lower lip spreading, 3-lobed, with middle lobe notched; calyx sticky-glandular, short-tubular, with short, sharp-tipped teeth, slightly inflated in fruit. **Leaves** opposite, lanceolate, sharply toothed, the upper ones reduced.

HABITAT In damp thickets, on river banks and shores.

FLOWERING June to September.

OBEDIENT-PLANT

COMMON SELF-HEAL

Common Self-Heal

Prunella vulgaris L.

Hairy to nearly smooth, simple or sparsely branched perennial with at least the tips of the square stems erect. **Flowers** violet, lavender, bluish, pinkish, or whitish, borne in 3-flowered clusters subtended by a leafy bract, the clusters forming dense terminal heads which elongate and may be 5–10 cm long in fruit. Corolla ascending, 10–16 mm long, strongly 2-lipped, the upper lip erect, entire, arching over the stamens, the lower lip spreading or drooping, 3-lobed, the middle lobe the largest, rounded and finely toothed on the margin; stamens 4, the filaments forked at the top. **Leaves** opposite, oblong, ovate, or lanceolate, entire or somewhat toothed.

HABITAT In fields, woods, waste places. Native, but a common weed.

FLOWERING May to October.

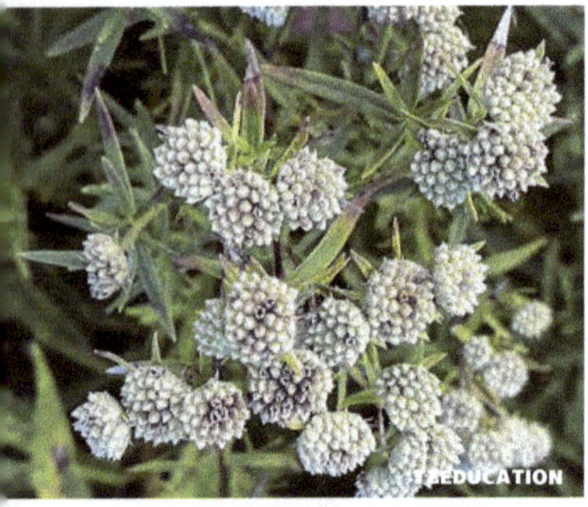

VIRGINIA MOUNTAIN-MINT

Virginia Mountain-Mint

Pycnanthemum virginianum (L.) T. Durand & B.D. Jacks. ex B.L. Rob. & Fernald

Pungent perennial up to 1 m tall, freely branching above the middle, finely hairy on the angles of the stems, smooth on the 4 sides. **Flowers** small, whitish or purplish, the lips dotted with purple, borne at ends of branches or in the axils of the leaves in dense crowded whorls which usually form heads. Corolla short, slightly irregular, 2-lipped, the upper lip straight, nearly flat; calyx tubular, nearly regular, the teeth all about the same length; stamens 4. **Leaves** numerous, crowded, sessile, smooth on the upper surface, often hairy on the veins beneath, linear-lanceolate, entire.

HABITAT In thickets, upland woods, and meadows, and on gravelly shores.

FLOWERING July to September.

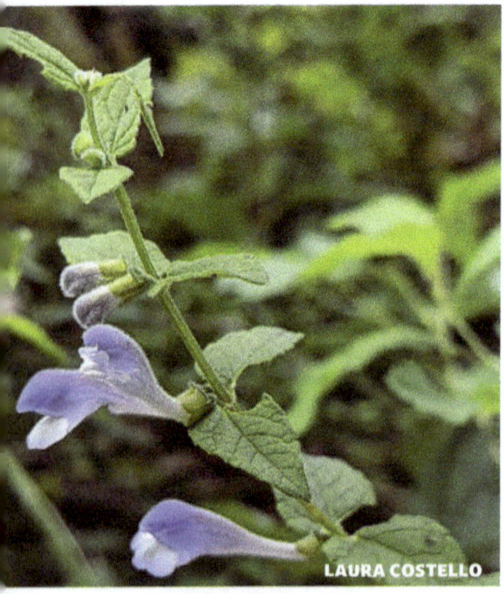

MARSH SKULLCAP

Marsh Skullcap

Scutellaria galericulata L.

SYNONYM *Scutellaria epilobiifolia* A. Ham.

More or less erect, non-aromatic perennial up to 7 dm tall, often freely branched, hairy on the ridges of the square stems. **Flowers** violet-blue, 15–25 mm

long, ascending, solitary on very short pedicels from the axils of the leaves. Corolla tubular, 2-lipped, the throat enlarged, the upper lip hooded, the lower lip broader and spreading, notched; calyx bell-shaped, 2-lipped, with a small but distinct protuberance on the upper side near the base, splitting to the base in fruit; stamens 4, in 2 pairs, the anthers bearded; nutlets 4. **Leaves** sessile or nearly so, having fine recurved hairs beneath, oblong-lanceolate, crenate, prominently veined.

HABITAT On rocky, sandy, or gravelly shores; in swamps, bogs, and wet thickets; along streams.

FLOWERING June to September.

NOTES This is a rather variable species, which, like so many other blue-flowered plants, sometimes has white or pink flowers.

MAD-DOG SKULLCAP

Mad-Dog Skullcap

Scutellaria lateriflora L.

Quite similar to the preceding species, but the flowers smaller (5–9 mm long) and borne in slender, compact, 2–24-flowered, 1-sided axillary racemes, the leaves petioled.

HABITAT In low, wet places, swampy woods and thickets, in meadows, along streams, frequently in a dense tangle of other vegetation,

FLOWERING July to September.

WOUNDWORT

Woundwort

Stachys palustris L.

Rank-smelling, simple or sometimes loosely branched perennial up to 1 m tall, having underground stolons which terminate in whitish tubers. **Flowers** rose-purple, mottled, borne in usually 6-flowered whorls in leafy terminal spikes which grow up to 2.5 dm long. Corolla 1.2–1.5 cm long, the lower lip 3-lobed, much longer than the nearly entire upper lip; calyx bell-shaped, hairy, the lobes about as long as the tube; stamens 4, ascending under the upper lip, the anthers in pairs, the upper pair shorter than the lower pair. **Leaves** opposite, lanceolate to narrowly ovate, pale green, firm and obscurely veined, hairy on both surfaces, slightly toothed; sessile or nearly so, rarely with elongate petioles.

HABITAT In damp ground in waste places, on shores, in fields and meadows. A variable, wide-ranging, circumpolar species.

FLOWERING late June to September.

SMOOTH HEDGENETTLE

AMERICAN GERMANDER

CARPET BUGLE

SIMILAR SPECIES The extensively creeping **Smooth Hedgenettle** (***Stachys tenuifolia*** Willd.) also occurs in Michigan. It has thin, conspicuously veined leaves which are dark green and glabrous above, and the marginal teeth are prominent. The flowers are pale rose and the calyx is glabrous, but sometimes has stiff hairs along the veins. Another Hedge-nettle, ***Stachys hyssopifolia*** Michx., is a delicate, linear-leaved Coastal Plain species which is known from the southwestern part of the Lower Peninsula.

■ American Germander

Teucrium canadense L.

Erect, hairy, very aromatic perennial up to 1 m tall. **Flowers** rose-purple to cream, the separate whorls of about 6 flowers forming slender, elongate spikes. Corolla tube short, the limb irregularly 5-lobed, split at the top so that the 4 small, nearly equal upper lobes and the larger lower lobe appear to form a single lip; calyx bell-shaped, with a dense, felt-like covering, 5-toothed; stamens 4, in 2 unequal pairs, protruding through the split in the corolla; ovary 4-lobed only at the apex, the style 2-lobed, protruding with the stamens; **fruit** consisting of 4 nutlets. **Leaves** opposite, short-petioled, light grayish-green, often tinged with purplish-red, soft-hairy, ovate to lance-oblong.

HABITAT In moist, usually open, soil, sometimes in open woods, common mud flats.

FLOWERING July and August.

SIMILAR SPECIES Another genus which appears to have a single lip, *Ajuga,* is represented in Michigan by 2 introduced species which sometimes escape from cultivation: **Carpet Bugle** (*Ajuga reptans* L.) and **Blue Bugle** (*Ajuga genevensis* L.), are low-growing and mat-forming, with stems up to 3 dm tall. The blue flowers are borne 4–6 in a cluster in the axils of the leaves, and have a very short (but not split) upper lip.

LENTIBULARIACEAE *Bladderwort Family*

This small but widely distributed family includes 3 genera, 2 of which occur in Michigan. Worldwide, about 400 species are reported. Its members are important elements of marsh and aquatic vegetation.

The flowers superficially resemble those of the Figwort family, but the plants are insectivorous and grow in water or wet places. The bladderworts have small, inflated bladders with a lid which traps insects. Butterwort catches insects on the sticky leaves. The spurred corolla and the calyx are 2-lipped; there are 2 stamens, and the fruit is a capsule.

KEY TO LENTIBULARIACEAE (BLADDERWORT FAMILY) SPECIES

1 Flowers lilac to purple, solitary; leaves broad and simple .... *Pinguicula vulgaris*

1 Flowers yellow, 2 or more on a stalk; leaves tiny or finely dissected .... *Utricularia*, 2

  2 Plants terrestrial, the minute leaves and bladders hidden in the mud .... *U. cornuta*

  2 Plants growing in water, having noticeable, finely dissected leaves.................. 3

3 Plants coarse, free-floating, the bladders borne on leafy branches .... *U. macrorhiza*

3 Plants slender, creeping, the bladders borne on separate leafless branches *U. intermedia*

Common Butterwort

Pinguicula vulgaris L.

Ground-hugging, rosette-forming perennial up to 20 cm tall. **Flowers** pale violet to reddish-purple, white in the throat, solitary on long peduncles. Corolla 2-lipped, the lips unequal, the tube short, the spur long and slender; calyx small, 2-lipped; stamens 2; ovary superior; **capsule** subglobose, splitting at the apex. **Leaves** all in a basal rosette, fleshy, yellowish-green, greasy in appearance and slimy to the touch, spatulate, elliptic, or oblong, the margin entire and usually inrolled.

HABITAT In bogs, on wet rocks or shores, chiefly where calcareous; rare in northern Michigan.

FLOWERING June and early July.

NOTES Small insects caught on the slimy leaf surfaces are digested by the plant.

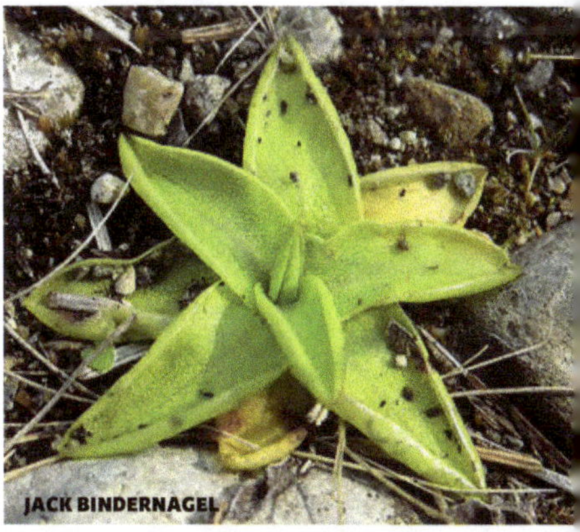

COMMON BUTTERWORT

Horned Bladderwort

Utricularia cornuta Michx.

Terrestrial plant of wet places, the erect, wiry scapes up to 3.5 dm tall, the branches and leaves few, delicate, hidden in the substratum. **Flowers** fragrant, bright yellow, usually 1–3 in a terminal raceme but sometimes up to 9, the freshly opened lower flower overlapping the buds above. Corolla strongly 2-lipped, the upper lip erect, the lower larger and spurred; calyx 2-lipped; stamens 2; pistil 1, the ovary superior; **cap-**

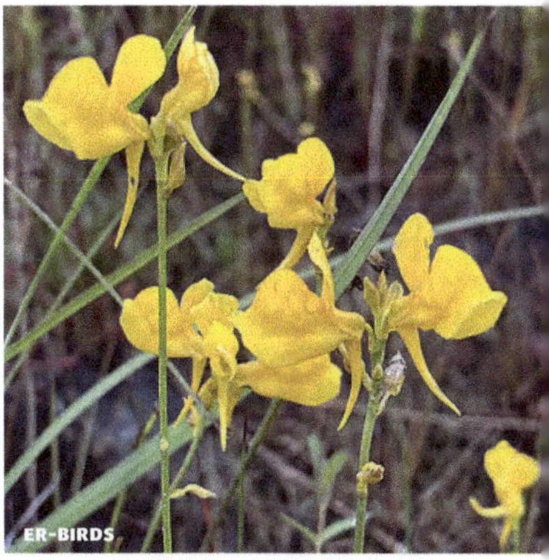

HORNED BLADDERWORT

sule covered by the beaked calyx, yellowish, persistent, upright. **Leaves** simple, threadlike, not readily seen, the minute **bladders** borne along the margin.

HABITAT On wet peaty, sandy, or muddy shores, in marly flats or bogs.

FLOWERING July to September.

NOTES This species often grows in great profusion, and may turn an area several hundred meters long into a golden sea. When growing in such profusion, as it does in spots along the Lake Michigan shore of the Upper Peninsula, it perfumes the air with an odor suggestive of locust trees or Scotch Broom.

Intermediate Bladderwort

Utricularia intermedia Hayne

INTERMEDIATE BLADDERWORT

Plants usually creeping on the bottom in shallow water and sending up rather stout, erect, bracted scapes up to 30 cm tall. **Flowers** yellow, 2–5 in a terminal raceme. Lower lip of the corolla about twice as long as the upper lip; **leaves** numerous, fan-shaped in outline, usually forked 3 times into very slender segments; **bladders** borne on separate leafless branches.

HABITAT Creeping at the bottom of shallow pools, on shores, and in quagmires, rarely free-floating.

FLOWERING May to September.

Common Bladderwort

Utricularia macrorhiza Leconte

SYNONYM *Utricularia vulgaris* L. p.p.

COMMON BLADDERWORT

Aquatic perennial with prolonged stems bearing plumose branches of foliage 2–11 cm in diameter, floating just below the surface of the water. **Flowers** 6–20, bright yellow, borne above the water on long, coarse scapes. Corolla lips about the same size, the spur longer than the lips. **Leaves** mostly alternate, sessile or short-petioled, elliptic to ovate in outline, very finely cut, the **bladders**, scattered among the leaves, relatively large and easily seen with the naked eye.

HABITAT In quiet waters.

FLOWERING May to September.

LINACEAE *Flax Family*

This cosmopolitan family of 9 genera includes about 250 species; a single genus is known from Michigan. Its principal economic importance is flax for linen and the linseed oil produced by some species. A few species are used as ornamentals.

Common Flax

Linum usitatissimum L.

Slender, erect, pale-green annual usually with a single stem up to 1 m tall. **Flowers** sky-blue, saucer-like, 2–2.5 cm broad, borne on very slender, erect pedicels in loose, few-flowered corymbs. Petals 5, separate or barely united at the base, broadly obovate, and flattened or shallowly notched at the apex, soon falling, closing in the evening and in cloudy weather; sepals 5, persistent, in 2 series, the inner ones bearing marginal hairs; stamens 5, the filaments broad, flat, united at the base; pistil 1, the ovary superior, the styles 5, long, slender, spreading, the stigmas linear; **capsule** globose, 5-celled. **Leaves** alternate, numerous, linear-lanceolate.

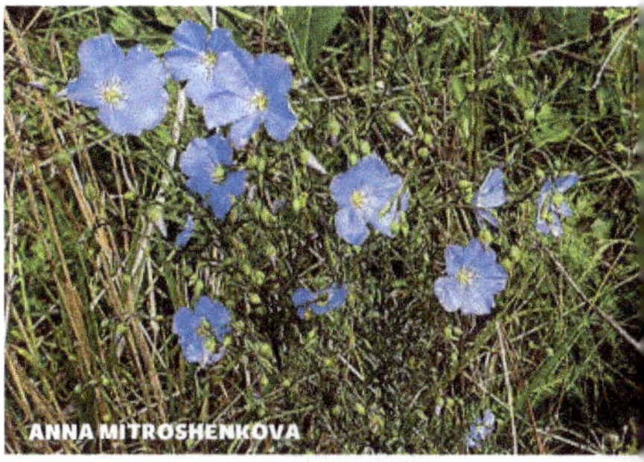

ANNA MITROSHENKOVA

COMMON FLAX

HABITAT In waste places and fields, along roads and railways. Introduced from Europe.

FLOWERING June to September.

NOTES Cultivated since prehistoric times, this species often grows wild here. It is important for the stem fibers, which are used in making linen and for linseed oil derived from the seeds.

SIMILAR SPECIES A similar but perennial species, **Prairie Flax**, ***Linum lewisii*** Pursh, is often grown for ornament. It usually has several stems spreading from the crown; the pedicels are arched and spreading, the stigmas capitate, and the sepals without marginal hairs. Five small native species, all with yellow or yellowish flowers, also occur in Michigan.

LYTHRACEAE *Loosestrife Family*

This widely distributed family is most abundant in the American tropics. It includes about 28 genera and about 620 species; 4 genera occur in Michigan. Several members are used for ornamentals, mostly in the southern part of the country.

The flowers are axillary or whorled. They have a well-developed hypanthium, i.e., a cuplike receptacle on which the calyx, corolla, and stamens appear to be borne. The ovary is superior. The stamens are usually the same number as the petals or twice as many. The style and stamens are often of 2 or 3 lengths.

Swamp Loosestrife

Decodon verticillatus (L.) Ell.

Herbaceous to slightly shrubby perennial with recurved stems up to 2.5 m long, the stems 4-sided, the bark of the submersed portion spongy, the arched branches rooting at the tip to form new plants. **Flowers** magenta, on short pedicels in axillary clusters. Petals 5; calyx with 5–6 erect teeth and a like number of longer, spreading, hornlike processes; stamens 10, of 2 lengths; pistil 1, the styles of 3 lengths; **capsule** globose, 3–5 celled. **Leaves** opposite or whorled, lanceolate to narrowly elliptic, nearly sessile, smooth, entire to somewhat wavy.

HABITAT Along margins of pools and small lakes, in bogs and swamps.

FLOWERING early June to September.

Purple Loosestrife

Lythrum salicaria L.

Somewhat hairy perennial, usually with many erect stems up to 1.5 m tall. **Flowers** showy, magenta to purple, cymose in axils of reduced upper leaves, and forming elongate, leafy slender, interrupted spikes. Petals usually 6 (5–7); calyx cylindrical, streaked, 5–7-toothed, having small appendages in the sinuses; stamens twice as many as the petals, of different lengths, inserted low on the throat of the calyx tube; pistil 1; **capsule** nearly cylindric, 2-celled. **Leaves** opposite or in 3's, sessile, lanceolate, heart-shaped or rounded to the base, 5–7.5 cm long; upper leaves reduced.

HABITAT In wet places, ditches, swamps, on lakeshores and low or marshy banks. Introduced from Europe; now common.

FLOWERING mid-June to early September.

NOTES Patches several acres in extent make a beautiful display when in bloom; however this noxious weed grows so vigorously that it tends to crowd out our native species.

SIMILAR SPECIES The native **Winged Loosestrife**, ***Lythrum alatum*** Pursh, is much less showy than Spiked Loosestrife. It is somewhat wand-like, simple or branched above, often becoming leafless below, the stems 4-angled and winged. Flowers reddish-purple, solitary and sessile or short-stalked in the axils of the upper leaves.

SWAMP LOOSESTRIFE

PURPLE LOOSESTRIFE

MALVACEAE *Mallow Family*

This family is distributed over much of the earth but is particularly abundant in the American tropics. There are about 246 genera and 4,225 species; 11 genera are known from Michigan. Economically the most important member is the cotton plant. The seeds produce long fibers which are used for cloth, cottonseed oil is pressed out of the seeds, and a meal is made from the remainder. Okra belongs to this family, and a European species of Althaea produces a mucilage which was the original basis for marshmallows. There are several well-known ornamentals: Hollyhock, Rose of Sharon, Poppy Mallow, Hibiscus, and Althaea.

Our members are easily recognized by the numerous stamens which are united around the style by their filaments and resemble a bottle brush in the center of the flower. The flowers are regular and perfect. There are 5 petals and 5 sepals; the ovary is superior and compound. The leaves are alternate and simple, and have stipules.

KEY TO MALVACEAE (MALLOW FAMILY) SPECIES

1 Flowers solitary .... *Hibiscus moscheutos*
1 Flowers in clusters 2
  2 Upper leaves deeply divided into 5–7 narrow segments .... *Malva moschata*
  2 Upper leaves shallowly lobed, nearly round or kidney-shaped .... *Malva neglecta*

Swamp Rose Mallow

Hibiscus moscheutos L.

SYNONYM *Hibiscus palustris* L.

Grayish, hairy perennial up to 2.5 m tall, having a mouse-like odor. **Flowers** large, showy, solitary, pink to purple or creamy-white. Petals 5, usually red or purplish at base; calyx 5-cleft, subtended by an involucre of numerous small bracts; stamens numerous, united by their filaments; pistils several; ovary and style united, the style 4–5 cm long, the upper half extending beyond the flower, with 5 densely hairy branches; **capsule** sub-globose, 2–2.5 cm long. **Leaves** alternate, broadly ovate to rounded, usually 3-lobed, those about the middle of the stem 7–18 cm long and 4.5–11 cm broad, but sometimes broader than long, dark green and glabrous on upper surface, velvety and grayish-hairy beneath.

HABITAT In salty, fresh, or brackish marshes.

FLOWERING August to October.

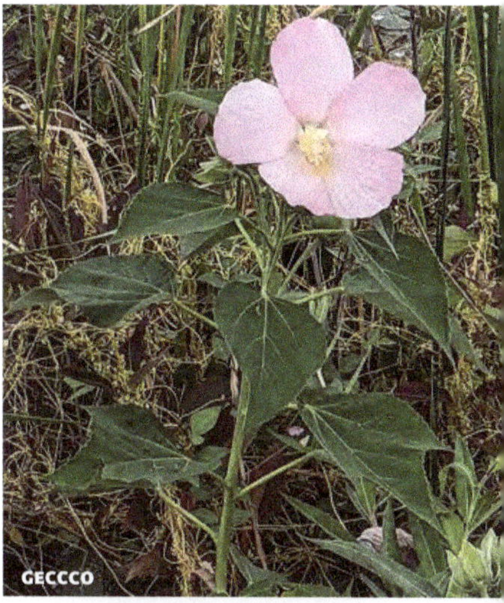

SWAMP ROSE MALLOW

Musk Mallow

Malva moschata L.

Erect or somewhat reclining, many-branched, hairy to nearly smooth, faintly musk-scented perennial 3–6 dm tall. **Flowers** white, pale pinkish, or rose, about 5 cm in diameter, slightly fragrant, borne in raceme-like clusters at the ends of branches, each flower subtended by an involucre of 3 small, linear, green bracts just below the calyx. Petals 5, triangular to broadly heart-shaped at apex, 2.4–3 cm long, attached to the stamen column at the base; calyx deeply 5-lobed, spreading; stamens numerous, the filaments united into a tube around the styles, the anthers pinkish or pale lilac; ovary densely hairy, of 15–30 carpels in a circle; **mature fruits** rounded on the back, sep-

MUSK MALLOW

COMMON MALVA

arating readily. **Basal and lower stem leaves** rounded, shallowly cleft or crenate or with 5 broad lobes; **upper stem leaves** usually 5–7-parted, the divisions deeply pinnately cut.

HABITAT Along roads and in waste places. Native of Europe; escaped from cultivation here.

FLOWERING July to September.

Common Mallow

Malva neglecta Wallr.

Low, sprawling, prostrate, or ascending biennial with stems up to 1 m long. **Flowers** pale lilac to white, in groups in the leaf axils. Petals heart-shaped at apex, twice as long as the calyx; calyx having a cluster of 3 small leaves at the base; **fruits** in a ring, usually 12–15. **Leaves** nearly round or kidney-shaped, often heart-shaped at the base, shallowly 5–9-lobed.

HABITAT In gardens, waste places, barnyards. A common weed; naturalized from Europe.

FLOWERING April to October.

MENYANTHACEAE *Buck-Bean Family*

A small family of of aquatic and wetland plants, with 6 genera and more than 60 species. Two genera in Michigan, the native **Buck-Bean** (*Menyanthes*) and the introduced **Floating-heart** (*Nymphoides peltata* (Gmel.) Kuntze), sometimes planted in water gardens and potentially invasive.

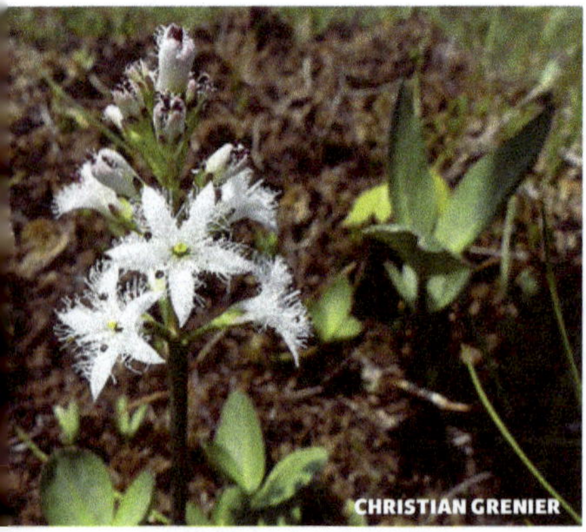

BUCK-BEAN, BOGBEAN

Buck-Bean, Bogbean

Menyanthes trifoliata L.

Low-growing smooth perennial with scapes and petioles up to 3 dm tall from a long rootstock. **Flowers** white or rose-tinged, borne in short, thick, compact racemes. Corolla with a short tube and 5 densely white-bearded lobes; calyx deeply 5-parted, the lobes shorter than the corolla tube; stamens 5, inserted on and shorter than the corolla tube; ovary 1, superior, the stigma 2-lobed; **capsule** ovoid, 6–10 mm long at maturity, bursting irregularly, many-seeded. **Leaves** smooth, long-petioled, crowded toward the base of the flowering stem, 3-foliolate, the leaflets oval, oblong or obovate, 2.5–7 cm long, entire to slightly crenate, pinnately veined.

HABITAT In shallow water, in pond margins, quagmires, bogs, marshes.

FLOWERING late April to mid-July.

MONTIACEAE *Candy-Flower Family*

A small family of 16 genera and about 230 species, includes members formerly placed in the Purslane family (Portulacaceae). Michigan representatives of the family (genus *Claytonia*) are characterized by their fleshy leaves, flowers with 5 petals, 2 sepals, and a 1-celled ovary.

Virginia Springbeauty

Claytonia virginica L.

Fleshy perennial with 2–40 weak unbranched stems up to 20 cm tall from a round, deep-seated tuber up to 5 cm thick. **Flowers** pinkish or nearly white, borne on pedicels in loose terminal racemes with an herbaceous bract at the base of the lowest pedicel. Petals 5, pinkish or white with rose to dark pink veins; sepals 2, ovate, separate; stamens 5, pink to rose, attached to the claws of the petals; style 3-cleft above the single ovary; **fruit** an ovoid capsule, opening along 3 sutures to show the 3–6 shining dark seeds. **Leaves** a single opposite pair on each flowering stalk and a few basal ones, fleshy, linear to linear-lanceolate or linear-oblanceolate, up to 17 cm long and 1 cm broad.

VIRGINIA SPRINGBEAUTY

HABITAT In rich, rather open woods and thickets, along roads in woods and clearings.

FLOWERING mid-April to late May.

NOTES This species often forms veritable carpets of bloom in the open woods of southeastern Michigan. In contrast to most members of the family, the flowers last for several days. The flowers of both of our Spring-beauties tend to face and follow the sun. They open only when the sun is shining but do not seem particularly sensitive to temperature changes. They close when picked and when the sun is hidden. Both species are satisfactory for the garden. They are especially effective in natural settings. The tubers are edible but too small to be of much importance as food.

Carolina Springbeauty

Claytonia caroliniana Michx.

Greatly resembling the preceding species, but differing in having fewer flowering stems (1–8), slightly smaller flowers, broader leaves (about a third as broad as long), and the bract at the base of the lowest pedicel dry and membranous.

HABITAT In sandy or rich woods, along streams in hardwood or mixed woods, alluvial thickets, or upland slopes.

FLOWERING late April to mid-June.

CAROLINA SPRINGBEAUTY

NELUMBONACEAE *Lotus-Lily Family*

A small family of aquatic flowering plants, with one genus, *Nelumbo,* containing Nelumbo lutea, native to North America and found in Michigan, and the Sacred Lotus (*Nelumbo nucifera* Gaertn.), widespread in Asia. Nelumbonaceae were previously included in the Waterlily family, Nymphaeaceae.

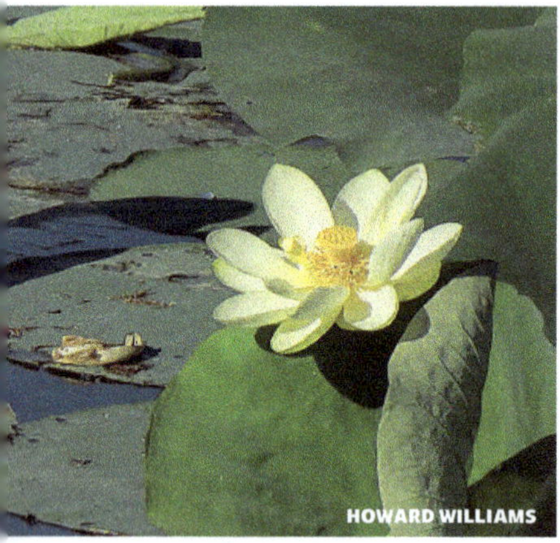

AMERICAN LOTUS

American Lotus

Nelumbo lutea (Willd.) Pers.

Large aquatic perennial, the petioles and peduncles up to 2 m long from the buried rootstock, the flowers and leaves usually held well above the surface. **Flowers** pale yellow, fragrant, up to 2.5 dm broad. Sepals and petals quite similar, 20 or more; stamens numerous; pistils numerous, embedded in pits in the enlarged top of the receptacle, which becomes dry and hard; up to 10 cm in diameter in fruit, the globular, nut-like fruits soon free. **Leaves** smooth, circular in outline, cup-shaped or depressed in the center where the petiole is attached, 3–6 dm in diameter, the veins prominent.

HABITAT In quiet streams, ponds, and lakes. Known in Michigan from only a few localities.

FLOWERING in August.

USES American Indians believed that the American Lotus had mystic powers; they often kept tubers about as a protection against witches. They also dried and stored the tubers for winter use with boiled meat, corn, or hominy. The leaves were also eaten. The dried fruiting heads are often used in winter bouquets.

NYCTAGINACEAE *Four-O'Clock Family*

This small family of 32 genera and about 290 species is distributed mostly in the tropics and subtropics. One genus occurs in Michigan. The only economic importance of the family is for ornamentals: Four-o'clock, Sand Verbena, and the subtropical woody vine, Bougainvillea, are the most familiar here.

Members of the family have the stems swollen at the joints; the calyx funnel-shaped or tubular, corolla-like.

Four-O'clock, Umbrella-Wort

Mirabilis nyctaginea (Michx.) MacMill.

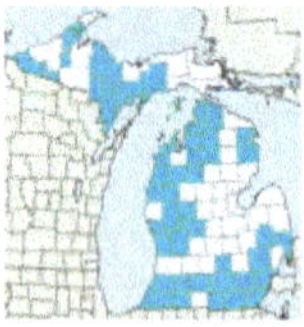

Glabrous, freely forking perennial up to 1.5 m tall. **Flowers** pink to rose-purple, 1–5 in a broad, open, 5-lobed, calyx-like, membranous involucre up to 2 cm in diameter; pedicels slender, short at first, elongating. Petals lacking; calyx corolla-like, with a very short tube flaring above and united to the ovary below, opening in the morning, soon falling; stamens 3–5; pistil 1, the style threadlike, the ovary superior; **achene** angled. **Leaves** opposite, broadly ovate to oblong and heart-shaped at base, petioled, entire.

HABITAT Usually in rich soil, along old roads and railways, on sandy and gravelly shores, on prairies. Native further west in the United States.

FLOWERING June to October.

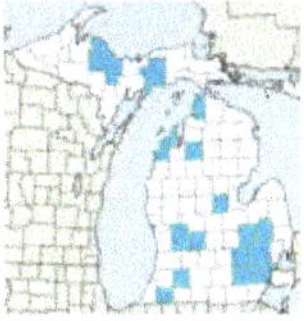

SIMILAR SPECIES *Mirabilis albida* (Walter) Heimerl, native to the Great Plains, is generally smaller; it is glandular-hairy, with oblong to lanceolate sessile or short-petioled leaves. The cultivated **Four-o'clock**, *Mirabilis jalapa* L., has considerably larger white, yellow, purple-red, or variegated flowers, which open in the late afternoon.

FOUR-O'CLOCK, UMBRELLA-WORT

NYMPHAEACEAE *Water-Lily Family*

This aquatic, cosmopolitan freshwater family includes 5 genera and about 70 species; 2 genera occur in Michigan. These plants are used to some extent for ornamentals.

The water lilies have submerged, horizontal rootstocks. They give rise to long petioles and peduncles which permit the leaves and flowers to float on the surface or to rise above the surface of the water. The solitary, showy flowers have few sepals, many petals (which in some species are so modified that they are difficult to distinguish from the stamens), numerous stamens, and either a solitary, many-celled ovary or several separate pistils.

Another aquatic plant, **Watershield** (*Brasenia schreberi* J.F. Gmel.), right, has a leaf with no sinus or cut and the petiole is attached to the center. It is in the Cabombaceae (Watershield Family), occurs statewide, and has small, dull purple flowers with 4–18 separate pistils, and oval, floating leaves up to 10 cm wide. The stems, petioles, and lower surfaces of the leaves are coated with a thick jelly.

KEY TO NYMPHAEACEAE (WATER-LILY FAMILY) SPECIES

1 Flowers white . . . . *Nymphaea odorata*
1 Flowers yellow 2
  2 Leaf blades nearly round . . . . (see *Nelumbo lutea*, p. 158)
  2 Leaf blades broadly ovate, with deep sinus extending nearly to center . *Nuphar advena*

Yellow Pond-Lily, Spatterdock

Nuphar advena (Ait.) Ait. f.

Aquatic perennial with erect leaves (floating only in very deep water) held well above the water on nearly round petioles rising from a rootstock creeping in the mud. **Flowers** bright yellow, solitary, usually raised above the water on round peduncles. Petals small, flattish, resembling and usually hidden by the stamens; sepals usually 6, large, concave, leathery, in 2 rows, deep yellow or greenish-yellow; stamens very numerous, strap-like, borne in several circles; ovary large, many-celled, the stigma a broad circular disk with 7–25 rays; **fruit** berrylike, many-seeded. **Leaves** broadly ovate with a broad V- or U-shaped sinus at the base, the sinus 5–15 cm wide, the 2 basal lobes nearly triangular; petioles round or nearly so.

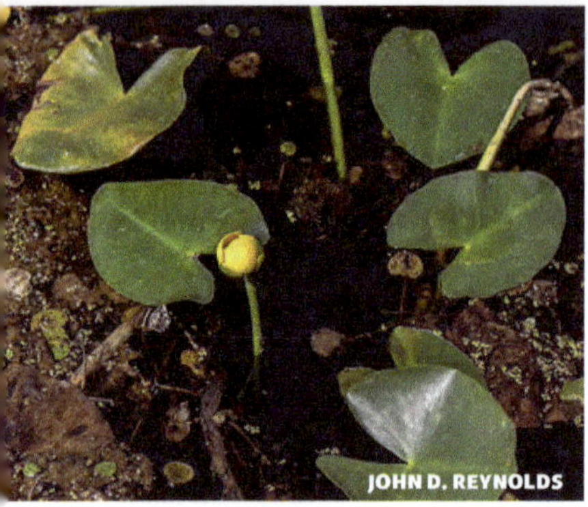

YELLOW POND-LILY, SPATTERDOCK

HABITAT In shallow, slow-running or stagnant water, in swamps and along the margins of ponds or lakes.

FLOWERING chiefly in June and July.

SIMILAR SPECIES Another common yellow pond-lily, **Bullhead-Lily**, ***Nuphar variegata*** Dur., is quite similar in general aspect but tends to be smaller; the sepals are tinged with purple at the base; the leaves are usually floating and have a deep, narrow closed sinus; the basal lobes are broad and rounded; and the petiole, somewhat flattened, is ridged or winged. It grows in slow streams or ponds and occurs throughout Michigan, but is more common in the northern part. It flowers chiefly in July and August. Tubers of this species were harvested by Indian women or were obtained by raiding the hoards collected by muskrats. The tubers were eaten raw, roasted, boiled with meat, or used to thicken soups. The seeds were ground into meal for bread, used for porridge, or parched and eaten like popcorn.

AMERICAN WHITE WATER-LILY

American White Water-Lily

Nymphaea odorata Ait.

Aquatic perennial from an elongate continuous rootstock with branches that are neither constricted nor easily detached. **Flowers** solitary, showy, floating, white or rarely the outer petals rose, very fragrant, 5–15 cm broad, open from early morning till noon for 3–4 days. Petals 17–32, in several rows, narrowly elliptic, the inner ones more like the stamens; stamens 36–100, large, conspicuous, variable, often somewhat petal-like; ovary large, globose, the stigmas radiating from the center; **fruit** globular, maturing under water. **Leaves** floating or in shallow water above the surface, green above, usually purple beneath, nearly round, with a narrow sinus, up to 25 cm in diameter, petiole purplish-green to red.

HABITAT Along the margins of ponds and lakes, in marshes and along slow streams (a dwarf form also occurs in bog pools). Most common in the southern part of Michigan.

FLOWERING June to September.

SIMILAR SPECIES The **Tuberous White Water-Lily**, *Nymphaea odorata* subsp. *tuberosa* (Paine) Wiersema & Hellq., is quite similar in general aspect, but the rootstock has numerous readily detachable knotty tubers; the flowers are odorless or only faintly fragrant; the petals are broader and more rounded at the apex; the sepals are green outside; the leaf blades are usually green beneath, rarely purple; and the petioles are usually striped. This species also grows in quiet or slow-moving water, pond margins, shallow bays and protected coves. It is more common in northern Michigan and flowers chiefly in July and August.

ONAGRACEAE *Evening-Primrose Family*

This family of 22 genera and 650 species is worldwide in distribution; 6 genera occur in Michigan. A number of species in several genera are cultivated as ornamentals. Evening-primrose, Sundrops, Fuchsia, and Clarkia are familiar examples.

There are 2 or 4 petals and calyx lobes; the stamens are the same number, or twice as many, as the petals. The ovary is inferior, and the calyx tube is attached to it; this may be very slender and is sometimes mistaken for the pedicel. The petals and stamens are attached at the top of the calyx tube.

KEY TO ONAGRACEAE (EVENING-PRIMROSE FAMILY) SPECIES

1 Petals 4 2
1 Petals 2 .... *Circaea canadensis* or *C. alpina*
 2 Flowers usually magenta .... *Epilobium angustifolium*
 2 Flowers white or yellow 3
3 Flowers yellow .... *Oenothera biennis* or *O. fruticosa*
3 Flowers white, becoming pink .... *Oenothera gaura*

Broad-Leaf Enchanter's-Nightshade

Circaea canadensis (L.) Hill

SYNONYM *Circaea quadrisulcata* var. *canadensis* (L.) H. Hara

Erect perennial up to 1 m tall, having threadlike stolons and slender rootstocks. **Flowers** small, white to roseate, borne in racemes that elongate in fruit. Petals 2, obovate; calyx tube slightly prolonged, the end filled with a cup-like disk, deciduous, the calyx lobes 2, reflexed; stamens 2; style 1; **fruit** small, bur-like, corrugated, covered with strong, hooked bristles, borne on reflexed pedicels. **Leaves** opposite, oblong-ovate, rounded at the base, shallowly wavy-toothed, firm, dark green above, 4–15 cm long, usually more than twice as long as broad, petioled.

HABITAT In rich deciduous woods, in dryish or moist soil.

FLOWERING July and August.

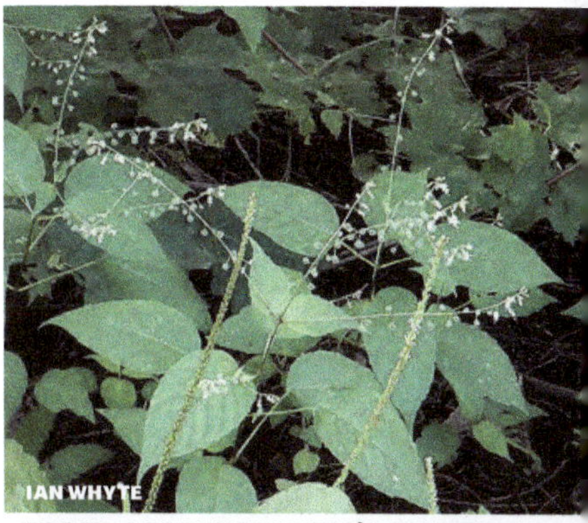

BROAD-LEAF ENCHANTER'S-NIGHTSHADE

Small Enchanter's-Nightshade

Circaea alpina L.

Smaller and weaker than the preceding species, up to 3 dm tall; the **leaves** ovate, usually less than twice as long as broad, pale green; the disk inconspicuous; the stigma deeply cleft; the **fruit** not corrugated.

HABITAT In moist or wet woods and mossy bogs.

FLOWERING June to September.

SMALL ENCHANTER'S-NIGHTSHADE

FIREWEED

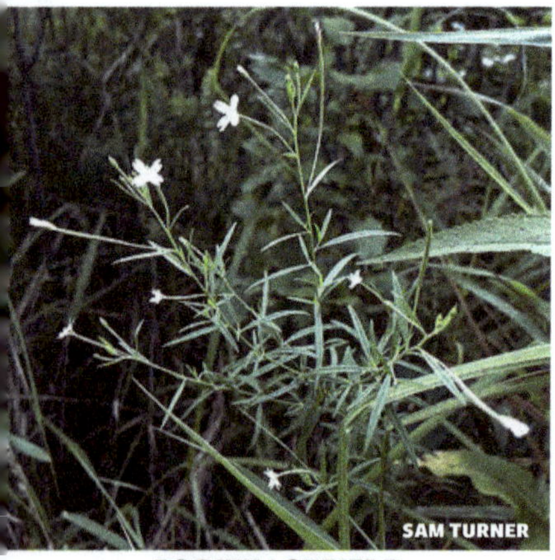

BOG WILLOWHERB

Fireweed

Epilobium angustifolium L.

SYNONYM *Chamaenerion angustifolium* (L.) Scop.

Erect perennial with solitary or few very leafy, simple or branched stems up to 3.5 m tall. **Flowers** magenta, purplish, or rose-purple, rarely white, borne in an elongate, simple raceme on densely hairy pedicels. Perianth borne at the top of the elongate calyx tube; petals 4, obovate, spreading, slightly unequal, entire; calyx tube cleft to the top of the ovary, the 4 lobes linear, colored about like the petals; stamens 8, curved downward; ovary inferior, elongate; style hairy at the base, longer than the stamens; stigma white, conspicuously 4-lobed; **capsule** very slender, 5–7 cm long, the seeds numerous, smooth, bearing a tuft of long whitish hairs at the summit. **Leaves** alternate, lanceolate, entire or finely toothed, green above, pale beneath, the lateral veins forming conspicuous marginal loops; petioles short.

HABITAT In dry soil in burned-over land, ravines, and recent clearings, along roads and railways, and in fields.

FLOWERING June to September.

NOTES This is a very variable species with a number of varieties and forms.

USES The fresh or moistened dried leaves were used by American Indians as poultices for bruises. The young shoots and leaves may be used as a substitute for asparagus. In spite of its attractive flowers, this species is not desirable in the garden as it tends to grow rank and straggly and to spread too rapidly.

SIMILAR SPECIES An introduced species, ***Epilobium hirsutum*** L., is becoming established in the state. This is a more spreading plant with a loose raceme; the flowers are fewer and a little smaller; the petals are notched at the tip; and all parts of the plant are quite hairy. Some much smaller species are quite common. **Bog Willowherb**, ***Epilobium leptophyllum*** Raf. usually has white flowers not over 5 mm wide, and the leaves and stem have tiny, incurved hairs. It grows in bogs, low ground, marshes, and wet meadows. ***Epilobium strictum*** Muhl. ex Spreng. is similar in appearance and habitat but is more sparingly branched. The flowers are pinkish and the hairs stand out at right angles. The common and variable ***Epilobium ciliatum*** Raf. has the leaves and lower part of the stem glabrous or nearly so.

Common Evening-Primrose

Oenothera biennis L.

Stout, erect, sometimes branched biennial up to 2 m tall, the ridged, often purplish stems produced from the previous year's rosette. **Flowers** yellow, 2–3 cm wide, opening in the evening, closing during the day, borne in dense terminal spikes which become greatly elongated, each flower subtended by a leaf. Petals 4 (15–25 mm long), in age often purplish, broadly ovate, attached with the stamens to the top of the calyx tube; calyx tube slender, greatly elongated beyond the ovary, 4-lobed, the lobes at first closely converging, later reflexed and the tips pointing directly back; stamens 8, equal in length; pistil 1, the ovary inferior; style very long, slender, the stigma with 4 linear lobes; **capsule** longer than the subtending leaf, oblong-cylindric, narrowed toward the blunt apex. **Rosette leaves** numerous, compact, elongate; **stem leaves** lanceolate to narrowly elliptic, 5–15 cm long, the margin slightly uneven or with a few small teeth; upper leaves sessile, lower leaves petioled.

HABITAT In dry or sandy soil, in fields, along roads, and in waste ground.

FLOWERING late June to fall.

NOTES Common Evening-Primrose varies greatly in abundance in a given locality; in some years it may be very abundant, in other years quite rare.

SIMILAR SPECIES Northern Evening-Primrose, ***Oenothera parviflora*** L., has slightly smaller flowers (petals 10–15 mm long); the tips of the calyx lobes are separate at base, not forming a tube in bud; and the plants tend to be smaller.

Cutleaf Evening-Primrose, ***Oenothera laciniata*** Hill, has fewer, large, yellow or whitish flowers (which become reddish) borne in the axils of the upper leaves and not forming a distinct spike. The leaves are decidedly wavy, toothed, or pinnately cut.

COMMON EVENING-PRIMROSE

NORTHERN EVENING-PRIMROSE

Sundrops

Oenothera fruticosa L.

Erect or ascending clustered perennial up to 1 m tall, usually covered with incurved hairs. **Flowers** blooming in the daytime, yellow, showy, up to 3.5 cm broad, borne in the axils of the upper leaves. Calyx tube threadlike, 5–15

SUNDROPS

SMALL SUNDROPS

mm long; **capsule** 4-angled, club-shaped, narrowed at base to a definite stalk. **Rosette leaves** (mostly lacking when the plant is in flower) ovate to spatulate, petioled; **stem leaves** alternate, lanceolate to narrowly elliptic or linear, tapered to a short petiole or sessile, hairy, entire or with a few scattered teeth; upper leaves often bearing short branchlets in the axils.

HABITAT In dry soil, in fields and open woods. Locally common throughout the state.

FLOWERING June and July.

SIMILAR SPECIES Small Sundrops, *Oenothera perennis* L. looks quite similar, but the tip of the flowering stem and the buds are nodding; the flowers usually open singly and become erect; and the petals are up to 9 mm long.

BIENNIAL EVENING-PRIMROSE

Biennial Evening-Primrose

Oenothera gaura W.L. Wagner & Hoch

SYNONYM *Gaura biennis* L.

Long-hairy biennial up to 1.5 m tall from a basal rosette. **Flowers** numerous, white, becoming pink or red, borne in slender, wand-like spikes. Petals 4, somewhat unequal, oblanceolate, clawed, about 5 mm long; calyx tube slender, 5–7 mm long, the 4 lobes reddish, reflexed in pairs; stamens 8, hanging downward, each filament with a small scale at the base; stigma 4-lobed, surrounded by a cup-like border; **fruit** hard and nut-like, 4-ribbed, hairy, tapered to both ends, sessile. **Rosette leaves** oblanceolate, 1–3 dm long; **stem leaves** alternate, tapering at both ends, sessile, 3–10 cm long, sparsely toothed.

HABITAT In dry open ground, sandy or waste soil, or in damp places.

FLOWERING June to October.

OROBANCHACEAE *Broom-Rape Family*

This is a family of 99 genera and more than 2,000 species, and Michigan hosts 13 genera. No species are now economically important, although some were formerly used medicinally. Many current members of the family were formerly included in the **Scrophulariaceae** (Figwort Family).

The plants in this family have no chlorophyll and are parasitic or saprophytic on the roots of green plants. The stems are usually yellowish, brownish, purplish, or whitish, and the leaves are reduced to scales. The flowers have a tubular, 2-lipped corolla, and the 4 stamens are borne on the corolla tube; the ovary is superior and becomes a many-seeded capsule.

KEY TO OROBANCHACEAE (BROOM-RAPE FAMILY) SPECIES

1 Plants lacking chlorophyll, not green 2
1 Plants having normal green color 4
2 Stems freely branched; flowers of 2 kinds, only upper ones having a well-developed corolla .... *Epifagus virginiana*
2 Stems usually unbranched; flowers all alike 3
3 Flowers numerous, in thick, compact spikes somewhat resembling cones of the White Pine .... *Conopholis americana*
3 Flowers solitary, borne on slender pedicels .... *Orobanche uniflora*
4 Leaves alternate or basal, usually lobed or pinnately dissected 5
4 Leaves opposite, not deeply cut 6
5 Leaves sessile, 3–5 lobed; bracts surrounding flowers scarlet .... *Castilleja coccinea*
5 Leaves mostly petioled, finely dissected; floral bracts not scarlet . *Pedicularis canadensis*
6 Flowers solitary in axils of normal leaves .... *Melampyrum lineare*
6 Flowers in clusters .... *Agalinis purpurea* or *A. tenuifolia*

Purple False Foxglove

Agalinis purpurea (L.) Pennell

SYNONYM *Gerardia paupercula* (Gray) Britt.

Erect, nearly smooth annual, simple or with ascending branches, up to 8 dm tall, the stem 4-angled. **Flowers** pink, rose-purple, or (rarely) white, borne in elongate, very loose terminal racemes, the pedicels shorter than the calyx and rising from the upper axils. Corolla inflated-tubular, open, with 5 spreading lobes, the upper lobes the smaller, the throat hairy inside, with darker spots and 2 yellow lines; calyx regular, 5-toothed; stamens 4, hairy; ovary superior. **Leaves** opposite, often with clusters of smaller leaves in the axils, linear, the upper leaves reduced to bracts subtending the pedicels, becoming blackish on drying.

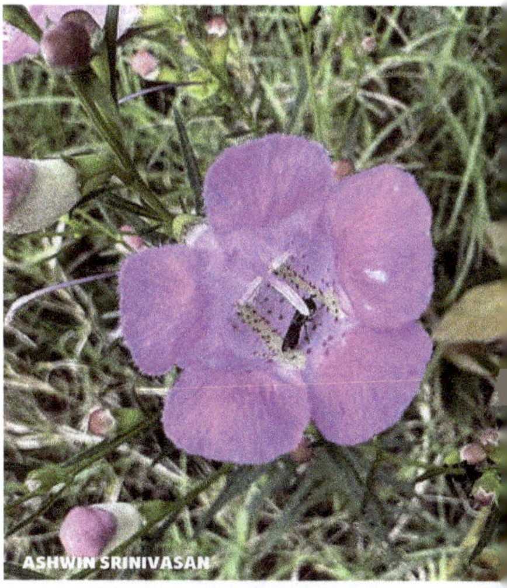

PURPLE FALSE FOXGLOVE

HABITAT Along marshy lakeshores and pond margins, in bogs, on damp open ground, and in grassy ditches.

FLOWERING August and September.

SLENDER FALSE FOXGLOVE

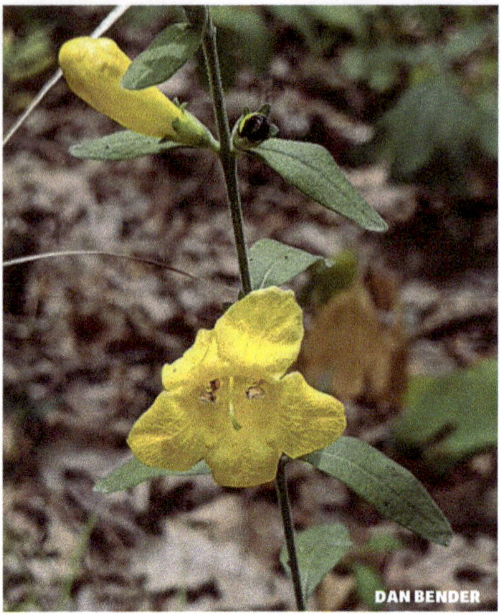

DOWNY FALSE FOXGLOVE

Slender False Foxglove

Agalinis tenuifolia (Vahl) Raf.
SYNONYM *Gerardia tenuifolia* Vahl

Quite similar to the preceding species in general aspect, but the leaves are narrower, the upper lobes of the corolla are longer than the lower, the corolla tube is smooth inside, and the pedicels are mostly longer than the flowers.

HABITAT In dry woods, thickets, and fields.

FLOWERING August to October.

SIMILAR SPECIES Downy False Foxglove, ***Aureolaria virginica*** (L.) Pennell (synonym *Gerardia virginica* (L.) B.S.P.), has large yellow flowers, 3–4.5 cm long; the lower leaves are ovate-lanceolate and usually have 2 pairs of large lobes below the middle; the capsule is hairy.

Two other species with conspicuous yellow flowers are ***Aureolaria flava*** (L.) Farw. (synonym *Gerardia flava* L.), with stem and capsule glabrous, and ***Aureolaria pedicularia*** (L.) Raf. ex Farw. (synonym *Gerardia pedicularia* L.), a glandular-hairy annual.

INDIAN PAINTBRUSH, PAINTED CUP

Indian Paintbrush, Painted Cup

Castilleja coccinea (L.) Spreng.

Stiffly erect, hairy annual or biennial with unbranched, usually solitary stems up to 6 dm tall from a basal rosette. **Flowers** yellowish or greenish, enclosed by conspicuous scarlet-tipped

3-lobed bracts, and borne in dense terminal spikes which become elongate in age. Corolla flattened laterally, 2-lipped, the upper lip (the larger) arching and forming a hood over the lower lip; calyx tubular, split along the lower and sometimes along the upper side, hairy, often tipped with scarlet or yellow; stamens 4, attached to the corolla tube; style curved, protruding from the corolla; **capsule** many-seeded. **Basal leaves** forming a rosette, hairy, linear to narrowly obovate or oblong, usually entire; **stem leaves** sessile, alternate, varying from entire (rarely) to 3–5-cleft, the segments narrow, the terminal one the longest.

HABITAT In damp sands and gravels, in peaty meadows, moist prairies, grassy thickets, and margins of woods.

FLOWERING May and June to September.

American Squawroot

Conopholis americana (L.) Wallr. Short, thick, fleshy, usually unbranched, often clustered parasite up to 3 dm tall from large rounded knobs on tree roots, yellowish to pale brown, changing to brown when bruised, in general aspect somewhat resembling slender elongate cones of White Pine. **Flowers** sessile or nearly so, each subtended by a basal bract or scale, borne in a dense spike. Corolla tubular, swollen at the base, curved, 2-lipped; calyx tubular, split down the upper side, irregularly toothed; stamens 4, attached to the corolla tube and extending beyond it; pistil 1, the ovary ovoid, superior, the stigma a flat disk. Scales fleshy at first, becoming dry and hard, persisting, the upper ones subtending the flowers.

HABITAT In rich woods, mostly under oaks.

FLOWERING April to July.

AMERICAN SQUAWROOT

Beechdrops

Epifagus virginiana (L.) W. Bart.

Low, slender, purplish to yellowish-brown, many-branched herbs up to 4.5 dm tall, the stems striped or flushed with brown or madder. **Flowers** nearly sessile, in elongate spikes which may form large panicles, the flowers of two kinds. Corolla of upper flowers tubular, somewhat 2-lipped, whitish and brown or madder, that of the lower flowers cap-like, not opening; calyx cup-shaped, 5-toothed; stamens 4, borne on corolla tube; ovary superior, forming a many-seeded capsule.

HABITAT Saprophytic or parasitic on roots of beech.

FLOWERING August and September.

BEECHDROPS

AMERICAN COW-WHEAT

NAKED BROOMRAPE

American Cow-Wheat

Melampyrum lineare Desr.

Simple or, more commonly, densely bushy-branched annual up to 5 dm tall, the stems somewhat 4-angled. **Flowers** white, small, borne in the axils of the leaves. Corolla tubular, 2-lipped, the lips closed by a yellow projection on the lower lip which is sometimes purplish-tipped; calyx bell-shaped, 5-cleft; stamens 4, ascending under the upper lip; ovary superior, flattish, the **capsule** flat. **Leaves** opposite, linear to elliptic-ovate, pointed, the lower ones entire, those in the inflorescence entire or with 2–6 bristle-tipped teeth.

HABITAT In dry, open, sandy woods and thickets, in bogs and damp peaty or rocky barrens.

FLOWERING July and August.

Naked Broomrape

Orobanche uniflora L.

SYNONYM *Aphyllon uniflorum* (L.) A. Gray

Whitish, pale purplish, or brownish parasite up to 3 dm tall. **Flowers** white, solitary on 1–4 hairy pedicels 6–20 cm long from the short, mostly underground stem. Corolla tubular, long, curved, 5-lobed; calyx bell-shaped, 5-lobed; stamens 4, borne on the corolla; pistil 1, the ovary superior, enlarged, and causing a swelling at the base of the corolla; **capsule** many-seeded, usually capped by the withered corolla, Scales smooth, oblong-ovate, blunt or pointed.

HABITAT In rich damp woods and thickets, growing on roots of various trees.

FLOWERING April to June.

SIMILAR SPECIES The uncommon (state threatened) **Clustered Broomrape**, *Orobanche fasciculata* Nutt., is somewhat similar, but the 3–10 pedicels are shorter (2–6 cm long); the stems are above ground and are 5–15 cm long; the scales are ovate-lanceolate and, at least the upper ones, sharply pointed. In Michigan, it is parasitic on **Wormwood** (*Artemisia caudata* Michx.), grows on dunes and other sandy ground, and flowers April to August.

Canadian Lousewort

Pedicularis canadensis L.

Low, coarse, clustered, densely hairy perennial up to 5 dm tall. **Flowers** yellowish to dull purplish-red or maroon, borne in short, dense, leafy, terminal spikes which become elongate in fruit. Corolla irregular, 2-lipped, 2–2.5 cm long, the upper lip hooded, flattened, longer than the lower, toothed; calyx oblique, entire except for a split along the front; stamens 4; pistil solitary, the ovary superior; **capsule** flattened, oblique. **Leaves** thick, mostly basal, oblong-lanceolate, pinnately cut almost to the midrib (the segments incised or sharply toothed), petioled, the stem leaves smaller, scattered, sessile.

HABITAT In dry woods and thickets and in clearings.

FLOWERING April to June.

SIMILAR SPECIES *Pedicularis lanceolata* Michx., with stem glabrous above and leaves opposite or nearly so, occurs in moist ground, bogs, and shores.

NOTES The common name Lousewort refers to the old belief that the presence of this species in fields caused lice in sheep.

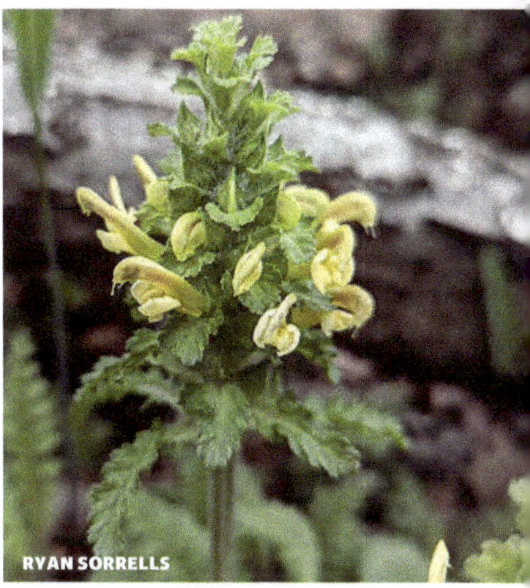
RYAN SORRELLS

CANADIAN LOUSEWORT

OXALIDACEAE *Wood-Sorrel Family*

This mostly tropical family of 5 genera includes about 570 species, most of which are in genus Oxalis; 4 species of *Oxalis* occur in Michigan. It is of little economic importance, but a few species of Oxalis are grown for ornament.

Our species are easily recognized by their regular flowers with 5 petals and sepals, 10 stamens, the 5-celled, superior ovary, and the shamrock-like leaves.

KEY TO OXALIS (WOOD-SORREL) SPECIES

1 Flowers white to purple; leaves all from base .... *Oxalis montana*

1 Flowers yellow; leaves on stem .... *Oxalis stricta*

Mountain Wood-Sorrel

Oxalis montana Raf.

Low perennial up to 1 dm tall, creeping by rootstocks. **Flowers** white with rose or purple veins, solitary, the flowering stalks long-hairy and bearing a pair of small bracts near the middle, the earliest flowering stalks longer than the leaves, the later ones shorter, recurving, and bearing small non-opening flowers. Petals 5, notched at apex, having a small yellow spot near the base; calyx deeply cut, the lobes lanceolate, often bordered with red; stamens 10, barely united at the base, of 2 lengths; pistil compound, deeply 5-lobed, the styles 5, distinct; **fruit** a capsule. **Leaves** all from the base, 3-foliolate, the leaflets broadly heart-shaped above, having a pleasant acid taste.

HABITAT In damp woods.

FLOWERING late May to August.

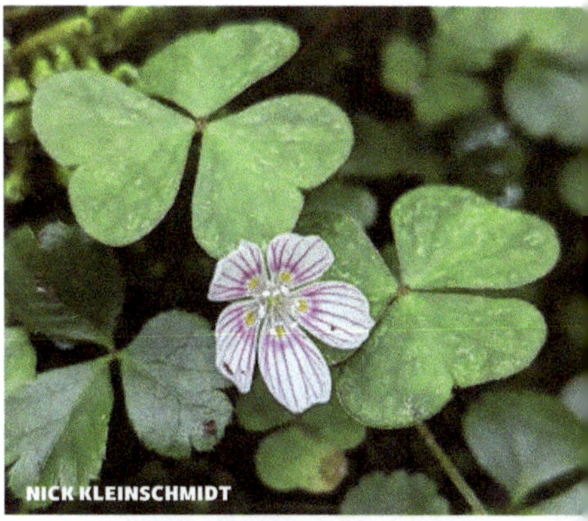
NICK KLEINSCHMIDT

MOUNTAIN WOOD-SORREL

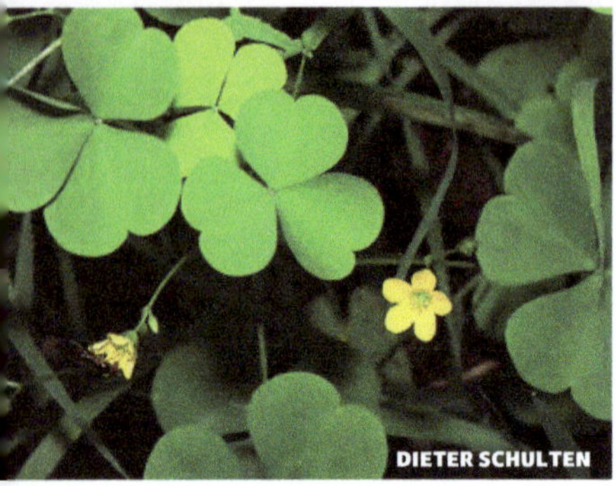

YELLOW WOOD-SORREL

Yellow Wood-Sorrel

Oxalis stricta L.

Erect, ascending, or partially reclining and matted, grayish-green perennial covered with appressed, whitish hairs, up to 5 dm tall (usually much shorter). **Flowers** yellow, 1–4 in an umbel. Petals sometimes red at the base; **capsules** erect on deflexed pedicels. **Leaves** alternate, the leaflets 1–3 cm broad; stipules oblong, firm.

HABITAT In woods, cultivated ground, lawns, gardens, fields, and along roadsides.

FLOWERING May to October.

PAPAVERACEAE *Poppy Family*

This group of nearly 50 genera and 570 species has many species, particularly poppies and bleeding-hearts, grown as ornamentals. Fifteen genera occur in Michigan. Opium is derived from capsules of certain species of poppy, and seeds of some kinds are used in cakes and on breads. The seeds do not contain the narcotic compounds.

The family in Michigan is characterized by regular or irregular flowers with 4 or more distinct or slightly united petals, 2 or 4 sepals, 6 to many stamens, and a single pistil which forms a capsule.

KEY TO PAPAVERACEAE (POPPY FAMILY) SPECIES

1 Plants with colored juice; flowers regular .... *Sanguinaria canadensis*

1 Plants with colorless juice; flowers irregular.. 2

2 Plants stemless, leaves all basal .... *Dicentra canadensis* or *D. cucullaria*

2 Plants with leafy stems .... *Capnoides sempervirens*

ROCK HARLEQUIN

Rock Harlequin

Capnoides sempervirens (L.) Borkh.

SYNONYM *Corydalis sempervirens* (L.) Pers.

Glabrous and glaucous, freely branched, erect annual or biennial up to 1 m or more tall. **Flowers** numerous, pink (rarely white) tipped with yellow, 1–2 cm long, borne in terminal clusters. Petals 4, in pairs, the uppermost of the outer pair with a spur at the base, the inner petals narrowed, keeled; sepals 2, small; stamens 6, in 2 unequal sets opposite the larger petals; pistil 1; **capsule** slender, 2–5 cm long. **Leaves** much divided into toothed or entire segments; basal leaves short-petioled, up to 12 cm long, the upper leaves smaller, sessile.

HABITAT In rocky places, on gravel slopes, and in recent clearings, often abundant on burnt-over areas.

FLOWERING May to September.

NOTES When crushed these plants have a nitrous odor.

SIMILAR SPECIES The related **Golden Corydalis**, ***Corydalis aurea*** Willd., has similar but golden-yellow flowers. It is a diffusely branched, often spreading, annual or biennial growing on rocky slopes, in open woods, and on the sandy shores of the Great Lakes. Both species are poisonous and are known to be fatal to sheep if eaten in quantity.

Dutchman's-Breeches

Dicentra cucullaria (L.) Bernh.

Fern-like, stemless perennial 1.2–2.5 flowering stem, often fragrant, white dm, tall from a cluster of small bulbs. or rarely pinkish, tipped with cream **Flowers** nodding in a row on the or yellow, with 2 conspicuous divergent spurs, resembling inverted, greatly inflated pantaloons. Petals 4, in 2 pairs, the outer pair longer than the pedicel, the inner pair at right angles and smaller, narrow, the tips enlarged, slightly crested, united to form an arch over the stamens; stamens 6, in 2 sets of 3; pistil solitary, the style slender, the stigma 2-lobed; **capsule** 1-celled, opening into 2 parts to the base, with 10–20 crested seeds. **Leaves** all basal, numerous, very finely dissected, the divisions in 3's, pale beneath.

HABITAT In rich woods.

FLOWERING in April, the plant usually disappearing by June.

DUTCHMAN'S-BREECHES

Squirrel-Corn

Dicentra canadensis (Goldie) Walp.

Greatly resembling and often growing with **Dutchman's-breeches**, but easily distinguished when in bloom by the heart-shaped rather than spurred base of the corolla and the more conspicuous crest on the inner petals. The small, grain-like, bright-yellow tubers are scattered instead of clustered. The sepals are often pinkish, tiny, and, as in the preceding species, so closely appressed to the corolla that they are easily overlooked.

HABITAT In rich woods.

FLOWERING in April, the plant disappearing by June.

NOTES This species tends to bloom a week or two later than the preceding one and thus is often in its prime when the other is fading. The flowers of Squirrel-corn resemble those of the common garden **Bleeding-Heart**, which is native to Japan, but they are smaller and paler.

SQUIRREL-CORN

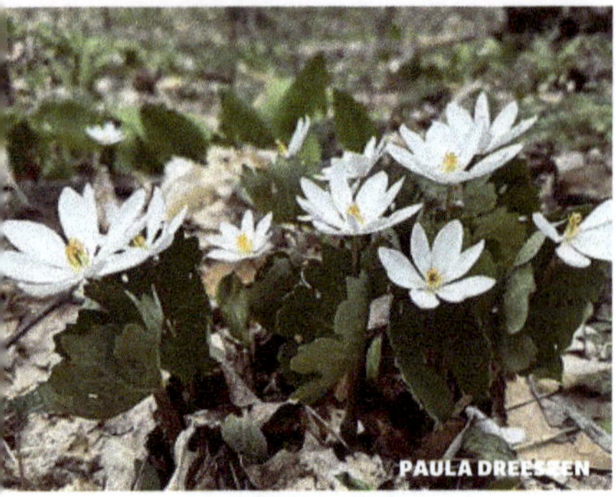

BLOODROOT

WOOD-POPPY

Bloodroot

Sanguinaria canadensis L.

Glabrous, stemless perennial seldom over 3 dm tall at flowering time. Juice (latex) red to reddish-orange. **Flowers** white (pink in one form), solitary, lasting a very short time, 2.3–3.5 cm wide. Petals 8–12 (more in the rare double form), borne in 2 or more rows; sepals 2, soon falling; stamens numerous; pistil 1, the **capsule** slenderly ellipsoid, 2 cm or more long, 1-celled, many-seeded. **Leaves** enfolding the base of the flowering stem at first, soon expanding and finally over-topping the capsules, roundish, 10–25 cm in diameter, palmately lobed or with merely a wavy margin, toothed or crenate, the veins (especially on the lower surface) prominent and orange-red.

HABITAT In rich open woods or thickets.

FLOWERING March to May (chiefly in April).

NOTES These attractive flowers are very susceptible to weather changes, opening in sunshine but closing up when it is cold or cloudy. Cold may cause the petals to fall even when the flower has been open only a few hours. The rootstock is very acrid and is distinctly poisonous.

USES American Indians used the juice for dyeing textiles, quills, and cane baskets. They also made a face paint and medicines from it.

SIMILAR SPECIES Two other species with colored juice may be found. Both have bright yellow flowers and yellow juice. **Wood-Poppy**, ***Stylophorum diphyllum*** (Michx.) Nutt., has large flowers (about 5 cm wide), solitary or in few-flowered clusters; there are 4 petals which are crushed in the bud, 2 sepals, many stamens, and a single pistil with a long, hairy style; the pistil produces a bristly capsule. **Celandine**, ***Chelidonium majus*** L., an introduced weed, has much smaller flowers with a very short style and smooth capsule.

PHRYMACEAE *Lopseed Family*

This family includes 15 genera and over 200 species; 3 genera are present in Michigan, the largest group being the Monkey-Flowers (*Erythranthe, Mimulus*).

Yellow Monkey-Flower

Erythranthe geyeri (Torr.) G.L. Nesom

SYNONYM *Mimulus glabratus* HBK.

Weak, prostrate or ascending, smooth to finely hairy perennial, rooting freely at the nodes. **Flowers** similar to the preceding species but the corolla yellow, 9–12 mm long; calyx irregular, becoming inflated and loosely enclosing the capsule, the **leaves** ovate or nearly round, wavy-toothed to nearly entire, the lower leaves petioled, the upper ones sessile or nearly so, palmately veined.

HABITAT In cool wet sand or soil, along lakes and streams, and in springy places.

FLOWERING June to September.

SIMILAR SPECIES *Erythranthe michiganensis* (Pennell) G.L. Nesom, is listed as **endangered** in the U.S and Michigan (and found only in Michigan), and similar to *E. geyeri*. It grows in marly springs, in cold streams through cedar swamps, on calcareous shores and ditches, usually near the shores of the Great Lakes.

Muskflower, *Erythranthe moschata* (Douglas ex Lindl.) G.L. Nesom, has a scent of musk and is sticky and hairy; the ovate leaves are pinnately veined; the calyx is nearly regular and does not become inflated.

YELLOW MONKEY-FLOWER

Allegheny Monkey-Flower

Mimulus ringens L.

Glabrous, erect perennial with square stems up to 1 m tall. **Flowers** blue-violet, pinkish, or rarely white, solitary on long pedicels from the leaf axils. Corolla tubular, 2–4 cm long, 2-lipped, the throat nearly closed; calyx 5-angled and 5-toothed; stamens 4; ovary superior; **capsule** covered by the calyx in fruit. **Leaves** opposite, sessile or clasping, lanceolate, narrowly oblong or oblanceolate, crenate.

HABITAT Shores, in wet meadows and swamps.

FLOWERING July to September.

SIMILAR SPECIES *Mimulus alatus* Ait., rare in southern Michigan, is similar, but the leaves are somewhat broader and petioled and the angles of the stems are somewhat winged.

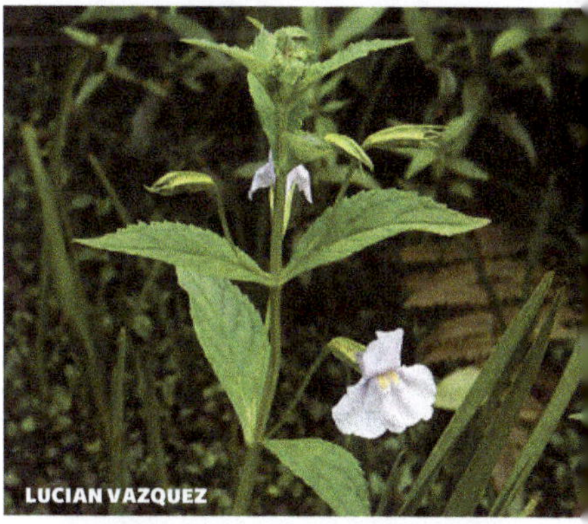

ALLEGHENY MONKEY-FLOWER

American Lopseed

Phryma leptostachya L.

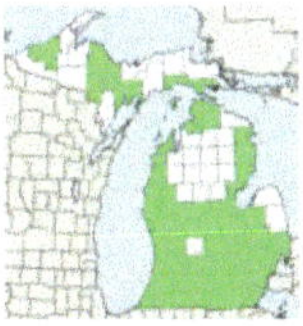

Slender perennial, simple or with a few divergent branches, up to 9 dm tall. **Flowers** purplish to rose or white, opposite, horizontal in slender, elongate terminal spikes (or from the upper axils). Corolla tubular, 2-lipped, the upper lip notched, the lower lip much larger, 3-lobed; calyx cylindrical, 2-lipped, the upper lip of 3 bristly teeth, the lower of 2 shorter, broader teeth, closing in fruit and becoming up to 1 cm long; stamens 4, borne on the corolla tube; pistil 1; ovary superior, forming a small 1-seeded nutlet reflexed against the stem. **Leaves** opposite, bright green, lower leaves long-petioled, upper leaves sessile or nearly so, ovate, coarsely toothed.

HABITAT In moist or wet woods.

FLOWERING July and August.

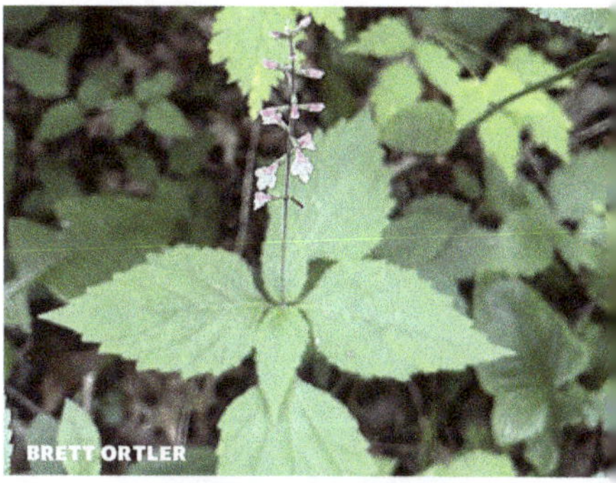

AMERICAN LOPSEED

PHYTOLACCACEAE *Pokeweed Family*

This mostly tropical and subtropical family includes about 4 genera and 33 species; only one genus and species is native to Michigan (see below). One or two genera are sometimes used for ornamentals, and Pokeweed is sometimes used for greens.

AMERICAN POKEWEED

American Pokeweed

Phytolacca americana L.

Stout, smooth, malodorous perennial with fleshy purple stems up to 4 m tall from a large poisonous root up to 1.5 dm in diameter. **Flowers** white, in racemes which appear to be opposite the leaves except at first. Petals lacking; sepals 5, petal-like, white or pinkish, rounded; stamens 10; pistils 10, united into a ring, the ovary superior, green when young, forming a dark purple, flattened, juicy berry with a seed in each of the 10 cells. **Leaves** large, alternate, entire, petioled, oblong-lanceolate to ovate.

HABITAT In rich low ground, open woods, or thickets, often along roads; especially abundant in recent clearings.

FLOWERING July to September.

USES Although the berries and roots are poisonous, the young leafy shoots are edible when well cooked and are sometimes used as greens. The plants are reportedly cultivated for food in France. The purplish-red juice of the berries gave rise to one common name "Inkberry." American Indians prepared a dull red or magenta pigment from the dried and pulverized berries and used it for stamping designs on baskets (using dies cut from potatoes). A yellow dye was made from the leaves.

PLANTAGINACEAE *Plantain Family*

As currently described, this is a large and diverse family of 107 genera and about 1,900 species that includes common flowers such as snapdragon and foxglove; the family is not related to the banana-like fruit called "plantain." The largest genus is *Veronica*, with about 450 species; 18 genera occur in Michigan. The drug digitalis, used in treating heart ailments, is derived from Foxglove (*Digitalis*).

KEY TO PLANTAGINACEAE (PLANTAIN FAMILY) SPECIES

1 Corolla having a spur at base .... *Linaria vulgaris*
1 Corolla not having a spur at base 2
  2 Stamens with anthers2 3
  2 Stamens with anthers4 4
3 Leaves mostly in whorls of 3–6; plants erect .... *Veronicastrum virginicum*
3 Leaves mostly opposite; plants sprawling .... *Veronica americana*
  4 Flowers blue and white, the corolla cut nearly to base .... *Collinsia verna*
  4 Not as in alternate choice 5
5 Flowers sessile or nearly so, borne in a rather compact cluster, corolla nearly closed at tip .... *Chelone glabra*
5 Not as in alternate choice; sterile stamen bearded .... *Penstemon hirsutus*

White Turtlehead
Chelone glabra L.

Stout, erect, leafy perennial up to 2 m tall, simple or with a few ascending branches, highly variable in shape of leaves and color of flowers. **Flowers** creamy-white to slightly pinkish, nearly sessile in compact terminal spikes. Corolla tubular, irregular, 2-lipped, the lips nearly closed, the upper lip arched, at first covering the bearded lower lip; calyx very deeply 5-parted; fertile stamens 4, the sterile stamen much shorter, slender, greenish; pistil 1, the ovary superior; **capsule** ovoid, splitting. **Leaves** opposite, sessile or with short, winged petioles, linear-lanceolate to ovate-lanceolate, sharply toothed.

HABITAT In low wet ground, along streams, in swales and swamps.

FLOWERING late July to October.

WHITE TURTLEHEAD

Spring Blue-Eyed Mary
Collinsia verna Nutt.

Weak, simple to many-branched, slightly hairy annual or biennial 1–6 dm tall. **Flowers** blue and white, in whorls of 4–6 or solitary in the upper axils. Corolla tubular, the tube swollen on the upper side near the base, 2-lipped, the upper lip erect, white, the middle lobe of the blue lower lip folded together forming a pouch which encloses the stamens; calyx bell-shaped, deeply 5-lobed; normal stamens 4, the filaments bearded, sterile stamen short, gland-like. **Lower leaves** ovate to nearly round, long-petioled, the **upper leaves** opposite or whorled, sessile or clasping by a heart-shaped base, ovate to lanceolate or linear, toothed or entire.

HABITAT In rich woods and thickets.

FLOWERING May to June.

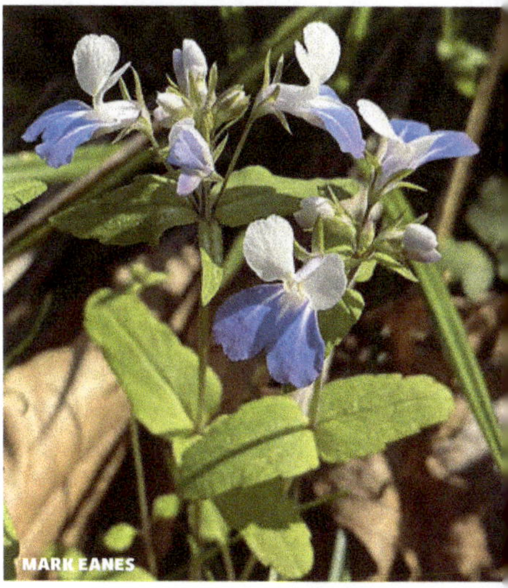

SPRING BLUE-EYED MARY

Butter-and-Eggs
Linaria vulgaris P. Hill

Erect perennial usually with many stems, up to 1.3 m tall from creeping rootstocks. **Flowers** yellow, in compact terminal racemes. Corolla tubular, 2-lipped, closed, strongly spurred; calyx 5-parted, the lobes overlapping; stamens 4, in 2 pairs of unequal length, attached to the base of the corolla tube; ovary 1, superior. **Leaves** very numerous, alternate, thickish, the midrib prominent, linear, tapering to both ends.

HABITAT In dry fields and waste places, along roadsides. Introduced from Europe.

FLOWERING May to October.

BUTTER-AND-EGGS

DALMATIAN TOADFLAX

NOTES Aberrant forms with white flowers, with regular corollas or with 3, 5, or no spurs also occur.

SIMILAR SPECIES A much coarser, introduced weedy species, **Dalmatian Toadflax**, *Linaria dalmatica* (L.) P. Mill., is less common. It may be up to 1 m tall, simple or branched. The corolla is yellow, and the flowers are 3.5–4 cm long. The leaves are ovate, ovate-lanceolate or oblong, clasping at the base. It is occasional statewide along roadsides and in fields.

Hairy Beardtongue

Penstemon hirsutus (L.) Willd.

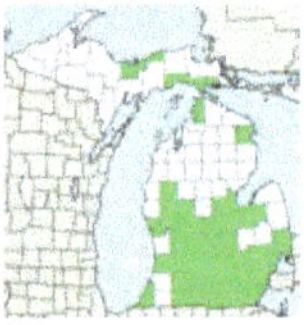

Stiffly erect, glandular-hairy perennial up to 1 m tall, with 1 to several stems which are usually purple, at least at the base. **Flowers** pale violet to dull purplish with white lobes, borne in stiff terminal racemes. Corolla tubular, flattened, the throat somewhat swollen, 2-lipped, closed, the upper lip 2-lobed, covering the lower one in bud; calyx deeply 5-lobed; fertile stamens 4 in 2 unequal pairs, not extending beyond the corolla tube, the sterile stamen flattened, densely yellow-bearded, longer than the fertile stamens; pistil 1, the ovary superior; **fruit** a capsule. **Leaves** many, opposite, sessile, somewhat clasping, sharply toothed, hairy at first, becoming smooth, the stem leaves lanceolate, the basal leaves lanceolate, oblanceolate, oblong, or elliptic.

HABITAT On grassy, sunny banks, along roads, and in dry or rocky ground.

FLOWERING June and July.

NOTES The name 'beardtongue' refers to the fifth, sterile, bearded stamen, which is found in all members of this genus.

SIMILAR SPECIES Three species of White Beardtongue are also found in Michigan: ***Penstemon digitalis*** Nutt. ex Sims, a glabrous, erect perennial up to 1.5 m tall, with white or faintly purple-tinged flowers, tubular corolla, the anthers bearded. ***Penstemon calycosus*** Small is quite similar, but the corolla is purplish outside, white inside, and the anthers are smooth. ***Penstemon pallidus*** Small is smaller, the flowers are white outside (but may have purple lines at base inside), the corolla is 1.7–2.2 cm long, flattened and strongly ridged within, the lower lip is longer than the upper, and the sterile stamen is densely bearded.

American Brooklime

Veronica americana (Raf.) Schwein. ex Benth.

Glabrous, somewhat fleshy, prostrate or sprawling perennial rooting at the lower nodes. **Flowers** blue to violet, in several lax racemes from the upper axils. Corolla wheel-shaped, the 4 lobes spreading; calyx 4-lobed; stamens 2, attached to the corolla tube; pistil 1, the ovary superior; capsule borne on threadlike pedicel 6–11 mm long. **Leaves** opposite, petioled, lanceolate, widest just above the base, pointed at the tip, sharply toothed.

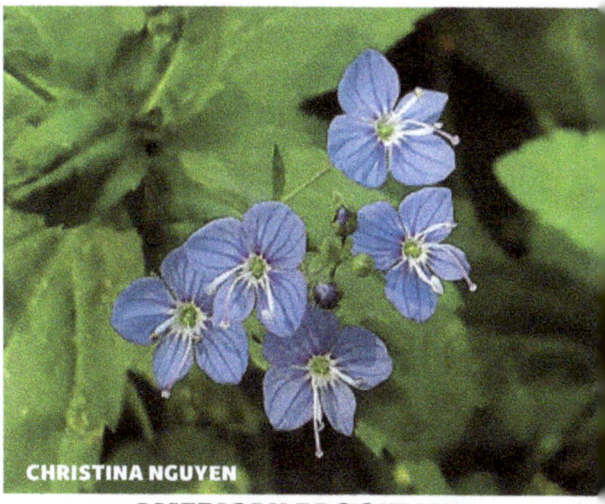

AMERICAN BROOKLIME

HABITAT In shallow water, wet places; frequent along streams and around springs.

FLOWERING June to September.

SIMILAR SPECIES The closely similar **European Brooklime,** *Veronica beccabunga* L., differs in having elliptic to ovate, crenate leaves which are widest near or above the middle and rounded at the tip. The fruiting pedicels are thicker and shorter (4–5 mm long). This species is uncommon in Michigan. Several other species of *Veronica* also occur in the state.

Culver's-Root

Veronicastrum virginicum (L.) Farw.

Tall, slender, erect, smoothish, many-stemmed perennial up to 2 m tall, branching only at the inflorescence. **Flowers** small, blue, purplish, or rarely white, borne in dense terminal spikes 5–25 cm long. Corolla tubular, nearly regular, the tube short, 4-lobed; calyx saucer-shaped, lobed; stamens 2, attached low on the corolla tube; ovary superior, 2-celled; **capsule** ovoid. **Leaves** short-petioled, sharply toothed, in whorls of 3–7, lanceolate to lance-ovate.

HABITAT In rich dry or moist woods and thickets, in meadows and on prairies.

FLOWERING late June to September.

HAIRY BEARDTONGUE

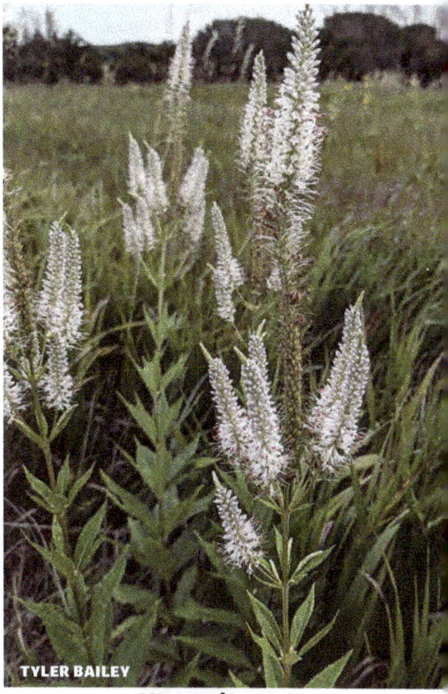

CULVER'S-ROOT

POLEMONIACEAE *Phlox Family*

This primarily American family includes 27 genera and about 300 species, the majority of which are native in western United States; 4 genera are present in Michigan. The family includes the garden phloxes, Jacob's-ladder, and Gilia. It is valued chiefly for such ornamentals.

The family is characterized by the regular flowers with both petals and sepals 5 and united, the 5 stamens borne on the corolla tube, and the superior, 3-celled ovary with 1 style and 3 stigmas.

WILD BLUE PHLOX

PRAIRIE PHLOX

Wild Blue Phlox

Phlox divaricata L.

Softly hairy perennial up to 5 dm tall from a partially reclining base, often with prostrate, evergreen rooting basal offshoots. **Flowers** blue to purplish-blue, rarely white, in terminal or axillary panicles which are at first compact, later becoming loose and spreading. Corolla salverform, the tube very slender, the 5 lobes notched at the apex; calyx 5-lobed, the lobes slender, pointed, longer than the tube, glandular; stamens 5, inserted at different heights in the corolla tube; pistil solitary, 3-celled, with single style and 3 stigmas; **capsule** ovoid. **Leaves** few, opposite, lanceolate or narrow-ovate, not sharp-pointed, the lower ones elliptic and evergreen.

HABITAT In damp to dry open woods or thickets, on rocky slopes and wooded dunes.

FLOWERING April to early June.

NOTES This attractive species is often very abundant. It blooms just after the first spring flowers have passed their peak and before the leaves are fully out on the trees. It makes an excellent plant for the garden, where it is attractive with tulips and other early flowering bulbs.

SIMILAR SPECIES **Moss-Pink**, *Phlox subulata* L., which is often cultivated, occurs infrequently in Michigan on sand dunes, forming dense evergreen mats. It has small, sharp-pointed, needle-like persistent leaves in tight clusters. In the spring it may be completely covered with white, roseate, pinkish, or lavender flowers. **Perennial** or **Garden Phlox**, *Phlox paniculata* L., is native in the southeast and is often cultivated here, rarely escaping.

Prairie Phlox

Phlox pilosa L.

Quite similar to **Wild Blue Phlox**, but the cymes more compact, leafy-bracted, the flowers smaller

(1.5–2 cm wide), usually red-purple, and the lobes of the corolla unnotched; the stems are erect from a crown and if offshoots are produced they are also erect; the leaves are linear to lanceolate and narrowed to a stiff, hard tip.

HABITAT In dry open woods and thickets, openings, sand dunes, waste places, and on prairies.

FLOWERING May to early June.

NOTES This species is quite variable, and several varieties have been described. It is said that it is more susceptible to garden pests, and thus more difficult to cultivate than Blue Phlox. It tends to bloom just after Blue Phlox has passed its prime.

■ Jacob's-Ladder, Greek-Valerian

Polemonium reptans L.

Spreading perennial up to 5 dm tall. **Flowers** deep blue, open-bell-shaped, borne in few-flowered corymbs. Corolla 10–15 mm long, divided to below the middle, the lobes rounded, somewhat spreading; calyx bell-shaped, 5-lobed; stamens 5, attached to the corolla tube, the filaments all bent toward one side of the corolla, shorter than the corolla; pistil solitary, the ovary superior, the stigma 3-parted; **capsule** ovoid. **Leaves** alternate, pinnately compound, the leaflets lanceolate or narrowly oval, 8–16 pairs plus a terminal one.

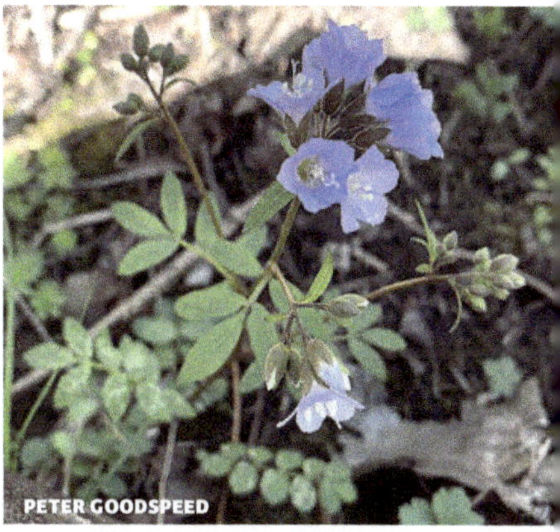

JACOB'S-LADDER, GREEK-VALERIAN

HABITAT Uncommon in rich woods and bottoms, thickets, and margins of meadows.

FLOWERING mid-April to June.

STATUS Threatened in Michigan.

NOTES Some of the numerous western species of this genus have a very disagreeable odor and are known as "skunkweed." In contrast to the phloxes, which are mostly pollinated by butterflies, the species of *Polemonium* are pollinated mainly by bees. Jacob's-ladder is said to be easy to grow in the garden.

■ POLYGALACEAE *Milkwort Family*

This widely distributed family has 29 genera and about 900 species; 2 genera in Michigan. It is of economic importance only for the few species cultivated as ornamentals.

The irregular flowers resemble those of the Pea Family, but their structure is not the same. There are 5 sepals, of which 2 or 3 are small and green and 2 (wings) are larger and petal-like. The 3 petals are connected with each other and with the stamen tube; one petal is often keel-like and crested with hairs. The filaments of the 8 stamens (6 in one species) are united into a split tube.

KEY TO POLYGALACEAE (MILKWORT FAMILY) SPECIES

1 Plants creeping; flowers showy, 1.5–2.5 cm long, few .... *Polygaloides paucifolia*

1 Plants erect; flowers small (not over 6 mm), numerous 2

  2 Flowers white; stems several from a tough, thick crown .... *Polygala senega*

  2 Flowers purple, pink, or green 3

3 Racemes headlike, rounded at apex .... ***Polygala sanguinea***

3 Racemes elongate 4

4 Flowers borne in loose racemes, leaves alternate .... ***Polygala polygama***

4 Flowers borne in very compact, spikelike racemes; leaves mostly whorled .... ***Polygala verticillata***

RACEMED MILKWORT

Racemed Milkwort

Polygala polygama Walt.

Glabrous, erect, ascending, or partially reclining, many-stemmed perennial up to 4.5 dm tall. **Flowers** of 2 kinds. Petaled flowers rose-purple to pink, 3–5 mm long, borne in loose, somewhat 1-sided, terminal racemes 2–12 cm long. Petals equal to or shorter than the wings, the keel fringed at the end; sepals 5, the 2 largest forming wings, the upper sepal appressed against the ovary; stamens 8, the filaments united in 2 sets, the anthers free; ovary 1, superior, the stigma 2-lobed; **capsule** flattened, 2-celled, each cell with a single seed. Self-pollinating, non-opening flowers small, borne in 1-sided racemes on white underground branches, forming plump fruits. **Leaves** numerous, alternate, the lower ones small and spatulate, the upper ones larger and linear to narrowly oblanceolate, entire, sessile.

HABITAT In sandy woods and dry open soil.

FLOWERING June to August.

PURPLE MILKWORT

■ Purple Milkwort

Polygala sanguinea L.

Annual with solitary, simple or branched, very leafy stems up to 4 dm tall. **Flowers** pink to rose-purple, greenish, or white, on short pedicels, overlapping in very dense, thick-cylindric to headlike racemes, which are rounded at the summit, 6–14 mm thick. Wings broadly rounded above, twice as long as the keel, 9-nerved; **fruits** soon dropping. **Leaves** numerous, alternate, linear or narrowly elliptic, 1–4 cm long, sessile.

HABITAT In moist acid soil, in fields, meadows, and open woods.

FLOWERING late June to October.

Seneca Snakeroot

Polygala senega L.

Erect perennial with several simple stems up to 5 dm tall from a thick crown. **Flowers** white or whitish, in dense, solitary terminal racemes 6–7 mm, thick. Wings nearly round, 3–3.3 mm long, equaling or exceeding the petals in length; **capsule** rounded, persistent. **Leaves** numerous, alternate, linear-lanceolate to ovate, irregularly toothed, 1.3–7 cm long, the upper leaves the largest.

HABITAT In dry or rocky places, chiefly in calcareous areas.

FLOWERING May to July.

USES American Indians carried its roots as a charm to ensure safety and health on their journeys.

Whorled Milkwort

Polygala verticillata L.

Slender annual with solitary stems up to 4 dm tall. **Flowers** white, greenish, or purplish, borne in slender tapering racemes, the lower branches of the inflorescence opposite or whorled. Wings shorter than or about equaling the capsule in length; keel yellow or yellowish. **Leaves**, at least those of the lower nodes, in whorls of 3–5, linear to narrowly lanceolate, pointed at apex, 1–2 cm long.

HABITAT In sterile open places, in moist to dry, usually sandy, soil, in grasslands and woods.

FLOWERING June to October.

Fringed Polygala

Polygaloides paucifolia (Willd.) J.R. Abbott

SYNONYM *Polygala paucifolia* Willd.

Creeping perennial up to 1 dm tall. **Flowers** irregular, of 2 kinds, the showy ones rose-purple to whitish, up to 2.3 cm, long, 1–4 on long pedicels from the end of the stem. Petals 3, united into a tube split along the top, the lower (middle) one keeled and conspicuously fringed at the end; sepals 5, the 2 lower ones small and bractlike, the upper sepal helmet-shaped, the lateral sepals petal-like and forming wings; stamens 6, in sets of 3, the filaments united into a split tube connected with the petals; pistil solitary, the style long, the ovary superior, flattened, 2-celled; **capsule** 2-seeded, flattened. Self-pollinating, non-opening flowers on underground branches, their fruits small, globular. **Leaves** few, mostly crowded near the top of the flowering stem, ovate, petioled, the lower leaves scattered, small and scalelike.

HABITAT In light soil in moist woods.

FLOWERING May and early June.

SENECA SNAKEROOT

WHORLED MILKWORT

FRINGED POLYGALA

POLYGONACEAE *Buckwheat Family*

This family, which grows in the temperate regions but primarily in the northern hemisphere, includes 56 genera and more than 1,200 species; 9 genera are found in Michigan. The family is of limited economic importance. Buckwheat and rhubarb (*Rheum*) are used for food, and a few species, such as Silver-Lacevine, are grown as ornamentals. Several species are common troublesome weeds.

The members of this family are characterized by the stipules in the form of cylindrical sheaths (called *ocreae*) above the swollen joints of the stems. The regular flowers have no petals, but the 3–6 sepals may be petal-like; there are 6–9 stamens, and the single pistil has a superior ovary.

KEY TO POLYGONACEAE (BUCKWHEAT FAMILY) SPECIES

1 Flowers greenish-yellow, often tinged with red, borne in terminal panicles; at least basal and lower leaves having spreading lobes; foliage acid to taste .... *Rumex acetosella*

1 Not as in alternate choice 2

2 Flowers solitary from axils of scales, which give a jointed appearance to ends of branchlets .... *Polygonum articulatum*

2 Not as in alternate choice 3

3 Plants erect; leaves broadly triangular; calyx shriveling to expose seed .... *Fagopyrum esculentum*

3 Plants usually partly decumbent, twining or creeping; leaves lanceolate, arrow-shaped, or heart-shaped; seeds enclosed by calyx 4

4 Stem prickly, flowers borne in short heads .... *Persicaria sagittata*

4 Stem not prickly; flowers in elongate clusters 5

5 Plants having elongate creeping stems .... *Persicaria amphibia*

5 Plants lacking elongate creeping stems............ 6

6 Stems twining .... *Fallopia ciliinodis*

6 Stems not twining 7

7 Summit of sheath fringed with bristles .... *Persicaria maculosa*

7 Summit of sheath having few or no bristles 8

8 Racemes erect .... *Persicaria pensylvanica*

8 Racemes arching or nodding at tip .... *Persicaria lapathifolia*

COMMON BUCKWHEAT

Common Buckwheat

Fagopyrum esculentum Moench

SYNONYM *Fagopyrum sagittatum* Gilib.

Erect annual 2–6 dm tall. **Flowers** white, in dense compound racemes. Petals lacking; calyx petal-like, 5-parted; stamens 8, alternating with yellow, honey-bearing glands; styles 3; **achenes** 3-sided, smooth and shining. **Leaves** alternate, triangular with heart-shaped or hastate base, the sheaths semicircular.

HABITAT In waste places and old fields, roadsides. Spreading from or persistent after cultivation.

FLOWERING June to September.

NOTES This species is the source of buckwheat flour.

Fringed Black Bindweed

Fallopia ciliinodis (Michx.) Holub

SYNONYM *Polygonum cilinode* Michx.

Freely branching, twining, high-climbing perennial which often covers other plants in extensive patches. **Flowers** small, borne in loose, slender racemes in numerous panicles; calyx white or pink-tinged, 4–5 mm long in fruit, not winged. **Leaves** heart-shaped-ovate, long-pointed at apex; sheathing stipules with a ring of deflexed bristles at base.

HABITAT Along roadsides, in dry thickets and borders of woods.

FLOWERING July and August.

SIMILAR SPECIES Two other, quite similar, species occur in Michigan, but neither has bristles at the base of the sheath. **Black Bindweed**, ***Fallopia convolvulus*** (L.) Á. Löve, an introduced species, has the calyx greenish, 4–5 mm long in fruit and barely or not at all winged. **Climbing False Buckwheat**, ***Fallopia scandens*** (L.) Holub, a native species, has thicker racemes; the fruiting calyx is 8–10 mm long, and has paper-like pinkish to yellowish wings. It grows in damp thickets and river bottoms, along shores, and in low woods.

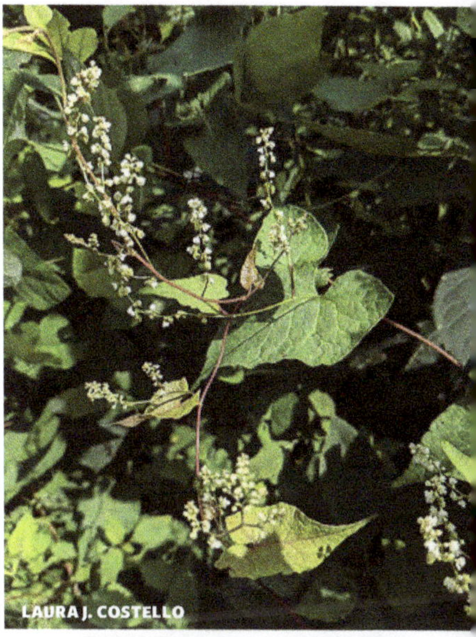

FRINGED BLACK BINDWEED

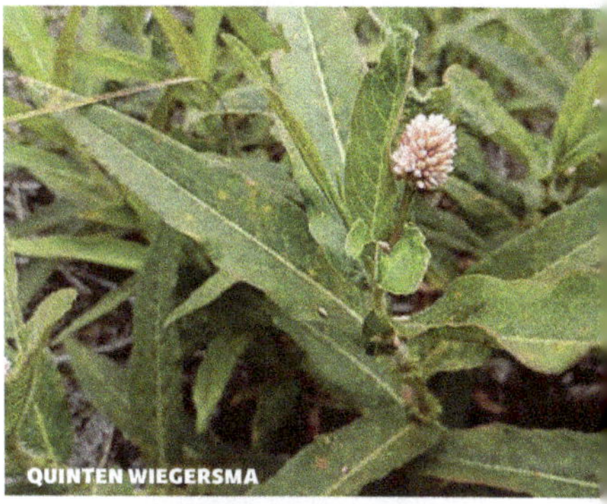

WATER SMARTWEED

Water Smartweed

Persicaria amphibia (L.) Delarbre

SYNONYM Polygonum amphibium L.

Aquatic or terrestrial, smooth or hairy, creeping perennial with tough branching stems up to 7 m long, leafy to the summit, floating, submersed, or creeping. **Flowers** pink, borne in 1–4 very dense, oblong or ovoid, erect spikes 1–4 cm long and 1–2 cm thick. Peduncles usually solitary (sometimes 2–4), holding the spike well above the water when in aquatic habitats. **Leaves** alternate, oblong-elliptic to narrowly lanceolate, often shining on upper surface and sometimes purplish or reddish beneath, the veins pinnate and prominent. The plants are nearly smooth in the aquatic forms and hairy in the terrestrial forms.

HABITAT Along the edges of lakes, in ditches, ponds, meadows, swamps. Common.

FLOWERING June to mid-September.

Dock-Leaf Smartweed

Persicaria lapathifolia (L.) Delarbre

SYNONYM *Polygonum lapathifolium* L.

Quite similar to the 2 preceding species in general aspect, the sheaths without bristles on the margin, the flower spikes usually arching or drooping at the tip. A highly variable species.

DOCK-LEAF SMARTWEED

SPOTTED LADY'S-THUMB

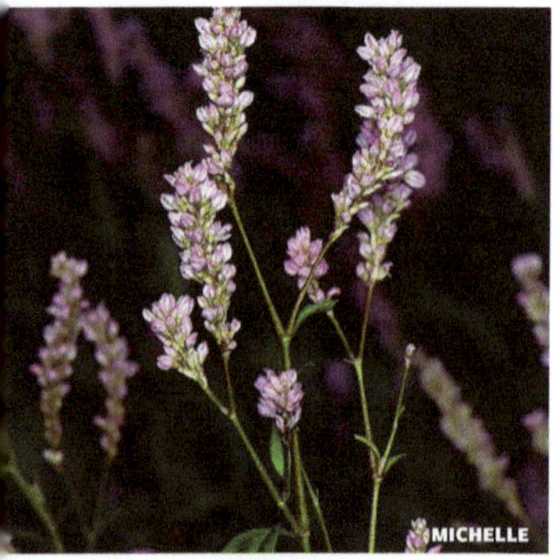

PINKWEED

HABITAT On wet shores, in thickets and clearings. Introduced from Europe.

FLOWERING July to November.

Spotted Lady's-Thumb

Persicaria maculosa Gray

SYNONYM *Polygonum persicaria* L.

Nearly glabrous, erect or partly reclining annual with simple or branching stems up to 10 dm, tall. **Flowers** pink or purplish to pink and green or nearly white, borne in 1 to several dense, cylindric spikes 1.5–4.5 cm long, the spikes usually in panicles. Petals lacking; sepals usually 5 (4–6), petal-like; stamens 3–9; **achenes** shiny. **Leaves** lanceolate, the well-developed ones 3–15 cm long, usually firm, often blotched with purple, sometimes thin in plants that are submerged part of the time; sheathing stipules fringed with bristles.

HABITAT Along roadsides and railroads, in cultivated ground and waste places; usually in damp soil. Introduced from Europe; now an ever-present weed.

FLOWERING June to October.

Pinkweed

Persicaria pensylvanica (L.) M.Gómez

SYNONYM *Polygonum pensylvanicum* L.

Erect to ascending annual quite similar in general aspect to the preceding species, but the sheaths lacking bristles on the margin, the flower spikes dense, erect, pink to purplish. This is a highly variable species with several recognized varieties.

HABITAT On damp shores, in wet thickets, clearings, and in cultivated soil.

FLOWERING late May to October.

Arrow-Leaf Tearthumb

Persicaria sagittata (L.) H.Gross

SYNONYM *Polygonum sagittatum* L.

Prickly, weak, usually leaning perennial with backward-pointing barbs on the ridges of the 4-angled stems. **Flowers** pink to white, borne in dense heads on long, glabrous peduncles.

Leaves narrowly arrow-shaped, 3–10 cm long, barbed on the under side of the midrib.

HABITAT In low ground and marshy places, and on wet sandy shores.

FLOWERING July to October.

Coastal Jointweed

Polygonum articulatum L.

SYNONYM *Polygonella articulata* (L.) Meisn.

Erect, branching, heath-like, jointed annual up to 6 dm tall, the stems and leaves quite similar in appearance. **Flowers** tiny, rose to white, solitary on short pedicels which rise from the sheathing stipules. Corolla lacking; calyx petal-like, 5-parted, persistent around the achene, stamens 8, styles 3. **Leaves** alternate, needlelike, jointed at the base, and readily falling.

HABITAT In dry sands, open ground, jackpine plains, and along lakeshores.

FLOWERING July and August.

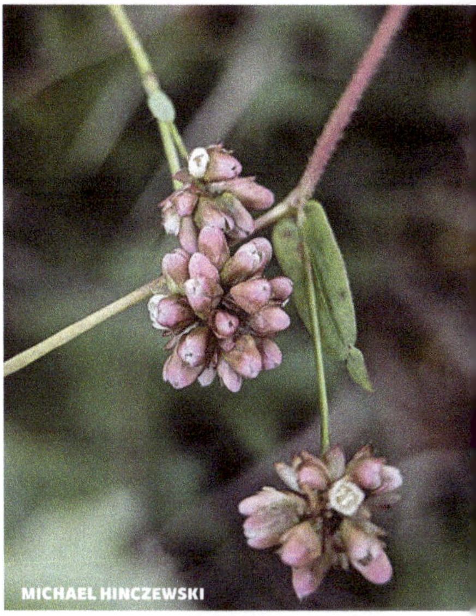

ARROW-LEAF TEARTHUMB

Sheep Sorrel

Rumex acetosella L.

Low, hairy, many-branched perennial up to 3 dm tall. **Flowers** usually unisexual, tiny, reddish, yellowish, or greenish, borne in slender panicles. Petals lacking; sepals 6, united at base, the 3 outer ones spreading at least in fruit, the inner ones convergent over the ovary. **Leaves** mostly basal, arrow-shaped with divergent lobes, petioled, the upper stem-leaves sessile and clasping.

COASTAL JOINTWEED

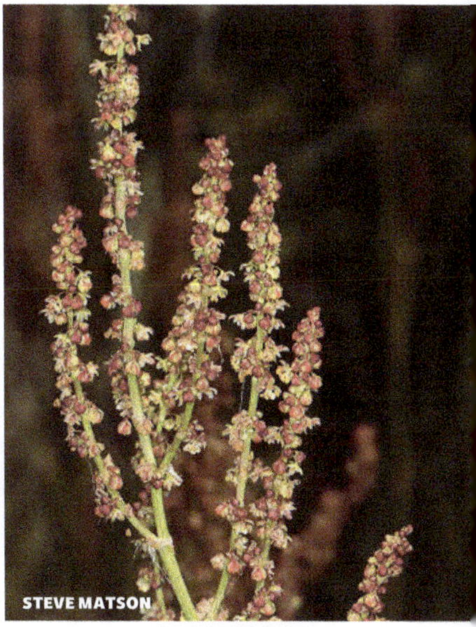

SHEEP SORREL

HABITAT In poor, dry or sandy, usually acid soil, in abandoned fields, along roadsides, and in disturbed ground of all kinds, often making an extensive ground cover. Introduced from Europe.

FLOWERING May to August.

NOTES This is a difficult weed to eradicate. The stems and leaves have a refreshing acid taste. Leaves of the larger **Curly Dock**, *Rumex crispus* L., are used for greens in the spring. Several native species also occur here.

PRIMULACEAE *Primrose Family*

This family includes 57 genera and about 800 species; 4 genera in Michigan. It is most abundant in north temperate regions but is widely distributed and occurs on all continents. It is of importance here only for a few ornamentals such as primroses, cyclamens, and Scarlet Pimpernel.

All members of this family are herbaceous. The petals are united, and the calyx is toothed, lobed, or deeply divided. The calyx and corolla lobes are usually 5 but may be 4–9 in some species. The stamens are the same number as the petals and opposite them. The fruit is a many-seeded capsule.

KEY TO PRIMULACEAE (PRIMROSE FAMILY) SPECIES

1 Leaves all basal, forming a rosette .... *Primula mistassinica*
1 Leaves borne on stem 2
  2 Flowers white .... *Lysimachia borealis*
  2 Flowers yellow 3
3 Plants creeping .... *Lysimachia nummularia*
3 Plants erect 4
  4 Flowers all or mostly borne in a terminal raceme .... *Lysimachia terrestris*
  4 Flowers borne from axils of leaves........ 5
5 Flowers borne in dense racemes .... *Lysimachia thyrsiflora*
5 Flowers solitary from axils........ 6
  6 Leaves opposite, petioles fringed; corolla lacking spots or streaks .. *Lysimachia ciliata*
  6 Leaves chiefly whorled, petioles lacking or very short; corolla dotted or streaked with black or red .... *Lysimachia quadrifolia*

STARFLOWER

Starflower

Lysimachia borealis (Raf.) U.Manns & Anderb.
SYNONYM *Trientalis borealis* Raf.

Erect, unbranched, fragile perennial up to 2.5 dm tall from long, creeping rootstocks, spreading by slender, elongate stolons. **Flowers** white, 8–14 mm wide, 1–4 on very slender, wiry peduncles from the axils of the leaves at the top of the stem. Corolla flat, star-shaped, cut almost or quite to the base, the lobes usually 7 (5–9), oblong, and pointed; calyx deeply lobed, the lobes usually 7 (5–9), very narrow, pointed, spreading flat; stamens the same number as the petals and opposite them, the filaments white, united into a ring at the base; pistil solitary, the style long and slender, the ovary 1-celled; **capsule** globular. **Leaves** mostly alternate and clustered at the top of the stem, sessile, lanceolate or narrowly ovate, 4–10 cm long, shallowly crenate or nearly entire, the leaves below the whorl few and scalelike.

HABITAT In moist woodlands and thickets, on peaty slopes, and in bogs.

FLOWERING May to late July.

Fringed Loosestrife

Lysimachia ciliata L.

Erect, simple or branched perennial up to 1.2 m tall, leafless below. **Flowers** yellow, borne on long, threadlike peduncles in whorls from the upper leaf axils. Corolla unspotted, 1.5–3 cm broad, the 5 lobes much longer than the stamens, irregularly toothed; calyx 5-toothed, the teeth tipped with a sharp point; stamens 5, separate. **Leaves** opposite, ovate to ovate-lanceolate, rounded to somewhat heart-shaped at base, 3–15 cm long; petioles very hairy, appearing fringed.

HABITAT In moist woods and thickets, on flood plains and in low, wet ground.

FLOWERING July and August.

MARK EANES

FRINGED LOOSESTRIFE

Creeping-Jennie

Lysimachia nummularia L.

Creeping or trailing perennial rooting freely at the nodes, often forming mats. **Flowers** bright yellow, showy, about 2 cm in diameter, solitary on peduncles from the axils of the leaves. Corolla wheel-shaped, dotted with red, so deeply 5-lobed that it appears to be made up of separate broadly ovate petals; calyx deeply 5-parted, red-dotted, spreading in flower but closing around the developing capsule; stamens 5, the filaments united into a tube at the base; ovary superior, red-dotted. **Leaves** opposite, short-petioled, broadly oval or orbicular, 10–25 mm long, entire.

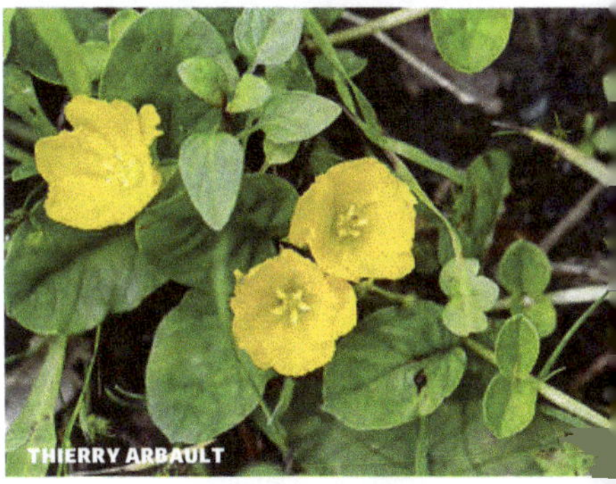

THIERRY ARBAULT

CREEPING-JENNIE

HABITAT Along damp roadsides, in grassy places, and along shores. Introduced from Europe; now often a weed in lawns.

FLOWERING June to August.

NOTES When flowers are absent, this species can appear like one of our native plants, such as *Chrysosplenium* or *Mimulus;* the orange to black glandular dots on the undersides of the leaves help identify Moneywort.

WHORLED LOOSESTRIFE

SWAMP-CANDLES

■ Whorled Loosestrife

Lysimachia quadrifolia L.

Erect, simple, slender perennial up to 8 dm tall. **Flowers** yellow, dark-streaked or spotted, borne on threadlike peduncles (2–5 cm long) from the axils of the whorled leaves. Corolla wheel-shaped, 5-lobed, the lobes lanceolate, entire; calyx deeply 5-parted; stamens 5, united into a ring at the base, the filaments unequal. **Leaves** usually in whorls of 4 or 5 (the lower ones sometimes opposite), sessile or nearly so, lanceolate to narrowly ovate, 3–9 cm long, 1–2.5 cm broad.

HABITAT In dry or moist, open or partially shaded ground, in woods or along shores.

FLOWERING May to August.

■ Swamp-Candles

Lysimachia terrestris (L.) B.S.P.

Erect, simple or sparingly branched perennial up to 1 m tall. **Flowers** yellow, dotted or streaked with red or purple, about 1 cm wide, borne in terminal racemes, the pedicels slender, horizontal or ascending. Corolla wheel-shaped, deeply 5-parted; stamens 5, the filaments united at the base, unequal in length. **Leaves** opposite, nearly sessile, thick, narrow-lanceolate, often black-dotted, entire, obscurely veined, 5–10 cm long; lower leaves smaller.

HABITAT Along wet shores and on low wet ground, in thickets, marshes, and swamps.

FLOWERING July and August.

■ Tufted Loosestrife

Lysimachia thyrsiflora L.

Erect, usually unbranched perennial up to 1 m tall, the stems clustered, red at least below. **Flowers** yellow, often spotted with red, purplish, or black dots, borne from the axils of the middle leaves in dense, short spikes, the peduncles stout, 2–4 cm long. Corolla tube very short, the limb deeply lobed, the lobes narrow, much shorter than the stamens; calyx cleft into 6 (5–7) linear segments, light yellowish-green, dotted with red; stamens rising from a ring at the base of the ovary; pistil solitary, the ovary green, speckled; **capsule** about as long as the calyx lobes. **Leaves** opposite, sessile, lanceolate to elliptic, 5–12 cm long, and up to 2.5 cm wide, the lower ones smaller (some merely ovate and scalelike).

HABITAT In swamps, wet meadows, low woods, springy marshes, and bogs.

FLOWERING May to July.

TUFTED LOOSESTRIFE

MISTASSINI PRIMROSE

Mistassini Primrose

Primula mistassinica Michx.

SYNONYM *Primula intercedens* Fern.

Slender perennial with naked flowering stems up to 2.5 dm tall from a basal rosette. **Flowers** lilac, pale pink, or white, with a conspicuous yellow center, borne in terminal umbels, on nearly equal, stiffly ascending pedicels. Corolla salverform, the 5-lobed limb flat, up to 2 cm broad, the lobes heart-shaped at apex; calyx tubular, 5-cleft; stamens 5; pistil 1; **capsule** cylindric, longer than the closely appressed calyx, often crowned with the withered corolla; seeds rounded, nearly smooth. **Leaves** all basal, nearly erect, oblanceolate, long-tapering to the base, irregularly and shallowly toothed, the margin somewhat inrolled.

HABITAT On wet, calcareous shores and rocks. Chiefly in northern Michigan.

FLOWERING May and early June.

RANUNCULACEAE *Buttercup Family*

This is a fairly large family and particularly important in the cool, temperate regions of the Northern Hemisphere. It includes 51 genera and more than 2,000 species; 22 genera are found in Michigan. The family is of some importance for ornamentals. Buttercups, Globe Flower (*Trollius*), Peony (*Paeonia*), Larkspur (*Delphinium*), Christmas Rose (*Helleborus*), Winter Aconite (*Eranthis*), Clematis, Columbine, and Meadow-rue are among the best-known of the garden species.

Most members of the family are herbaceous. The petals, sepals, stamens, and pistils are all separate; the petals are often lacking, in which case the sepals may be petal-like. The stamens are usually numerous and spirally arranged; there are generally several to many one-celled pistils. The leaves are often deeply divided or compound, but in some species they are simple and merely toothed or shallowly lobed; the petiole is usually enlarged at the base, and there are seldom stipules present. The juice is acrid and, in many species, poisonous.

KEY TO RANUNCULACEAE (BUTTERCUP FAMILY) SPECIES

1 Both petals and sepals present 2
1 Petals lacking, or if present so modified as to be scarcely recognizable as such; sepals often petal-like and colored 5
 2 Sepals green or yellowish green; often pale and soon falling 3
 2 Sepals petal-like in color and texture 4
3 Flowers few to several; fruit an achene . . . . ***Ranunculus***
3 Flowers numerous, in a dense, thick raceme; fruit a berry . ***Actaea pachypoda*** or ***A. rubra***
 4 Flowers white; leaflets 3 . . . . ***Coptis trifolia***
 4 Flowers scarlet and yellow; leaflets more than 3 . . . . ***Aquilegia canadensis***
5 At least some of flowers unisexual 6
5 All flowers having both stamens and pistils 7
 6 Vines . . . . ***Clematis occidentalis*** or ***C. virginiana***
 6 Erect plants . . . . ***Thalictrum dasycarpum*** or ***T. dioicum***
7 Plants low and stemless 8
7 Plants with leafy stems 9
 8 Leaves with 3 lobes . . . . ***Hepatica americana***
 8 Leaves with 3 leaflets . . . . ***Coptis trifolia***
9 Leaf margins with shallow, rounded teeth . . . . ***Caltha palustris***
9 Leaf margins deeply lobed or leaves compound; flowers not both showy and yellow . . 10
 10 Leaflets having a few rounded lobes 11
 10 Margin of leaflets or lobes having numerous sharp teeth 12
11 Flowers few, borne above a whorl of leaves; fruit an achene . . . . ***Thalictrum thalictroides***
11 Flowers solitary from axils of alternate stem leaves; fruit a 2–3-seeded follicle . . . . ***Enemion biternatum***
 12 Fruit berrylike; stamens much more conspicuous than tiny sepals 13
 12 Fruit an achene; sepals conspicuous . . . . ***Anemonastrum, Anemone***
13 Flowers numerous, borne in racemes . . . . ***Actaea pachypoda*** or ***A. rubra***
13 Flowers solitary . . . . ***Hydrastis canadensis***

WHITE BANEBERRY

White Baneberry

Actaea pachypoda Ell.

Perennial up to 1 m tall from a creeping rootstock. **Flowers** whitish, borne on thick red pedicels in a sub-cylindric raceme 5–20 cm long, the pedicels becoming thicker in fruit, nearly as thick as the peduncle. Petals 4–10, small, flat, with slender claws, resembling modified stamens; sepals 4–5, soon falling when the flowers open; stamens numerous, the filaments slender; pistil 1, the stigma sessile, broad and cap-like; **berry** white (red in one form), shining, subglobose, about 1 cm long, the stigma persisting as a red or purple "eye-spot." **Leaves** large, 2–3 times pinnately compound, the leaflets whitish beneath, ovate, sharply cleft and toothed.

HABITAT In rich woods and thickets.

FLOWERING May and June. Fruit ripe July to Oct.

NOTES The berries are reported to be poisonous, and the rootstock is a violent purgative.

Red Baneberry

Actaea rubra (Ait.) Willd.

Like the preceding species, but the pedicels are thin and remain so, the leaflets are less sharply toothed, the petals taper to the apex, and the berries are red (white in one form).

HABITAT In woods and thickets.

FLOWERING May and early June.

NOTES Both species of baneberry may have either red or white berries, but, regardless of the color of the berries, the species with thick red pedicels is *A. pachypoda*, or **White Baneberry**, and that with thin pedicels is **Red Baneberry**.

RED BANEBERRY

KEY TO ANEMONASTRUM AND ANEMONE SPECIES

1 Leaves cut into numerous long, narrow segments, flowers usually red; plants of sandy or gravelly shores of the Great Lakes . . . . *Anemone multifida*

1 Not as in alternate choice 2

2 Plants usually 1–2 dm tall, unbranched, having a single flower; leaves compound . . . . *Anemone quinquefolia*

2 Usually much taller and coarser, branched, and having more than 1 flower; leaves deeply cut, but not compound 3

3 Leaves of flowering stem sessile or nearly so; fruiting heads nearly globose . . . . *Anemonastrum canadense*

3 Leaves of flowering stem petioled; fruiting heads ovoid to cylindric 4

4 Stem leaves typically 3; plants green . . . . *Anemone virginiana*

4 Stem leaves typically 5–9 in a whorl, of 2 sizes; plants grayish green . . . . *Anemone cylindrica*

Meadow Anemone

Anemonastrum canadense (L.) Mosyakin

SYNONYM *Anemone canadensis* L.

Robust perennial with 1–2 usually freely branching stems up to 7 dm tall, usually growing in clumps or patches. **Flowers** solitary on peduncles, white, 1.5–5 cm wide. Petals lacking; sepals 5, petal-like, white, oblong or ovate, unequal; stamens numerous, conspicuous; pistils numerous, glabrous or somewhat hairy; fruiting heads globose; achenes flat, broadly wing-margined, beaked with a long, hairlike, pointed style. Primary involucre large, of 3 sessile, deeply 2–3-lobed, coarsely toothed leaves; secondary involucres of 2 smaller but similar leaves. **Basal leaves** 5–15 cm broad, long-petioled, 5–7-parted nearly to the base, the divisions usually 3-cleft and coarsely toothed.

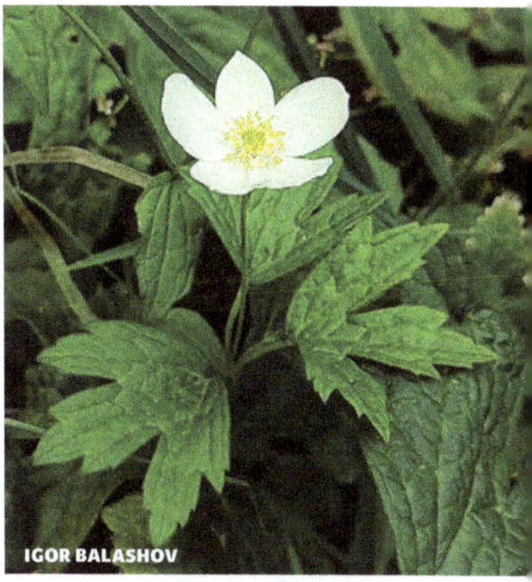

MEADOW ANEMONE

LONG-HEAD THIMBLEWEED

HABITAT In damp thickets, open woods, and meadows, along gravelly, usually calcareous, shores, and in shallow ditches.

FLOWERING late May to early August.

NOTES This species is more attractive than our other white-flowered anemones because of its more conspicuous, yellow-centered, brilliantly white flowers and its habit of growing in clumps. It makes a satisfactory garden flower, though it tends to spread too rapidly.

USES American Indians used the crushed rootstocks on wounds and hemorrhages and in infusions for treating infections. They also used this plant for a tea to relieve lung congestion.

Long-Head Thimbleweed

Anemone cylindrica Gray

Similar to the preceding species, but the plants grayish-green; the involucre usually of 5–9 leaves, 3 of which are noticeably larger than the others; the secondary peduncles usually lacking involucres; the fruiting heads cylindric (typically more than twice as long as thick); the styles pointed, crimson, the tips recurved.

HABITAT In dry open soil, along roadsides, on dry and rocky lakeshores, and in dryish open woods.

SIMILAR SPECIES A third thimbleweed, ***Anemone virginiana*** L., resembles the two preceding species in size and habit and is difficult to distinguish from them. It has 3 involucral leaves, with the segments wedge-shaped in general outline and the margins straight or nearly so toward the base. The styles at maturity are curved upward and the mature fruiting heads are 7–11 mm thick.

RED WINDFLOWER

Red Windflower

Anemone multifida Poir.

Freely branching, silky-hairy perennial, seldom over 3 dm tall. **Flowers** solitary on erect peduncles, red, purplish, or yellowish inside and yellowish, greenish, purplish, or red outside. Petals lacking; sepals 5 (14–16 in one form), petal-like; stamens many; pistils numerous, the styles threadlike, erect, often falling off in fruit; fruiting head short-cylindric (2 cm long and 1 cm thick) or nearly globose, woolly; **achenes** pubescent to woolly. **Basal leaves** long-petioled, 2–3 times divided or cleft into long, narrow, rather pointed segments; leaves of the involucre 3, like the basal leaves, petioled; the main peduncle leafless, the secondary peduncles with small petioled leaves resembling the basal leaves.

HABITAT In dry, slaty or calcareous gravel, on ledges, and in sand, frequently on lakeshores. Common on the sandy beaches of Lakes Michigan and Huron at the tip of the Lower Peninsula; also on Isle Royale.

FLOWERING May to September (more profusely early in the season).

SIMILAR SPECIES Another colorful and very beautiful anemone, the **Pasque-Flower**, ***Pulsatilla nuttalliana*** (DC.) Bercht. & J.Presl, has large blue, purple, or lavender flowers, which are open and bell-shaped at first, resembling a crocus. The entire plant is covered with silky hairs, and the leaves are finely divided.

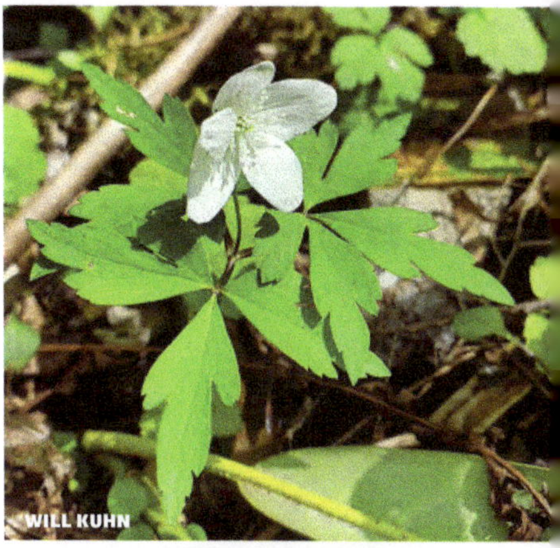

WOOD ANEMONE

Wood Anemone

Anemone quinquefolia L.

SYNONYM *Anemonoides quinquefolia* (L.) Holub

Small, unbranched perennial with a solitary slender stem 1–2 dm tall. **Flowers** solitary, white, 1.5–2 cm wide. Petals lacking; sepals usually 5 (4–9), petal-like, often tinged with pink or crimson; stamens numerous; pistils few, hairy; fruiting head globose; **achenes** tipped with hooked styles. **Basal leaf** solitary, long-petioled, palmately compound, with 3–5 sharply toothed or cut leaflets (or those with 3 leaflets having the lower leaflets so deeply cut that there appear to be 5 leaflets); involucre usually of 3 (2–4) petioled leaves similar to the basal leaves.

HABITAT In low or moist openings, thickets, open woods, and rich partially shaded ground.

FLOWERING late April to mid-June.

Tall Thimbleweed

Anemone virginiana L.

Stiffly erect, greenish perennial up to 1 m tall. **Flowers** solitary on long peduncles, white, greenish-yellow, or red-tinged, 1.5–3.8 cm wide. Petals lacking; sepals 5, petal-like (sometimes leathery); stamens numerous, the styles strongly spreading at maturity, persistent; fruiting heads 12–15 mm thick, woolly, ovoid or somewhat thimble-shaped, the **achenes** woolly. **Basal leaves** long-petioled, hairy, deeply palmately lobed and cleft, the divisions convex, the margins slightly rounded on the sides at the base, the involucre usually of 3 (2–5) short-petioled smaller leaves; the main peduncle leafless, the secondary peduncles usually with a pair of small leaves near the middle.

TALL THIMBLEWEED

RED COLUMBINE

HABITAT In dry or rocky open woods, thickets, and clearings.

FLOWERING late June to September.

Red Columbine

Aquilegia canadensis L.

Slender, graceful, many-branched perennial up to 1 m tall, **Flowers** 2.5–5 cm long, scarlet and yellow, solitary, nodding at the ends of slender branches. Petals 5, bright red to pinkish outside, yellow inside, having a short spreading lip and a slender spur nearly 2 cm long with a nectar-filled terminal knob; sepals 5, petal-like, reddish or greenish-yellow, 1 cm or more long; stamens very numerous, small, projecting as a column well beyond the perianth; pistils 5, the styles long and thin; **follicles** erect, the seeds numerous, shiny. **Leaves** basal or alternate on the stem, 2–3 times compound, the divisions in 3's; basal leaves long-petioled, 10–20 cm broad, the leaflets 2–5 cm broad and sessile or with short stalks; upper leaves similar but smaller.

HABITAT In sandy or rocky, open or partially shaded ground; frequent in open woods and thickets, on dry slopes and ledges with scanty soil; sometimes on springy slopes and in peat bogs.

FLOWERING late April to mid-July.

NOTES This handsome plant, easy to grow and often cultivated, is visited by hummingbirds and large moths that can reach the nectar at the base of the long spurs. It is the only native Michigan columbine, but it has several color variations. The common **Garden Columbine**, *Aquilegia vulgaris* L., often escapes and may grow wild for a time. It has blue, purple, pink, or white flowers with short, thick, recurved spurs. Several long-spurred varieties are also cultivated.

USES Wild Columbine was used in many ways by American Indians. Men rubbed it on the hands as a love charm. The pulverized seeds were also used as a "man's perfume" by some tribes. An infusion was made of the pounded seeds for treating headache and fever, and a remedy for stomach trouble was derived from the roots. The seeds were a commodity of intertribal commerce.

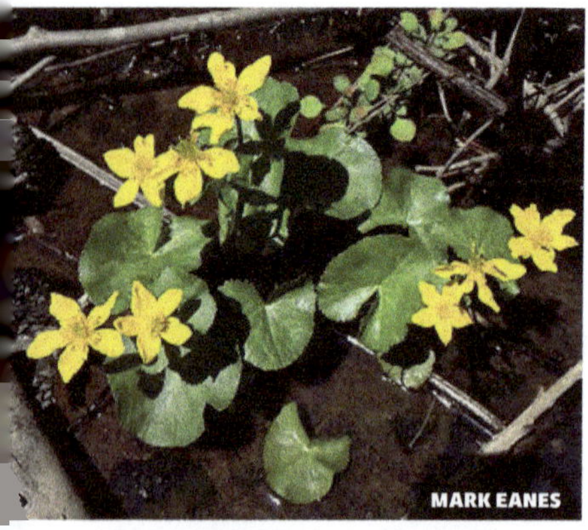

MARSH MARIGOLD, COWSLIP

Marsh Marigold, Cowslip

Caltha palustris L.

Glabrous, decumbent or erect perennial often forming low rounded mounds, the stems clustered, branching, short at flowering time but later up to 7 dm tall. **Flowers** bright yellow, 3–4 cm wide, borne in loose cymose clusters mostly above the leaves. Petals lacking; sepals 5–10, petal-like, green at first, becoming bright yellow to yellow-orange; stamens numerous; pistils 5–10, separate; **follicles** compressed, many-seeded, spreading. **Leaves** nearly kidney-shaped or round, with a deep sinus, margin crenate, toothed or nearly entire; upper leaves smaller and short-petioled or sessile.

HABITAT In wet places, low meadows, along streams, in swamps and wet woods. Often occurring in great profusion.

FLOWERING late April to mid-June.

NOTES It is said that this species has more than 25 common names. The leaves and stems are edible when cooked, but are poisonous when raw. The plant is so acrid that animals usually avoid it. American Indians used the boiled, mashed roots for treating running sores.

Virgin's-Bower

Clematis virginiana L.

Perennial vine with freely branching stems 2–3 m long, trailing or climbing over shrubs, trees, and fences by twisting the petioles around the support. **Flowers** unisexual, creamy white, borne in large cymose clusters from the leaf axils, about 2 cm wide. Petals lacking; sepals 4, separate; staminate flowers with numerous stamens; pistillate flowers with a cluster of silky, long-styled pistils surrounded by sterile stamens and the sepals; **achenes** brown or reddish with long, whitish, feathery, somewhat curly styles 2 cm or more long, forming conspicuous fluffy masses. **Leaves** opposite, 3-foliolate, the leaflets thin, ovate, often heart-shaped at the base, toothed or lobed.

HABITAT In moist places, lowlands, damp thickets, borders of woods and along streams.

FLOWERING July to September. Fruits ripe in late summer and early fall.

NOTES Various species of *Clematis* are cultivated; some have very showy flowers which may be as much as 10–15 cm in diameter.

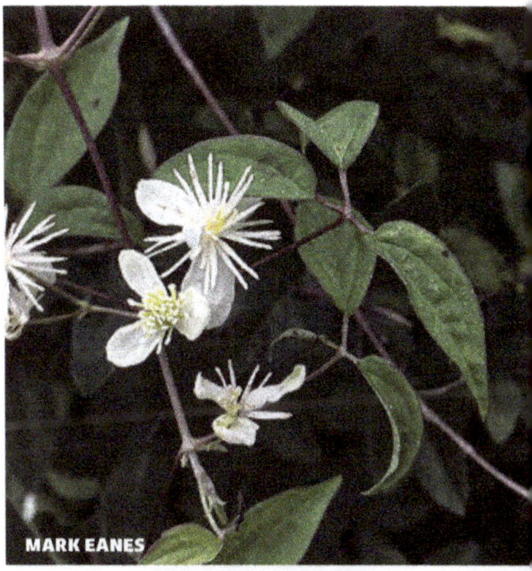

MARK EANES

VIRGIN'S-BOWER

Purple Clematis

Clematis occidentalis (Hornem.) DC.

SYNONYM *Clematis verticillaris* DC.

Woody-stemmed climber differing from the preceding species by having large (5–8 cm broad) blue, mauve, or purple solitary **flowers** nodding in the axils of the leaves; the sepals covered with long soft hairs; the **achenes** with long, softly hairy styles, borne in a dense cluster.

HABITAT In rocky open woods and on slopes.

FLOWERING May to June.

MFEAVER

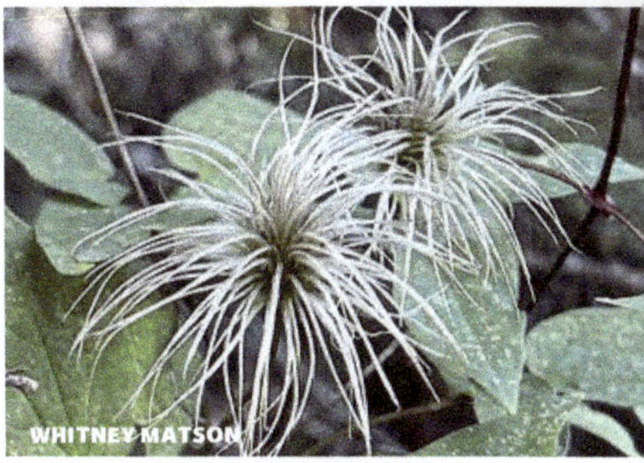

WHITNEY MATSON

PURPLE CLEMATIS

THREE-LEAF GOLDTHREAD

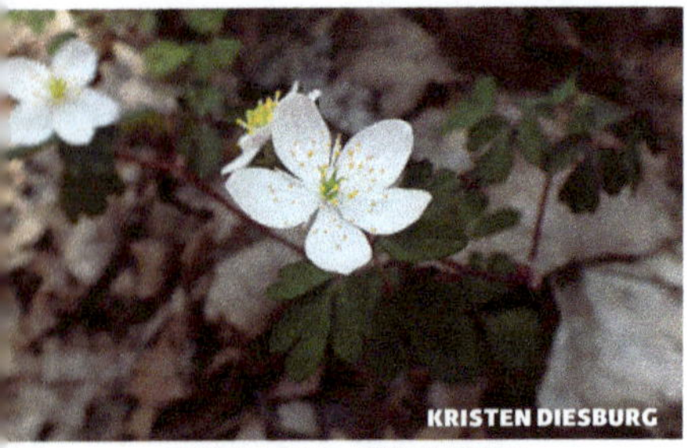

FALSE RUE-ANEMONE

ROUND-LOBED HEPATICA

Three-Leaf Goldthread

Coptis trifolia (L.) Salisb.

SYNONYM *Coptis groenlandica* (Oeder) Fern.

Small, stemless, evergreen perennial up to 15 cm tall from a golden-yellow, threadlike rootstock. **Flowers** solitary, white, about 1 cm wide. Petals 5–7, small, thick and fleshy, club-shaped, dull greenish to yellow, and not always easily recognizable as petals; sepals 5–7, petal-like, white, sometimes tinged with green below; stamens 15–25, white, small but conspicuous; pistils 3–9, pale yellowish-green, long-stalked, the styles elongate and frequently curled downward; **follicles** divergent, pointed. **Leaves** all basal, 3-foliolate, shiny on upper surface, 2.5–5 cm wide, toothed, the leaflets obscurely 3-lobed.

HABITAT Usually in acid, peaty soil, in bogs, low wet woods, cedar swamps, mossy woods.

USES A bright yellow dye can be made by boiling the rootstocks. American Indians used a decoction of the rootstocks for treating sore gums and for lessening the pain of teething.

False Rue-Anemone

Enemion biternatum Raf.

SYNONYM *Isopyrum biternatum* (Raf.) Torr. & Gray

Slender, glabrous perennial up to 4 dm tall from fibrous or somewhat tuberous roots. **Flowers** white, solitary from the axils or terminal, 1–1.5 cm wide. Petals lacking; sepals 5, petal-like, soon falling; stamens up to 40; pistils 4 (3–6); **follicles** spreading at maturity, oblong, pointed, sessile, the seeds 2–5, smooth. **Leaves** alternate, 2–3 times compound, the leaflets roundish, 2–3-lobed.

HABITAT In rich moist woods or thickets.

FLOWERING April and May.

Round-Lobed Hepatica

Hepatica americana (DC.) Ker-Gawl.

Low, hairy perennial seldom over 1.5 dm tall, **Flowers** blue, lavender, purple, maroon, pinkish, or whitish, closing in cloudy weather and at night, solitary on hairy flowering stems with a calylike involucre of 3 round-tipped bracts just below the flower. Petals lacking; sepals usually 6

or 7, petal-like; stamens numerous, small, of unequal lengths; pistils several; **achenes** oblong, pointed, hairy. **Leaves** usually broader than long, heart-shaped at the base and having 3 broad, blunt to rounded lobes, appearing after the flowers and persisting until the following spring, the old leaves frequently purplish red.

HABITAT In dry woods, on sandy slopes and in thickets.

FLOWERING principally in April and May.

NOTES This plant is fairly common in oak woods, where the clusters of flowers stand out well above the old leaves. It is an excellent garden plant and is as attractive as many cultivated species, both in flower and in leaf after the flowers are gone. It blooms with or just after Crocuses.

USES American Indians used Hepatica medicinally and as a charm to put on traps for fur-bearing animals.

Sharp-Lobed Hepatica

Hepatica americana var. *acuta* (Pursh) Mabuchi & J. Massey

SYNONYM *Hepatica acutiloba* DC.

Very similar to the preceding species, but the leaf blades are usually longer than broad, the 3 divisions more sharply pointed and frequently notched or lobed; the involucral bracts are pointed, and the flowers tend to be showier.

HABITAT In rich, often calcareous, woods.

FLOWERING March to early May.

NOTES The two species of Hepatica seldom grow together naturally, probably because var. *acuta* does best in a less acid soil.

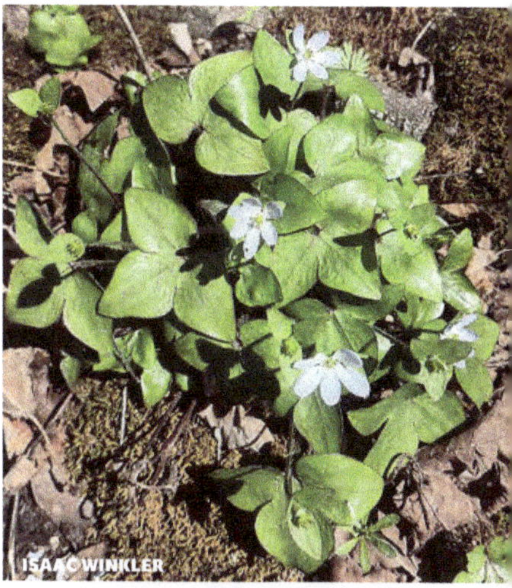

SHARP-LOBED HEPATICA

Goldenseal

Hydrastis canadensis L.

Perennial with a single unbranched stem up to 4 dm tall. **Flower** solitary, terminal, whitish. Petals lacking; sepals 3, petal-like, falling off as the flower opens; stamens numerous, conspicuous; pistils 12 or more; **fruit** a raspberry-like head of small, 2-seeded red berries. **Stem leaves** 2, alternate near the top of the stem, palmately cut, sharply toothed; **basal leaf** similar, solitary, long-petioled, becoming 13–20 cm broad.

HABITAT In woods in rich damp leaf mold.

FLOWERING in May.

STATUS Threatened in Michigan.

NOTES The rootstock and leaves of this species contain alkaloids which are poisonous but have medicinal value. Formerly quite abundant, the plant was so extensively collected for medical purposes that it is now rare, but it is sometimes grown commercially.

GOLDENSEAL

KEY TO RANUNCULUS (BUTTERCUP) SPECIES

1 Plants growing on ground 2
1 Plants growing in water 5
2 Plants producing trailing or creeping, rooting stems . . . . *R. carolinianus*
2 Plants not producing rooting stems 3
3 Flowers conspicuous, petals longer than sepals 4
3 Flowers inconspicuous, petals shorter than sepals . . . . *R. abortivus*
4 Flowers bright yellow; basal leaves palmately divided . . . . *R. acris*
4 Flowers pale yellow; basal leaves pinnately divided . . . . *R. fascicularis*
5 Flowers yellow . . . . *R. flabellaris*
5 Flowers white . . . . *R. longirostris*

KIDNEY-LEAF BUTTERCUP

TALL BUTTERCUP

Kidney-Leaf Buttercup

Ranunculus abortivus L.

Glabrous perennial with 1–3 stems up to 7 dm tall, branching above. **Flowers** few to many, less than 1 cm wide, pale yellow, the petals shorter than the sepals; **achenes** nearly round, shining. **Stem leaves** variable, simple or divided into lanceolate or linear segments, the upper leaves sessile, the lower stem leaves petioled; **basal leaves** simple and kidney-shaped to nearly round, or (less commonly) 3-lobed, cleft, or compound.

HABITAT In low woods, thickets, and clearings, and on damp slopes.

FLOWERING April to July.

Tall Buttercup

Ranunculus acris L.

Erect, branching, many-stemmed perennial, up to 1.3 m tall from a short, thickish, erect rootstock. **Flowers** bright yellow, borne on freely branching terminal peduncles. Petals twice as long as the sepals, glossy above, dull beneath; style short and curved, the stigma persistent, covering one side of the style; **achene** smooth, flattened, the beak nearly central. **Basal leaves** large, long-petioled, 5–7-divided to the base, the divisions cleft and toothed; **stem leaves** similar but less divided, sessile or with petioles sheathing at the base.

HABITAT In fields and clearings, along roadsides, in meadows and pastures. Introduced from Europe.

FLOWERING May to August.

NOTES This common species often colors whole fields yellow in late June and is reputedly the cause of most buttercup poisoning in cattle. Cows poisoned by buttercups may produce milk with a reddish color or bitter taste. The poison is broken down when the plants are dried, so that buttercups cause no trouble in hay.

Swamp Buttercup

Ranunculus carolinianus DC.

SYNONYMS *Ranunculus hispidus* var. *nitidus* (Chapm.) T. Duncan, *Ranunculus septentrionalis* Poir.

Coarse, fleshy perennial with soft, hollow, ascending or trailing stems up to 1 m long, rooting and sending up leaves at the nodes, the roots thick and fibrous. **Flowers** few, bright yellow, 2–3 cm wide. Petals 5, somewhat longer than the sepals; sepals 5, spreading, becoming reflexed; stamens and pistils numerous, the style long, nearly straight, the stigma at the tip; fruiting heads globose to ovoid; **achenes** pitted, wing-margined, beaked. **Basal and lower leaves** compound, the divisions in 3's, the leaflets stalked, cleft, sharply toothed, the stipules conspicuous, brown, and leathery.

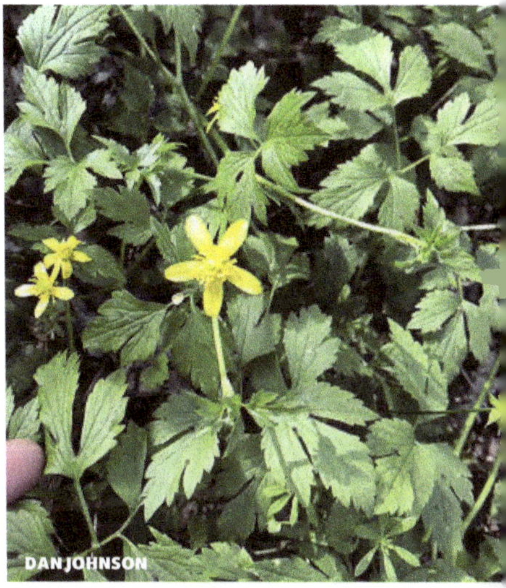

SWAMP BUTTERCUP

HABITAT In open or shady places, alluvial thickets, woods, meadows, and marshy shores, along stream banks.

FLOWERING April to July.

SIMILAR SPECIES **Creeping Buttercup**, *Ranunculus repens* L., a common and somewhat similar species, also has creeping stems, but the style is short and has the persistent stigma down one side. It was introduced from Europe.

Early Buttercup

Ranunculus fascicularis Muhl. ex Bigelow

Small perennial, with weak silky stems up to 2 dm, tall from thickened roots. **Flowers** 2–8 on a stem, 1–3 cm wide, pale yellow. Petals 5 or more; fruiting heads globose, 5–8 mm long; **achenes** slightly flattened, having a narrow wing and a long, straight or curved beak. **Basal leaves** pinnately 3–5-parted and having linear to oblong lobes.

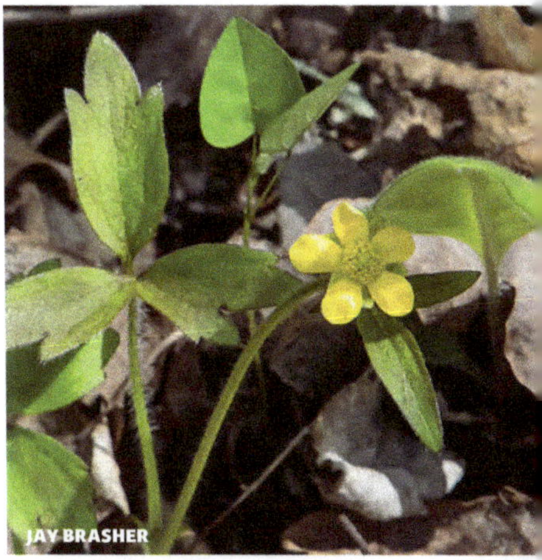

EARLY BUTTERCUP

HABITAT In thin soil in open woods and thickets, on exposed hills or calcareous ledges.

FLOWERING late April to early. June.

SIMILAR SPECIES The less common **Hairy Buttercup**, *Ranunculus hispidus* Michx., is quite similar, but it has palmately divided leaves and thicker fruiting heads (7–11 mm in diameter).

YELLOW WATER-CROWFOOT

Yellow Water-Crowfoot

Ranunculus flabellaris Raf.

Amphibious, submerged or floating species with stout, hollow, elongate stems. **Flowers** golden yellow, 1–7 on branching stems above the water, 1.5–2.5 cm wide. Petals 5–8, nearly twice as long as the sepals; sepals greenish to yellowish-green; fruiting head globose, the **achenes** having a corky thickening at the base and along the margin. **Leaves** twice compound, the divisions in 3's, narrow to almost threadlike, flabby, up to 7 cm long. Strand or shore plants have shorter stems and thicker leaves.

HABITAT In quiet water or on muddy shores.

FLOWERING May to June.

LONG-BEAK WATER-CROWFOOT

Long-Beak Water-Crowfoot

Ranunculus longirostris Godr.

Aquatic or amphibious, the stems elongate in water, shorter and rooting at the nodes in mud. **Flowers** white, solitary, 1–2 cm broad. Petals and sepals 5; **achenes** ridged, the beak long. **Leaves** very finely and deeply divided into stiff threadlike segments; stipules united to the petioles for at least 3/4 their length.

HABITAT In shallow, usually calcareous, waters of rivers, lakes, and ponds.

FLOWERING May to September.

NOTES The other species of white buttercup or crowfoot occurring in Michigan are distinguished by rather slight technical differences.

EARLY MEADOW-RUE

Early Meadow-Rue

Thalictrum dioicum L.

Delicate perennial up to 6 dm tall. **Flowers** greenish or straw-colored, unisexual, borne in axillary or terminal panicles, the individual flowers small and inconspicuous but the panicles graceful and attractive. Staminate flowers with 4–5 small, thin, oblong to oval, green, straw-colored, or purplish sepals; stamens numerous, at first erect and brownish but soon yellow and drooping on threadlike filaments. Pistillate flowers with 4–5 smaller, firm, green or purple sepals; pistils 8–20, the stigma threadlike; **achenes** ellipsoid with a short point, ridged. **Leaves** 1–3 below the flowers, 2–3 times 3-parted, the leaflets with rounded teeth or lobes, stalked; upper leaves with green, crescent-shaped stipules; not fully expanded at flowering time.

HABITAT In rich rocky woods and ravines and on slopes. More common in the southern part of the state.

FLOWERING late April to May.

NOTES This and other species of Meadow-rue are often cultivated for the attractive light-green foliage and the feathery flower clusters.

Purple Meadow-Rue

Thalictrum dasycarpum Fisch., C.A.Mey. & Avé-Lall

Considerably more robust than the preceding species, this plant may be 2 m or more in height, with many thick, often purplish stems, which give it the name Purple Meadow-Rue. **Flowers** chiefly unisexual, borne in large whitish panicles, usually well above the leaves. **Leaves** 3–4 times compound; stem leaves 3–7, the upper ones sessile or nearly so, their ovate stipules brown, the leaflets entire or with 2–3 lobes, well developed at flowering time.

HABITAT In damp soil, along streams, in swamps, damp thickets, and meadows.

FLOWERING June and July.

NOTES Though not colorful, this species is attractive because of the large, striking panicles. It can be easily grown in a damp location but requires lots of space. The ripening seeds have a delicate odor which American Indians used for perfume, rubbing the seeds over their clothing.

PURPLE MEADOW-RUE

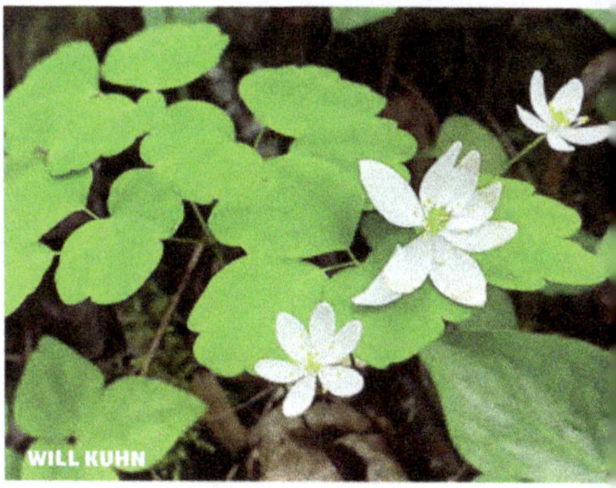

RUE ANEMONE

Rue Anemone

Thalictrum thalictroides (L.) Eames & Boivin

SYNONYM *Anemonella thalictroides* (L.) Spach

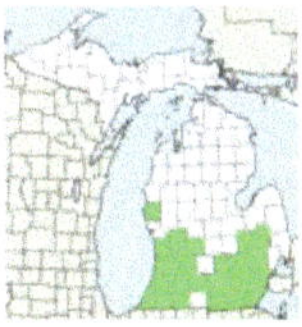

Delicate, glabrous, weak-stemmed perennial up to 3 dm tall. **Flowers** delicate pink or white, 2 or more in a loose umbel above a whorl of leaves (involucre), 1–2 cm broad. Petals lacking; sepals petal-like, 5–10 (usually 7–8), white or pink (in one rare form green and leaflike); stamens numerous, small; pistils 4–15, **achenes** ovoid with 8–10 ribs. **Leaves** on the stem in a whorl forming an involucre, the leaves compound, the divisions in 3's, the leaflets rounded, heart-shaped at the base, 3-lobed at the apex, not fully expanded at flowering time, petioled; basal leaves similar.

HABITAT In open woods and thickets, along roadsides.

FLOWERING late April to early June.

NOTES This species makes a charming addition to the wildflower garden. Forms in which there are extra sepals or in which the sepals are 3-lobed are particularly attractive. The small starchy tubers grow in a cluster and resemble miniature sweet potatoes. They are edible when cooked but are much too small to use conveniently.

SIMILAR SPECIES Because of the general similarity this species is sometimes confused with False Rue Anemone and Wood Anemone, but both of those species have solitary flowers. The latter also has sharply toothed leaves.

ROSACEAE *Rose Family*

A large, nearly cosmopolitan family of 108 genera and over 4,800 species, particularly abundant in Asia; 33 genera reported for Michigan. This family is of considerable economic importance in temperate regions. The principal fruits are apple, apricot, pear, cherry, plum, prune, peach, nectarine, quince, blackberry, raspberry, and strawberry. Such well-known ornamentals as Spiraea, Cotoneaster, Pyracantha, Ninebark, Flowering Quince, Japanese Cherry, Hawthorn, and Rose belong to this family.

Most members of the family are easily recognized as such. The flowers are usually showy, the petals and calyx lobes are 5 each, the calyx often with an extra row of bracts, the stamens in multiples of 5, and the pistils 1 to many. The petals and stamens appear to be borne on the edge of a saucer-shaped or cup-shaped receptacle (hypanthium), and the petals usually extend out horizontally from the base. The leaves are alternate, often compound, and usually have stipules.

KEY TO ROSACEAE (ROSE FAMILY) SPECIES

1 Leaves simple, palmately lobed .... *Rubus parviflorus*
1 Leaves compound 2
  2 Flowers rose-colored .... *Rosa palustris*
  2 Flowers not rose-colored 3
3 Plants stemless, leaves and flowering stalk from base of plant 4
3 Plants having leafy stems 5
  4 Flowers white .... *Fragaria virginiana* or *F. vesca*
  4 Flowers yellow .... *Waldsteinia fragarioides*
5 Flowers borne in very slender spikes; calyx bearing a band of stiff, hooked bristles .... *Agrimonia gryposepala* or *A. parviflora*
5 Flowers borne in a loose cluster; calyx not bristly 6
  6 Fruiting heads of achenes with long, feathery, persistent styles .... *Geum*
  6 Fruiting heads of achenes with short styles, soon falling off 7
7 Woody shrubs .... *Dasiphora fruticosa*
7 Not woody shrubs 8
  8 Flowers solitary on peduncles from nodes 9
  8 Flowers few to many, borne in cymes 10
9 Leaves pinnately compound .... *Potentilla anserina*
9 Leaves palmately compound .... *Potentilla simplex*
  10 Flowers purple or red-purple .... *Comarum palustre*
  10 Flowers not purple or red-purple 11
11 Principal leaves having 3 leaflets 12
11 Principal leaves having 5 or more leaflets 13
  12 Flowers yellow .... *Potentilla norvegica*
  12 Flowers white (rarely rose) .... *Sibbaldia tridentata*
13 Leaves palmately compound 14
13 Leaves pinnately compound .... *Drymocallis arguta*
  14 Petals pale yellow, much longer than lobes of calyx .... *Potentilla recta*
  14 Petals deep yellow, same length as, or shorter than, calyx lobes 15

15 Lower surface of leaves white-woolly .... *Potentilla argentea*
15 Lower surface of leaves slightly hairy .... *Potentilla intermedia*

Tall Hairy Agrimony

Agrimonia gryposepala Wallr.

Erect, branching, minutely glandular perennial, up to 1.8 m tall. **Flowers** yellow, rather small, borne in slender, spike-like terminal and axillary racemes, the peduncles short, each subtended by a deeply cleft, hairy-margined bract. Petals 5, inserted on the disk; calyx tube top-shaped, with a band of stiff, hooked bristles around the top, the 5 lobes ovate, pointed; stamens 5–15; ovary inferior; styles 2; **fruit** dry, the hardened calyx tube enclosing the 2 achenes, the hooked bristles hardened and persisting. **Leaves** mostly below the middle, alternate, widely separated, petioled, pinnately compound with small leaflets between the larger ones, the large leaflets usually 7 (or 5), thin, bright green, oblong-lanceolate to narrow-obovate, coarsely toothed, glabrous or with scattered long hairs on the veins beneath; having a disagreeable odor when crushed; stipules large and leaflike.

HABITAT In woods and thickets, and along the borders of woods.

FLOWERING July and August.

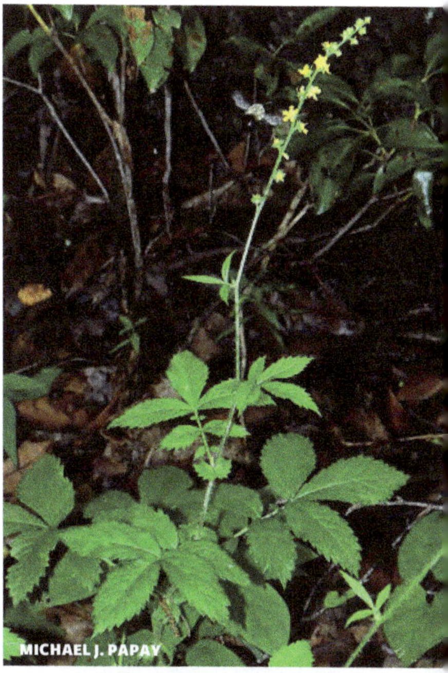

TALL HAIRY AGRIMONY

Swamp Agrimony

Agrimonia parviflora Ait.

Quite similar to the preceding species in aspect, but the larger leaflets 11–15 on the middle and upper leaves, the leaflets softly hairy beneath, the stem of the inflorescence densely hairy, and the fruits only 4–5 mm long.

HABITAT In damp thickets and on rocky slopes.

FLOWERING August and September.

SWAMP AGRIMONY

Marsh Cinquefoil

Comarum palustre L.

SYNONYM *Potentilla palustris* (L.) Scop.

Stout, erect perennial up to 6 dm tall from a partly reclining, somewhat woody base. **Flowers** purple, 1.8–3 cm broad, several in a loose terminal cluster. Petals ovate to lanceolate, much smaller than the calyx lobes and about the same length as the calyx bracts but broader, persistent; calyx deeply 5-lobed, the lobes large, purple at least inside, alternating with 5 smaller bracts; pistils numerous, borne on a superior, hairy, conical receptacle which becomes spongy and partly enclosed by the calyx. **Leaves** alternate, long-

MARSH CINQUEFOIL

SHRUBBY CINQUEFOIL

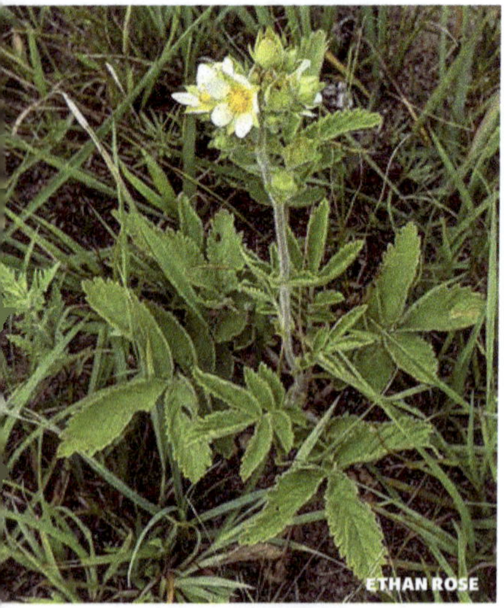

TALL CINQUEFOIL

petioled, pinnately compound; leaflets 5–7, oblong-lanceolate to oblanceolate, toothed, green on both sides, smooth or nearly so; upper leaflets close together.

HABITAT Around lakes, in swamps and bogs, and along streams.

FLOWERING June to August.

Shrubby Cinquefoil

Dasiphora fruticosa (L.) Rydb.

SYNONYM *Potentilla fruticosa* L.

Bushy-branched shrub up to 1 m, tall. **Flowers** bright yellow, 2–3 cm wide, solitary or few at the ends of branches. Petals 5, rounded to broadly elliptic; calyx flat, deeply 5-lobed, the lobes alternating with 5 narrower bracts; stamens numerous; pistils numerous, borne on a dry receptacle, the style soon falling; **achenes** long-hairy. **Leaves** numerous, short-petioled, pinnately compound, the 5–9 leaflets narrowly oblong.

HABITAT In wet or dry open ground, especially in calcareous soils, in meadows, bogs, and on shores.

FLOWERING June to October.

Tall Cinquefoil

Drymocallis arguta (Pursh) Rydb.

SYNONYM *Potentilla arguta* Pursh

Rather coarse, erect perennial up to 1 m tall, covered with clammy brownish hairs. The **flowers** whitish or creamy, borne in rather compact cymes. Petals broadly ovate to nearly round; stamens 30 (sometimes 25), borne in 5 groups on the glandular disk; style thickened at the middle. **Basal leaves** pinnately compound; leaflets 7–11, oval to ovate, toothed, downy beneath.

HABITAT In rocky or bushy places, on prairies, and in alluvial soils.

FLOWERING June to August.

Virginia Strawberry

Fragaria virginiana P. Mill.

Stemless, tufted perennial up to 3 dm tall, usually with several runners. **Flowers** white, in loose, few-flowered cymes. Petals 5, obovate or obcordate, inserted on the calyx; calyx deeply cleft into 5 ovate hairy lobes and with a smaller bract in each sinus; stamens numerous, inserted on the calyx; pistils many, on a rounded, somewhat hairy receptacle, the **achenes** numerous, borne in pits on the surface of the enlarged red pulpy receptacle (the strawberry). **Leaves** basal except for one subtending the cyme, 3-foliolate, the leaflets obovate, thick, firm, hairy, coarsely toothed.

HABITAT In fields, along roadsides, on open slopes, and in thin woods.

FLOWERING April to June. Fruits ripe in early summer.

NOTES This is a variable species, and several varieties have been described. It is probably our best-known and most delicious wild fruit. The domestic strawberry is a cross between this species and a South American species, but it seldom has the special flavor of the well-ripened wild berry.

VIRGINIA STRAWBERRY

Woodland Strawberry

Fragaria vesca L.

Similar to the preceding species but generally smaller; the **flowers** in a raceme; the calyx lobes spreading or reflexed; the **leaves** strongly veined above; and the **achenes** on the surface (not in pits) of the cone-shaped fruit.

HABITAT In rocky woods and on ledges.

FLOWERING May to August.

WOODLAND STRAWBERRY

KEY TO GEUM (AVENS) SPECIES

1 Petals white 2

1 Petals yellowish to purplish .... *G. rivale*

  2 Petals as long as, or longer than, calyx lobes; peduncles slender .... *G. canadense*

  2 Petals much smaller than calyx lobes; peduncles stout .... *G. laciniatum*

White Avens

Geum canadense Jacq.

Variable, smooth to slightly hairy, slender, erect perennial up to 1.2 m tall from a basal rosette. **Flowers** white, about 1 cm in diameter, solitary on threadlike, hairy or glandular peduncles. Petals spreading, equaling or exceeding the calyx lobes in length, 5–9 mm long, 2–4.5 broad; calyx tube saucer-shaped, deeply 5-lobed with bractlets at each sinus; stamens many; pistils several to many, borne on a broad, densely hairy receptacle; fruiting

WHITE AVENS

ROUGH AVENS

PURPLE AVENS, WATER AVENS

heads spherical; **achenes** 30–160; styles jointed, elongating after flowering, the bearded upper segment falling off, leaving the stiff, hooked lower segment attached to the achene. **Basal leaves** long-petioled, simple or with 3 (sometimes 5 or 7) leaflets; **lower stem leaves** short-petioled or sessile, mostly with 3 leaflets; **upper stem leaves** 3-cleft to simple, sharply toothed; stipules 1–2 cm long.

HABITAT In rich open woods, borders of woods, and thickets, along roadsides and in fields throughout Michigan.

FLOWERING June to early August.

Rough Avens

Geum laciniatum Murr.

Similar to the preceding species, but with the peduncles stout and hairy, the receptacle smooth or only sparsely hairy, and the petals much smaller than the calyx lobes, 2–5 mm long, 1–2 mm wide.

HABITAT In damp thickets, meadows, and along roadsides only in the southernmost part of Michigan.

FLOWERING June and July.

Purple Avens, Water Avens

Geum rivale L.

Hairy, spreading or ascending, sparsely branched perennial up to 1 m tall. **Flowers** several, nodding, yellow suffused and veined with purple, about 2 cm wide. Petals 5, obcordate, constricted at apex, contracted to a claw at the base; calyx hairy, purple, bell-shaped, with 5 erect lobes and a small bract in each sinus; **achenes** borne on a dry, cylindrical receptacle, the styles elongating greatly, the plumose upper section soon falling from the persistent, stiff basal section. **Basal leaves** pinnately compound, the terminal 1–3 leaflets much larger than the others; **stem leaves** 3-foliolate or 3-lobed, coarsely toothed; stipules leaflike, green or purplish.

HABITAT In wet meadows and bogs.

FLOWERING May to August.

USES American Indians made a beverage from the fragrant rootstocks.

SIMILAR SPECIES Two species of yellow avens occur in Michigan. The flower structure and characteristic tailed achenes are similar to those of the above species. ***Geum aleppicum*** Jacq. has basal leaves in which the terminal and side segments are about the same size and are narrowed to the base. In ***Geum macrophyllum*** Willd. the terminal segment of the leaf is conspicuously larger than the lateral ones and the base is heart-shaped or truncate. Both species grow in moist or wet soil in roadside ditches, thickets, and clearings, and flower from June to August.

Silverweed

Potentilla anserina L.

Low, tufted perennial with slender, arching, many-jointed runners up to 8 dm long. **Flowers** bright yellow, 1.8–2.7 cm across, solitary on erect, hairy, often reddish, axillary peduncles. Petals 5, nearly round, attached with the stamens to the flattish disk. **Leaves** all basal, 4–4.5 dm long, pinnately compound, nearly smooth on upper surface, densely hairy beneath, the leaflets 7–25, oblong or obovate, blunt at the apex, sharply toothed, the lower leaflets smaller than the upper.

HABITAT On sandy or gravelly lakeshores, in sandy fields or on banks. Very common on lakeshores in Michigan.

FLOWERING June to September.

SILVERWEED

Silvery Cinquefoil

Potentilla argentea L.

Erect or ascending, loosely clustered perennial up to 3 dm tall, branched above. **Flowers** yellow, 8–10 mm wide, borne in loose cymes. Pistils and achenes smooth, the style thickened, about the same length as the achene (or shorter). **Leaves**, except the uppermost, petioled, palmately compound; leaflets usually 5, green on upper surface, white-woolly beneath, sessile, oblanceolate to obovate, narrowed at the base and pinnately incised, the teeth long and sharp, the margins inrolled; withered stipules persistent.

HABITAT In dry open ground. A common weed; naturalized from Europe.

FLOWERING June to early August.

SILVERY CINQUEFOIL

Downy Cinquefoil (not illustrated)

Potentilla intermedia L.

Quite similar to the preceding species but tending to be larger; the inflorescence more freely branched; the leaflets grayish, hairy beneath, and less deeply cut; the plants shorter-lived. Along roadsides and in waste places.

FLOWERING May to August.

ROUGH CINQUEFOIL

SULPHUR CINQUEFOIL

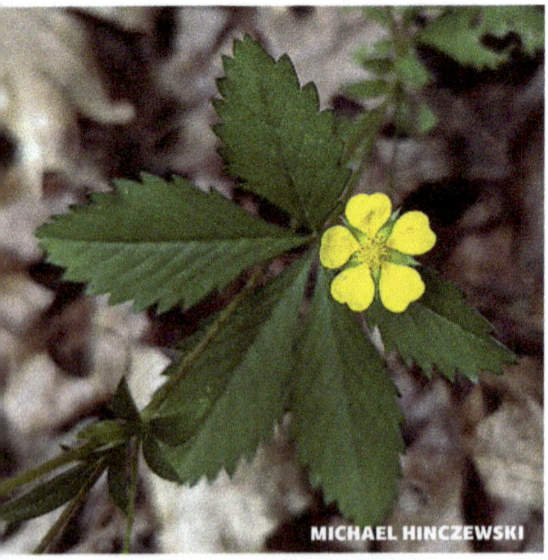

OLDFIELD CINQUEFOIL

Rough Cinquefoil

Potentilla norvegica L.

Stiffly hairy, erect or ascending annual, biennial, or (rarely) short-lived perennial 1–9 dm tall, the stiff hairs spreading. **Flowers** yellow, borne in leafy cymes, the petals obovate and usually shorter than the calyx lobes, calyx enlarging in fruit. **Lower leaves** 3-foliolate, long-petioled, the leaflets obovate to oblanceolate, coarsely toothed, green, having some long stiff hairs but not woolly; **upper leaves** sessile, the leaflets usually narrow.

HABITAT In waste places, along roads, in open meadows and clearings.

FLOWERING June to October.

Sulphur Cinquefoil

Potentilla recta L.

Stout, erect, hairy or hairy-glandular perennial 3–6 dm tall, many branched above. **Flowers** pale yellow, 1.32.9 cm wide, borne in loose terminal cymes. Petals broadly obcordate, with a deep, rounded notch; calyx lobes 5, broader than the bracts, spread out flat when the flower is in bloom, but upright and enclosing the bud at first and becoming so again to cover the maturing seeds; style terminal, shorter than the mature smooth achene. **Basal leaves** palmately compound, the leaflets 5–7, oblanceolate, often glandular, with 7–17 long-pointed teeth; petioles long and hairy.

HABITAT Along roadsides and in fields and waste places. A common weed; introduced from Europe.

FLOWERING late July through August.

Oldfield Cinquefoil

Potentilla simplex Michx.

Rather coarse, hairy perennial, erect and up to 5 dm tall at flowering time, but the stem soon elongating, forking, arching, and rooting at the tips, the rootstock irregularly enlarged, up to 10 cm long and nearly 2 cm thick. **Flowers** bright yellow, about 1 cm wide, 1 (sometimes 3) on slender, hairy peduncles from the axils or from a point opposite the origin of the petiole, the first flower usually from the axil of the second stem leaf. Petals obcordate. **Leaves** alternate, palmately compound, mostly well expanded at flowering time, the leaflets 5, green on both sides, hairy beneath, narrowly obovate or oblanceolate, sharply and coarsely toothed in the upper 3/4; basal leaves similar or with linear-lanceolate inrolled auricles.

HABITAT In dry or moist soil, in open fields, thickets, open woods, and waste places.

FLOWERING April to June.

Swamp Rose

Rosa palustris Marsh.

Freely branching, spiny shrub up to 2 m tall, the spines short, usually recurved. **Flowers** showy, fragrant, rose, 5–7.5 cm across. Petals 5, broadly ovate or somewhat obcordate; calyx tube urn-shaped, the 5 lobes very long, pointed; stamens numerous; pistils many, separate; fruit (hip) fleshy, the calyx tube enclosing the achenes, the sepals falling from the mature fruit. **Leaves** alternate, compound, odd-pinnate, the leaflets 5–9 (usually 7), finely toothed; stipules usually adherent to the base of the petiole.

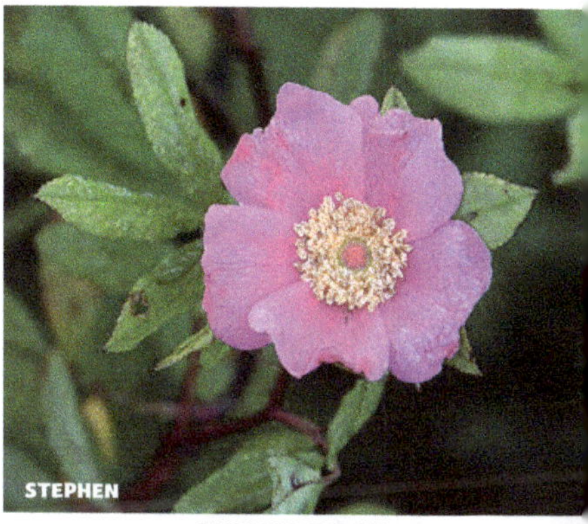

SWAMP ROSE

HABITAT Along swamps, stream borders, roadsides, on beaches, in wet thickets, dry open places, or in woods.

FLOWERING May to July.

NOTES Several species of wild rose, often difficult to distinguish, occur in Michigan. The species of this genus vary greatly and hybridize readily.

USES The hips of various species were often eaten by American Indians, and they may be used for jelly making. They constitute a source of vitamins, particularly vitamin C, for which they were used in England during the World War. Roses have been in cultivation since ancient times and are probably our best-known ornamental plant.

Thimbleberry

Rubus parviflorus Nutt.

SYNONYM *Rubus nutkanus* Moc. ex Ser.

Erect, branched shrub up to 2 m tall. **Flowers** showy, white, 3–5 cm across, borne in a large, few-flowered corymb. Petals 5, spreading, oval; calyx lobes long-hairy, with a long slender appendage; stamens numerous; pistils many, borne on an elongate, spongy receptacle, each forming a small, juicy, 1-seeded **fruit**, these together forming the tart, flattened thimble-shaped berry which separates readily from the receptacle. **Leaves** alternate, palmately 3–5-lobed, heart-shaped at base, the lobes all about the same length, irregularly coarse-toothed, sparsely hairy on both sides.

THIMBLEBERRY

HABITAT In rocky woods and thickets and on

shores. Often forming the ground cover in open woods in northern Michigan.

FLOWERING June and July.

USES The tart, juicy Thimbleberries are often made into jam or eaten fresh. American Indians used them fresh or pressed them into cakes, which were then dried. They also ate the young, tender shoots as a vegetable.

SIMILAR SPECIES The **Purple-Flowering Thimbleberry**, ***Rubus odoratus*** L., which occurs only in the Lower Peninsula, is quite similar but has rose-purple flowers and a dry, insipid berry.

THIMBLEBERRY

Three-Toothed Cinquefoil

Sibbaldia tridentata (Aiton) Paule & Soják

SYNONYM *Potentilla tridentata* Ait.

Evergreen creeping perennial with somewhat trailing woody branches which give rise to ascending flowering stems up to 3 dm tall. **Flowers** white (sometimes pinkish) in stiff, few- to many-flowered cymes. **Leaves** palmately compound, the 3 leaflets leathery, oblong wedge-shaped, tapering to base, entire except for 3 (sometimes 5) teeth at the apex, bright green, smooth.

HABITAT In dry open places, in rocky, gravelly, or peaty soil.

FLOWERING late May to October.

THREE-TOOTHED CINQUEFOIL

Barren Strawberry

Waldsteinia fragarioides (Michx.) Tratt.

SYNONYM *Geum fragarioides* (Michx.) Smedmark

Stemless tufted perennial up to 2 dm tall from a stout creeping rootstock. **Flowers** yellow, 3–8 on a bracted flowering stem. Petals 5, obovate, inserted on the calyx tube and longer than the calyx lobes; bractlets on calyx minute, soon falling; stamens 8; pistils 2–6, the styles slender; receptacle dry, not becoming greatly enlarged in fruit; **achenes** 2–6, hairy. **Leaves** from the base, long-petioled, 3-foliolate, the leaflets obovate, incised or crenate.

HABITAT In woods, thickets, and clearings.

FLOWERING late April to June.

BARREN STRAWBERRY

RUBIACEAE *Madder Family*

This large, mainly tropical and subtropical family has about 614 genera and over 6,500 species; 7 genera are reported for Michigan. The family is important as the source of coffee, ipecac, and quinine. *Gardenia* and other genera are grown as ornamentals.

The leaves are in whorls or are opposite and have stipules (which may be very small) between the petioles. The corolla is regular and 4-lobed; the calyx is adherent to the inferior ovary, and the 4 stamens are borne on the corolla tube.

Buttonbush

Cephalanthus occidentalis L.

Smooth, spreading shrub 1–3 m tall. **Flowers** small, white, borne in dense, globose, peduncled heads 2–2.5 dm in diameter. Corolla tubular, regular; calyx tubular, 4-toothed; stamens 4, attached to the corolla tube; the thread-like styles with their knobby stigmas protruding well beyond the corollas and forming a halo around the heads; **fruits** small, dry. **Leaves** opposite or in whorls of 3 or 4, petioled, oblong-ovate or elliptic.

HABITAT In wet places, often in water, in wooded swamps and flood plains; also in drier habitats such as jack-pine plains and the sandy beds of receding lakes.

FLOWERING July and early August.

BUTTONBUSH

KEY TO GALIUM SPECIES

1 Principal stem leaves in whorls of 6; plants prostrate .... *G. triflorum*

1 Principal stem leaves in whorls of 4; plants erect 2

  2 Flowers numerous, white, in dense clusters .... *G. boreale*

  2 Flowers few, yellowish at first, becoming maroon .... *G. lanceolatum*

Northern Bedstraw

Galium boreale L.

Rather stiff, erect perennial up to 8 dm tall, with hollow 4-angled stems covered with short stiff hairs and enlarged at the nodes. **Flowers** with heavy sweet fragrance, tiny, numerous, white, borne in conspicuous, dense, terminal panicles. Corolla saucer-shaped, 4-lobed; calyx very small, not lobed, adherent to the ovary; stamens 4, inserted on the corolla; pistil 1, the ovary inferior, 2-celled; **fruit** separating into 2 seed-like bristly parts. **Leaves** in whorls of 4, sessile, linear-lanceolate, strongly 3-nerved, hairy, and inrolled on the margin.

HABITAT Along roadsides, in rocky soil, on shores and gravelly banks, along streams.

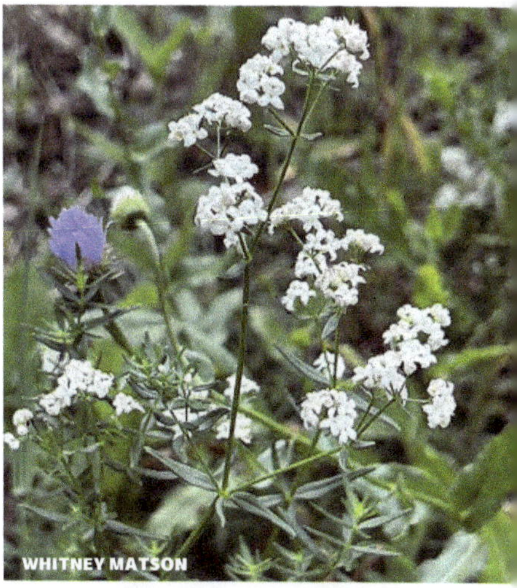

NORTHERN BEDSTRAW

STICKY WILLY

WILD LICORICE

FRAGRANT BEDSTRAW

FLOWERING June to August.

USES American Indians used roots of this species and of ***Galium tinctorium*** (L.) Scop. to make a red dye. *G. tinctorium* is a weak, matted, many-branched perennial with 3-flowered inflorescences, the principal leaves in whorls of 5 or 6, the corolla 3-lobed, and the fruit smooth.

SIMILAR SPECIES Another bedstraw, ***Galium aparine*** L., **Sticky Willy**, is a weak or reclining many-stemmed annual, with a clinging or sticky feel owing to the backward-pointing bristly hairs. The fruit is densely bristly; the leaves are mostly in whorls of 8 and are margined with curved bristles. This is often abundant in deciduous woods in the Lower Peninsula.

The cultivated **Scotchmist** or **Baby's-breath**, ***Galium sylvaticum*** L., has the lower leaves in whorls of 8, the upper leaves mostly in 4's and 6's; the numerous stems are erect or nearly so; the small white flowers are borne on threadlike ascending pedicels in loose, leafy racemes; and the fruit is smooth.

Another interesting introduced species is **Yellow Bedstraw**, ***Galium verum*** L., which has bright yellow flowers, the leaves in 8's or 6's. According to early Christian tradition, this is the bedstraw that filled the manger in Bethlehem.

Wild Licorice

Galium lanceolatum Torr.

Nearly smooth, erect perennial with slender 4-angled stems up to 6 cm tall. **Flowers** yellowish, becoming dull maroon with lines of cream-color, mostly sessile on 3-flowered forking peduncles in the axils of the leaves. Corolla spreading, about 4 mm wide; **fruit** bristly. **Leaves** in whorls of 4, lance-ovate to lanceolate, the largest ones about the middle of the stem.

HABITAT In rich dry woodlands, often under beech and maple; on sandy bluffs and in rocky or gravelly soil.

FLOWERING June and early July.

Fragrant Bedstraw

Galium triflorum Michx.

Smooth, weak perennial with simple or forked stems up to 10 dm long (sometimes longer), often sweet-scented in drying. **Flowers** tiny, greenish-white, borne on 3-flowered or 3-forked peduncles in the axils of the leaves; **fruits** densely bristly. **Leaves** mostly in whorls of 6, elliptic-lanceolate to linear, some greatly reduced.

HABITAT In woods and thickets.

FLOWERING mid-May to September.

Long-Leaf Summer Bluet

Houstonia longifolia Gaertn.

Smooth to hairy perennial with numerous stems, up to 2.5 dm tall. **Flowers** pale purple to lilac or white, borne in terminal or axillary cymes. Corolla funnel-shaped, with 4 equal lobes; calyx short, 4-lobed, the tube adherent to the ovary; stamens attached to the corolla tube; ovary inferior, 2-celled; **capsule** projecting beyond the calyx tube. **Stem leaves** usually 2–4 pairs below the branches, sessile, opposite, linear to narrow-oblong, 1-nerved, the stipules connecting the bases of opposite pairs; **rosette leaves** narrowly oblanceolate.

LONG-LEAF SUMMER BLUET

HABITAT In open, sandy or rocky fields and jack-pine plains, on sandy banks and exposed ledges.

FLOWERING mid-May to mid-July.

Partridge-Berry

Mitchella repens L.

Low, mat-forming evergreen herb, the slender, trailing stems freely branching and rooting at the nodes. **Flowers** somewhat fragrant, twin, united at the base, white, often tinged with rose-purple or scarlet. Corolla funnel-shaped, with 4 short, spreading lobes, densely long-hairy within; stamens 4, alternate with the lobes, attached to the corolla tube; pistil 1, the ovary inferior, the style long, with 4 stigmas; **fruit** a double, edible, scarlet berry crowned with the calyx teeth of the twin flowers (the "eyes"), persisting through the winter and often still present when the ensuing year's flowers are blooming. **Leaves** dark green, opposite, and with minute stipules between the opposite petioles, nearly round or broadly ovate, blunt at apex, often heart-shaped at base, entire, the veins often white, prominent on upper surface.

PARTRIDGE-BERRY

HABITAT In woods, on dry or moist knolls; particularly common in pine forests.

FLOWERING June and July.

NOTES The bright red berries make a nice contrast against the deep green leaves, and these plants are often grown in terraria. The berries have a pleasant, slightly aromatic flavor and are eaten by birds.

USES American Indians ate the berries and made a medicinal infusion from the plants.

SANTALACEAE *Sandalwood Family*

This family, of temperate and tropical regions, includes 41 genera and about 1,000 species; 3 genera and 3 species occur in Michigan. Current treatments of the family include the Viscaceae (mistletoes, *Arceuthobium*), previously considered distinct. The family is of no economic importance here, but the sweet-scented sandalwood of the tropics is prized for cabinet-making and for use in perfumes.

BASTARD-TOADFLAX

Bastard-Toadflax

Comandra umbellata (L.) Nutt.

SYNONYM *Comandra richardsiana* Fern.

Glabrous, erect perennial up to 30 cm tall, from a long, freely branching underground rootstock or the rootstock on or near the surface. **Flowers** white, about 5 mm wide, borne in small clusters on ascending or diverging branches of terminal corymbs. Petals lacking; calyx petal-like, slender, bell-shaped, 5-lobed at the apex and having a 5-lobed disk in the center; stamens 5, opposite the sepals and inserted in the edge of the disk between the lobes; pistil solitary, the style long and slender; **fruit** a dry, 1-seeded, roundish nut, covered by the lower part of the leathery calyx tube. **Leaves** very numerous, elliptic to oblanceolate, the upper leaves the largest, quite firm.

HABITAT In dry sandy soil, calcareous gravels, or marly soils.

FLOWERING May to August.

SARRACENIACEAE *Pitcher-Plant Family*

This family of bog-inhabiting plants comprises 3 genera and some 30 species in the Western Hemisphere; one species is native to Michigan. These plants are of considerable interest but are very difficult to grow. None is of economic value.

Purple Pitcher-Plant

Sarracenia purpurea L.

Low-growing, stemless perennial up to 7 dm tall from a rosette of leaves. **Flowers** showy, purplish or wine-red (sometimes yellowish-green), nearly globose, 3.7–5.5 cm broad, solitary and nodding at the end of the glabrous flowering stem. Petals 5, separate, incurved and arching over the style, fiddle-shaped, narrower than the sepals; sepals 5, ovate, usually greenish to deep purplish-red or maroon, often having 3 or 4 small, similarly colored bracts at the base; stamens numerous; pistil solitary, the ovary globular, 5-celled, the stalk of the style dilated at the top into a 5-rayed, greenish or yellowish umbrella-shaped structure, the rays protruding between the petals and bearing the small hooked stigmas on their lower surface. **Leaves** (often lasting more than one season) all basal, numerous, ascending, usually green, often veined with purple, somewhat pitcher-shaped or trumpet-shaped, curved, broadly winged, narrowed to a roundish petiole, hollow, with an upstanding or arching hood

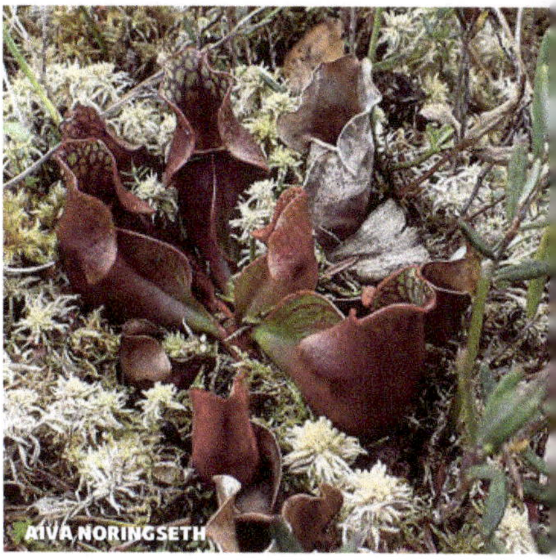

PURPLE PITCHER-PLANT

above the opening, 10–30 cm long, nearly smooth outside and bristly on the inner side of the lid, the hollow usually partly filled with liquid and decaying insects.

HABITAT In wet sphagnum, peat bogs, and tamarack swamps,

FLOWERING May and June (sometimes until July).

NOTES This species is one of our most attractive bog plants. An Indian name for it means "frog-leggings." The leaves and roots were used medicinally. It is now thought that the decaying insects found in the hollow of the leaf provide food, not for the plant, but for the larvae of the flies that cross-pollinate the flowers.

■ SAXIFRAGACEAE *Saxifrage Family*

This family of 40 genera and about 640 species has its best representation in the United States, principally in the West; 6 genera occur in Michigan. The family formerly included a number of ornamentals such as Its chief economic importance previously was for ornamentals: Mock-orange (*Philadelphus*), Deutzia, and Hydrangea, those genera now placed in Hydrangeaceae; Coral-bells (*Heuchera*) remains in the Saxifrage family. Currants and gooseberries (*Ribes*) are bush fruits of some importance, but are now placed in Grossulariaceae.

The flowers are perfect and regular. The 5 (rarely 4) petals and sepals with the stamens (the same number as the petals or twice as many) are borne on the edge of a cup-like or saucer-like structure (hypanthium). There are usually 2 pistils, which may be separate or united. The leaves are opposite or alternate, are usually simple, and usually lack stipules. This family has many characters in common with the Rose Family, with which it is easily confused.

KEY TO SAXIFRAGACEAE (SAXIFRAGE FAMILY) SPECIES

1 Petals with entire margins 2

1 Petals with finely fringed margins .... *Mitella diphylla* or *M. nuda*

  2 Basal leaves narrowly oblong to oblanceolate .... *Micranthes pensylvanica*

  2 Basal leaves cordate-ovate .... *Tiarella stolonifera*

SWAMP SAXIFRAGE

TWO-LEAF MITERWORT

Swamp Saxifrage

Micranthes pensylvanica (L.) Haw.
SYNONYM *Saxifraga pensylvanica* L.

Sticky-glandular perennial with stout flowering stems up to 10 dm tall. **Flowers** yellowish, yellowish-green, or greenish, small, up to 1 cm wide, borne on glandular pedicels in terminal panicles which are at first dense but soon elongate, becoming 1–6 dm long. Petals 5, spreading, inserted at the top of the calyx tube; calyx tubular, with 5 lanceolate reflexed lobes which are 2–4 times as long as the tube; stamens 10; pistils 2, the ovaries adherent to the calyx tube, the styles short and thick; **capsule** inflated, many-seeded. **Leaves** in a basal rosette, thick, narrow, oval, ovate, obovate, or oblanceolate, 1–2.5 dm long, obtuse at the apex and narrowed to a broad, clasping, often reddish petiole, shallowly crenate to entire, hairy, the midrib prominent beneath.

HABITAT In swamps, wet meadows, and boggy thickets and on seepage banks.

FLOWERING April to June.

Two-Leaf Miterwort

Mitella diphylla L.

Slender, stiffly erect, straight, stiffly hairy perennial with numerous unbranched stems which have a single pair of opposite leaves at or near the middle, rootstock stout, with stolons lacking. **Flowers** white, tiny, borne on very short, thick pedicels in a stiff, slender raceme. Petals 5, deeply and narrowly pinnately fringed or cut, alternate with the calyx lobes which appear petal-like; calyx cup-like with 5 lobes, whitish; stamens 10, inserted on the calyx tube, not extending out of the flower; pistil 1; **capsule** opening wide and exposing the numerous blackish seeds at maturity. **Stem leaves** sessile or nearly so, ovate, palmately 3–5-lobed; **basal leaves** petioled, wider and more heart-shaped at the base, pointed at the apex, hairy and prominently veined beneath.

HABITAT In rich, loamy and rocky, usually moist woods.

FLOWERING in April and May.

Naked Bishop's-Cap

Mitella nuda L.

Small, stiffly erect perennial up to 2 dm tall, the rootstocks and stolons threadlike, the flowering stalk usually leafless but occasionally having 1 or more petioled leaves. **Flowers** delicate, pale yellowish-green, 7–8 cm wide, borne in a simple, stiff, slender raceme. Petals deeply and very narrowly pinnately cut, the threadlike segments spreading; calyx short, attached to the base of the ovary. **Leaves** nearly round in outline, heart-shaped at the base, having scattered, short, stiff hairs on both sides; margin deeply crenate.

HABITAT In cool mossy woods or swamps. A common ground cover in swampy woods throughout Michigan.

FLOWERING May to August.

Creeping Foamflower

Tiarella stolonifera G.L. Nesom

SYNONYM *Tiarella cordifolia* var. *bracteata* Farw.

Low perennial up to 2 dm tall, the leaves all basal. **Flowers** small, white, borne in terminal racemes up to 14 cm long, the flowering stalk and pedicels glandular-hairy. Petals 5, soon falling, inserted on the calyx tube between its lobes; calyx white, bell-shaped, with 5 lobes which resemble petals; stamens 10; pistil 1, with 2 uneven sides; **capsule** of 2 uneven parts which separate readily, few-seeded. **Leaves** all basal, petioled, heart-shaped-ovate, 5–12 cm long when mature, smaller at flowering time, having stiff, spine-like hairs on both sides, the margin shallowly lobed and having rounded teeth.

HABITAT In swampy woodlands and drier rich woods, often making an extensive ground cover in woods.

FLOWERING in May and June.

NOTES There is considerable variation in this species, and several forms have been described.

SIMILAR SPECIES **Richardson's Alumroot,** *Heuchera richardsonii* R. Br., is somewhat similar in aspect but is taller (up to 1 m), the flowers are greenish or red-tinged, and there are only 5 stamens which, with the styles, extend well out of the flower. This species is also found in woods, primarily in southern Michigan.

NAKED BISHOP'S-CAP

CREEPING FOAMFLOWER

SCROPHULARIACEAE *Figwort Family*

This cosmopolitan family now includes 58 genera and more than 1,800 species. In the past it encompassed about 275 genera and over 5,000 species, but many species have been reassigned to other families, especially the Orobanchaceae and Plantaginaceae; 3 genera in Michigan.

The family is characterized by bisexual flowers with tubular corollas (fused petals) that are bilaterally symmetrical (two-lipped). Most species have 4 stamens, 2 of which are usually shorter than the other 2.

LANCE-LEAF FIGWORT

Lance-Leaf Figwort

Scrophularia lanceolata Pursh

Coarse, erect, strong-smelling, many-stemmed perennial up to 2 m tall, smooth or finely hairy, the stems square, the sides flat or shallowly grooved. **Flowers** numerous, small, 7–11 mm long, dull reddish brown except for the yellowish-green lower lobe, borne in large, loose, cylindric panicles, 1–5 dm long. Corolla lustrous, short, the tube wide, 2-lipped, the middle lobe of the lower lip drooping or turned back; calyx regular, deeply saucer-shaped, 5-lobed; normal stamens 4, the sterile filament yellowish green, usually wider than long; **capsule** ovoid, dull brown. **Leaves** opposite, lanceolate to ovate, wedge-shaped, tapering or sub-truncate at the base, coarsely toothed, the petioles wing-margined.

HABITAT In open woods, thickets, along roads, and in open fields. Common and often conspicuous in northern Michigan.

FLOWERING in June and July.

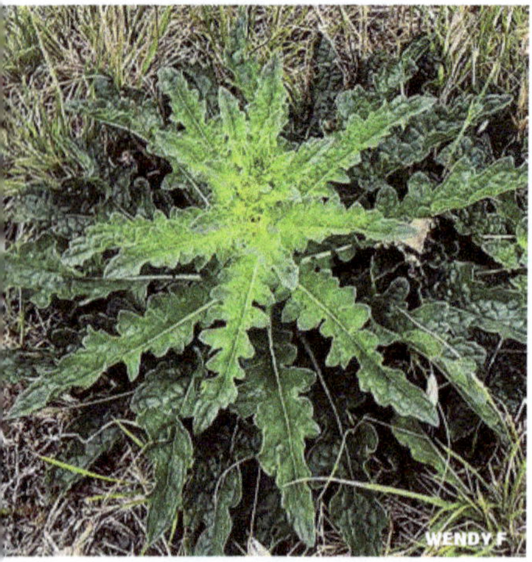

MOTH MULLEIN

Moth Mullein

Verbascum blattaria L.

Slender, green, smooth or slightly hairy, simple or branched biennial. **Flowers** yellow (white in one form), borne on pedicels 1–1.5 cm long in elongate, loose, terminal racemes; corolla 1.5–2.5 cm wide, the filaments all violet-bearded. **Leaves** alternate, not decurrent on stem, variable, narrowly triangular, oblong, lanceolate or oblanceolate, sometimes pinnately cut or lyrate, toothed to nearly entire; upper leaves sessile, clasping, lower ones petioled.

HABITAT In open pastures and old fields. Introduced from Europe.

FLOWERING late June through September.

Common Mullein

Verbascum thapsus L.

Robust, stiffly erect, unbranched, yellowish or grayish, densely woolly biennial up to 2.3 m tall, producing a rosette of leaves the first year, from which the stout strongly winged, flowering stem rises the second year. **Flowers** pale yellow, borne in a stiff, elongate, compact terminal spike. Corolla open-bell-shaped to wheel-shaped, the 5 rounded lobes only slightly unequal; calyx regular, deeply 5-lobed; stamens 5, all fertile, attached to the corolla tube, the 3 upper filaments densely white-bearded; ovary superior, with 5 spreading styles; **capsule** longer than the calyx, many-seeded. **Leaves** felted on both sides with many branched hairs, thick, oblong blade long-decurrent on the stem, entire or somewhat crenate, lower leaves having winged petioles.

COMMON MULLEIN

HABITAT In fields, along roadsides, and in waste places, sometimes in open forests. Introduced from Europe.

FLOWERING July to September.

SOLANACEAE *Nightshade Family*

This large family of 102 genera and some 2,700 species is most abundant in tropical America; 11 genera in Michigan. It includes food and drug plants, as well as ornamentals. Potato, tomato, eggplant, and green and red peppers are among the best-known foods. Belladonna, atropine, stramonium and henbane are important drugs. Petunia, *Salpiglossis, Datura, Solanum,* and others are grown for ornament. Tobacco also belongs to this family.

The foliage is rank-scented and often narcotic. The fruits are narcotic in some species, deadly in some, edible in others. The leaves are alternate; the regular corolla is often plaited (folded lengthwise), and consists of 5 united petals; the 5 sepals are united; the 5 equal stamens are borne on the corolla; the 2-celled superior ovary produces a capsule or berry.

Jimsonweed

Datura stramonium L.

Glabrous, stout, ill-scented, poisonous annual up to 1.5 m tall. **Flowers** showy, white to pale violet, solitary and erect at the forks of the stem, each subtended by a leaf and 2 young branches. Corolla funnel-shaped, 5-lobed, 7–10 cm long; calyx about half as long as the corolla, 5-toothed; stamens 5, erect; **capsule** globular to ovoid, 3–5 cm long, many-seeded, usually prickly.

JIMSONWEED

JIMSONWEED

HABITAT In fields, barnyards, and waste ground. Introduced from Asia.

FLOWERING July to October.

Large-Flowered Ground-Cherry

Leucophysalis grandiflora (Hook.) Rydb.

SYNONYM *Chamaesaracha grandiflora* (Hook.) Fern.

Erect, slightly hairy annual 1.5–9 dm 3–5 cm broad, 2–4 in the upper axils. tall. **Flowers** white, yellow-centered, Corolla wheel-shaped, nearly flat, 5-angled with a 5-lobed hairy spot in the center; calyx 5-lobed, becoming greatly inflated in fruit; stamens 5; **fruit** a globose berry which is enclosed by the calyx. **Leaves** glandular-hairy, sticky to the touch, wing-petioled, lance-ovate, the margin wavy or entire.

HABITAT In recent clearings, along roadsides, in open woods, and on sandy and rocky shores.

FLOWERING June to August.

LARGE-FLOWERED GROUND-CHERRY

Clammy Ground-Cherry

Physalis heterophylla Nees

Sticky-hairy, freely branching perennial with deeply buried thick rootstocks. **Flowers** yellow with a dark center, 1.5–2.5 cm broad, solitary, drooping. Corolla tube very short, the limb between wheel-shaped and bell-shaped, shallowly 5-lobed; calyx small, 5-lobed, becoming greatly inflated in fruit; stamens 5, erect, the anthers separate; **fruit** a small yellow globose berry enclosed by the inflated calyx. **Leaves** alternate, hairy, ovate, usually coarsely and irregularly toothed.

HABITAT In dry, open woods and orchards, along roads and railways, at the foot of sand dunes, in gardens and waste places.

FLOWERING June to September.

CLAMMY GROUND-CHERRY

SIMILAR SPECIES The related low, weak **Husk-Tomato**, *Physalis pubescens* L., is an annual that is often cultivated. It has larger, sweetish, edible berries which are used in preserves, pies, or sauce. American Indians ate the berries raw or cooked, and used the roots medicinally. The attractive **Chinese Lantern-Plant**, *Alkekengi officinarum* Moench (synonym *Physalis alkekengi* L.), native to Asia, has been introduced as an ornamental. It has a large, bright red to scarlet fruiting calyx about 5 cm long and is used extensively in winter bouquets.

Bittersweet Nightshade

Solanum dulcamara L.

Somewhat woody perennial vine, all parts of which are poisonous. **Flowers** violet or purple (rarely white) borne in several-flowered, loose cymes which arise between the nodes or opposite the leaves. Corolla wheel-shaped, about 1 cm wide, deeply 5-cleft; calyx 5-cleft; stamens 5, extending beyond the corolla, the anthers converging around the style, opening at the tip by pores; ovary superior, 2-celled; **fruit** an ovoid or ellipsoid bright red berry. **Leaves** ovate, with or without 1 or 2 basal lobes.

HABITAT In moist thickets and clearings. Naturalized from Europe.

FLOWERING mid-May to September.

SIMILAR SPECIES **Jerusalem Cherry**, ***Solanum pseudocapsicum*** L., is often grown indoors. It has white flowers and globose scarlet to yellow berries. **Green Nightshade** (***Solanum nitidibaccatum*** Bitter), is an erect, widely branching annual with white or very pale flowers and shiny black berries. The berries are poisonous at first, but it is said that they are nontoxic when fully ripe, and they are then sometimes used in cooking.

THIERRY ARBAULT

BITTERSWEET NIGHTSHADE

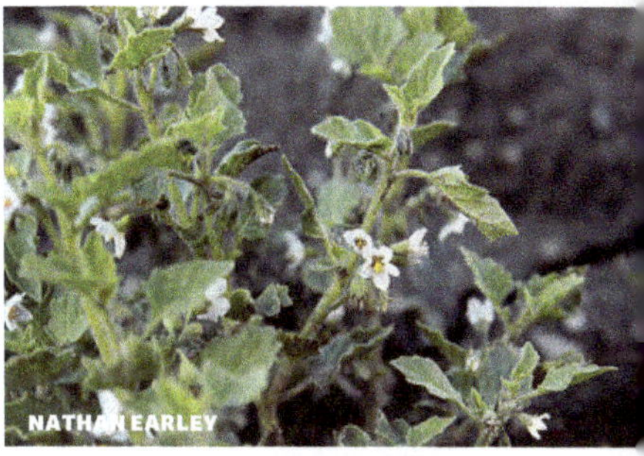

NATHAN EARLEY

GREEN NIGHTSHADE

VERBENACEAE *Vervain Family*

This predominantly tropical family has 30 genera and over 800 species; 3 genera present in Michigan. Economically the family is most important for teakwood from the East Indies. Verbena and, to a lesser extent, Callicarpa and Lantana are cultivated for ornamentals.

The slightly irregular flowers, each subtended by a bract, are borne in dense spikes; the 5 petals are united, and the 4 stamens are borne on the corolla tube; the ovary is somewhat 4-lobed and produces 4 nutlets; the stem is usually square; and the leaves are opposite.

KEY TO VERBENA SPECIES

1 Plants stiffly erect; leaves with regular teeth 2

1 Plants low, prostrate with branches radiating from base, leaves cut or lobed .... *V. bracteata*

2 Spikes numerous, stalked; plants of moist places .... *V. hastata*

2 Spikes solitary or few, sessile; plants of dry places .... *V. stricta*

Bigbract Verbena

Verbena bracteata Cav. ex Lag. & Rodr.

Prostrate perennial with numerous partially reclining or ascending hairy branches radiating from the base. **Flowers** small, purple to bluish, borne in elongate, thick, terminal, sessile spikes with conspicuous divergent bracts. **Leaves** opposite, hairy, pinnately cut or 3-lobed, narrowed at the base, 1–6 cm long.

BIGBRACT VERBENA

HABITAT In dry sunny places, in fields, on prairies.

FLOWERING June to September.

Blue Vervain

Verbena hastata L.

Stiff, erect, rough-hairy perennial up to 1.5 dm tall, freely branched above, the stems angled. **Flowers** small, violet-blue or blue, in numerous slender, erect, very compact spikes on fairly long peduncles. Corolla tubular, with 5 spreading lobes and a dense ring of hairs at the top of the tube; calyx tubular, hairy, 5-toothed; stamens 4, inserted on the corolla tube and not extending beyond it; pistil 1, the ovary superior, the style slender, terminal, the stigma 2-lobed; nutlets 4, readily separating. **Leaves** opposite, lanceolate to narrowly elliptic or ovate-lanceolate, coarsely sharp-toothed, 4–18 cm long, sometimes having 2 basal lobes.

HABITAT In damp thickets, swales, and moist meadows, along streams, in swamps and wet fields.

FLOWERING June to October.

USES This species is readily grown in a moist garden. American Indians made a tea from the leaves and also used the plant for treating nosebleed.

BLUE VERVAIN

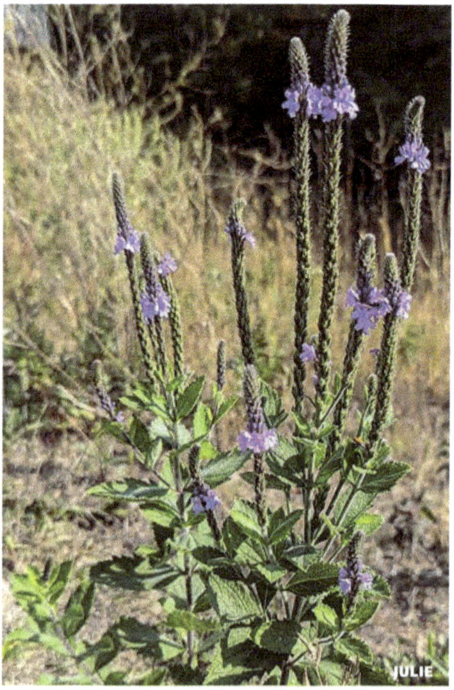

HOARY VERVAIN

Hoary Vervain

Verbena stricta Vent.

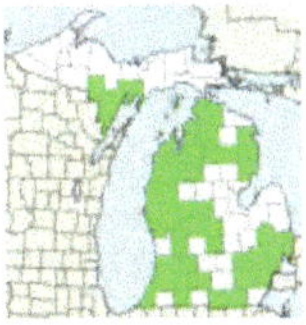

Coarser than **Blue Vervain**, the **flowers** larger (8–10 mm wide) and more irregular, and the color more roseate; the spikes larger, solitary or few, sessile; the entire plant covered with long, soft, white hairs; **leaves** nearly round, oval, or oblong-ovate, strongly veined, coarsely toothed.

HABITAT In dry soil, open sandy ground, on prairies, in barren fields, and along roads.

FLOWERING July and August.

NOTES The root system of this species extends down more than one meter. The plants show remarkable resistance to drought. The foliage is so bitter that cattle will not eat it even when forage is scarce.

VIOLACEAE *Violet Family*

This widely distributed family is found on all continents and includes 24 genera and about 1,000 species; 2 genera present in Michigan. Violets and pansies are well-known ornamental plants. Over 100 species of violet are cultivated.

Our members of this family are low herbs with irregular flowers. There are 5 petals, one of which is spurred, and 5 distinct sepals; the 5 stamens are often slightly united at the base and enclose the ovary; the fruit is a 3-sided capsule.

KEY TO VIOLA (VIOLET) SPECIES

1 Flowers white or yellow (sometimes having blue veins) 2
1 Flowers lilac, bluish, or lavender 6
  2 Flowers white 3
  2 Flowers yellow . . . . *V. pubescens*
3 Plants with leafy stems . . . . *V. canadensis*
3 Plants stemless 4
  4 Leaves broadly heart-shaped to kidney-shaped 5
  4 Leaves lanceolate . . . . *V. lanceolata*
5 Plants having stolons . . . . *V. minuscula*
5 Plants lacking stolons . . . . *V. renifolia*
  6 Plants with leafy stems 7
  6 Plants stemless 8
7 Lateral petals beardless, petals with a darker lilac-purple spot near center . . . . *V. rostrata*
7 Lateral petals bearded, petals lacking a darker center spot . . . . *V. labradorica*
  8 Leaves cordate-ovate, margin merely toothed . . . . *V. affinis*
  8 Leaves deeply cut into narrow segments . . . . *V. pedata*

Common Blue Violet

Viola affinis Le Conte

SYNONYM *Viola papilionacea* Pursh

Small, densely tufted, stemless, glabrous perennial. **Flowers** violet or bluish, with a white center, the earlier flowers borne above the leaves, solitary on peduncles from the base of plant. Style enlarged upward, capitate with a conical beak on the lower side, the stigma inside the tip of the beak; self-pollinating, non-opening flowers borne on prostrate peduncles. **Leaves** heart-shaped-ovate, long-petioled, crenate, glabrous.

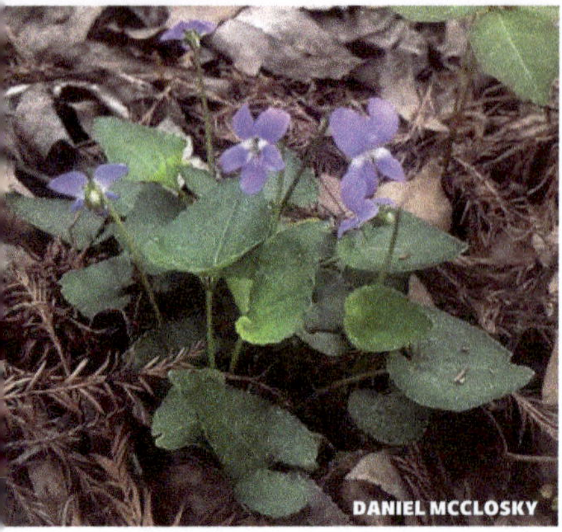

COMMON BLUE VIOLET

CANADIAN WHITE VIOLET

LABRADOR VIOLET

HABITAT In damp woods, thickets, meadows, and on shady ledges.

FLOWERING late March to early June.

SIMILAR SPECIES Among the other common blue-flowered violets, ***Viola sororia*** Willd. is very similar to this species. It also has self-pollinating, non-opening flowers on prostrate peduncles, but the leaves and petioles are downy.

Marsh Violet, *Viola cucullata* Ait., is smooth, the spurred petal has no beard, and there is a dark eye-spot in the center of the flower. It is frequently very difficult to distinguish these and some other species.

Canadian White Violet

Viola canadensis L.

Smooth perennial with one to many stems up to 5 dm tall from a thickset sub-woody rootstock. **Flowers** white, veined with purple, and soon violet-tinged on the back, about 2.5 cm wide, irregular, solitary on slender peduncles from the axils of the leaves. Lateral petals bearded, the spur on the lower petal short and rounded; sepals slender, separate, having basal auricles; stamens flattened, closely surrounding the ovary; pistil solitary, the ovary superior, the style capitate; **capsule** subglobose; seeds numerous, brown. **Leaves** heart-shaped-ovate, the upper ones narrower; basal leaves long-petioled; stipules narrowly lanceolate, thin, dryish, pointed.

HABITAT In deciduous woods.

FLOWERING late April to July, infrequently and sporadically to October.

American Dog Violet

Viola labradorica Schrank

SYNONYM *Viola conspersa* Reichenb.

Resembling **Long-spurred Violet** in general appearance and color of the flowers, but there is no darker spot in the center of the flower, the style is bent down at the tip and somewhat hairy, the lateral petals are bearded, and the spur is usually only 4–5 mm long.

White Bog Violet

Viola lanceolata L.

Stemless, freely stoloniferous perennial, often growing in dense mats. **Flowers** white, with purple veins; petals all beardless, the spur on lower petal shorter than the blade. **Leaves** lanceolate to oblanceolate, gradually tapering to the margined petiole, obscurely crenate.

HABITAT In damp, open ground or in light shade.

FLOWERING May to June.

LAUREN MCLAURIN

WHITE BOG VIOLET

Northern White Violet

Viola minuscula Greene

SYNONYM *Viola pallens* (Banks) Brainerd

Small, stemless, stoloniferous perennial. **Flowers** white, veined with purple, borne above the leaves, solitary on peduncles from the base, less than 1 cm broad, fragrant. Lateral petals beardless, the spur on the lower petal shorter than the blade. Self-pollinating, non-opening flowers on peduncles above the ground, evident at flowering time. **Leaves** all basal, glabrous, heart-shaped-ovate to kidney-shaped, crenate; petioles smooth or with a few short hairs.

HABITAT In wet or springy woods or thickets, on slopes, and in openings.

FLOWERING early April to July.

SIMILAR SPECIES *Viola incognita* Brainerd, perhaps our commonest stemless, white-flowered violet, occurs statewide, and differs in having the lateral petals bearded.

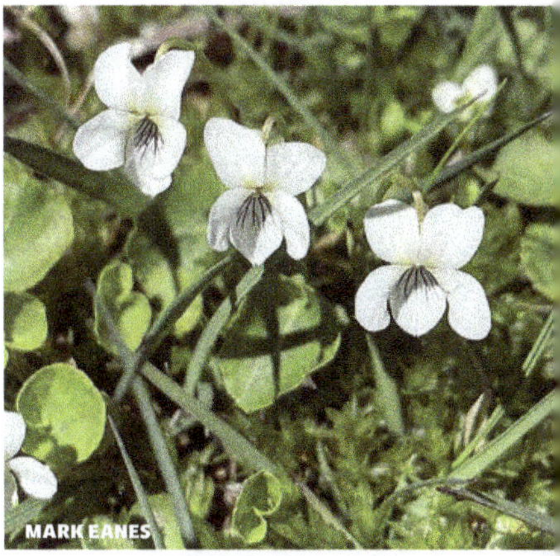

MARK EANES

NORTHERN WHITE VIOLET

Bird-Foot Violet

Viola pedata L.

Small, glabrous perennial up to 15 cm tall. **Flowers** lilac, lavender, or bluish, 2–4 cm wide, flat and wheel-shaped, borne on peduncles which hold them above the leaves. Petals all beardless, the lower one with a white spot at the base; style club-shaped, beakless, beardless; no cleistogamous, (self-pollinating) flowers produced. **Leaves** all from the base, fan-shaped, and divided into 3 main parts, the divisions cut nearly to the base into linear or narrowly oblanceolate segments, which may in turn be toothed or cleft near the apex; both early and late leaves less finely cut, smaller and thicker.

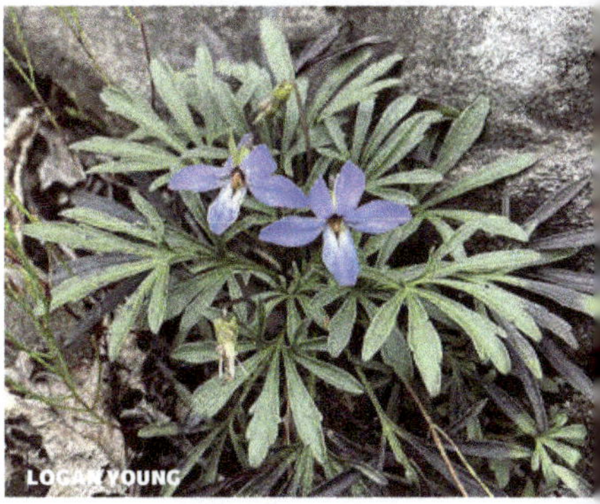

LOGAN YOUNG

BIRD-FOOT VIOLET

DOWNY YELLOW VIOLET

KIDNEY-LEAF WHITE VIOLET

LONG-SPUR VIOLET

HABITAT Usually in rather sterile soil, in open grassland, sunny woods, and openings, on rocky or sandy slopes.

FLOWERING April and early June.

NOTES Unlike most of the violets, this species is difficult to grow in the garden.

Downy Yellow Violet

Viola pubescens Ait.

Softly hairy perennial with 1 to a few leafy stems 3–4 dm tall, leafless at the base or with one leaf. **Flowers** yellow, with dark purplish or blackish veins, about 1–1.5 cm wide, solitary on peduncles from the axils of the leaves, borne either above or below the leaves. Lateral petals slightly bearded, the spur on lower petal short; **capsule** ovoid, white-woolly. **Stem leaves** few, borne near the summit of the stem, broadly ovate, petioled, densely downy, at least at first, the teeth coarse and shallow, 13–23 on each half; stipules green, semi-ovate.

HABITAT In rich deciduous woods, thickets, cool slopes, rocky places.

FLOWERING late-April to early June.

Kidney-Leaf White Violet

Viola renifolia Gray

Stemless, non-stoloniferous perennial with threadlike rootstocks. **Flowers** white, with purple veins, the petals all beardless. **Leaves** whitened with long silky hairs when young, kidney-shaped to round, heart-shaped at the base.

HABITAT In cool, mossy woods, and swamps, on calcareous slopes.

FLOWERING May and June.

Long-Spur Violet

Viola rostrata Pursh

Leafy-stemmed perennial up to 12 cm tall at flowering time. **Flowers** bluish-lilac, with a darker spot at the center, about 2 cm wide, solitary on peduncles from the axils of the leaves. Lateral petals beardless, the spur on the lower petal 10–16 mm long, and upcurved at the tip; style straight, slender, neither capitate nor bearded. **Leaves** mostly heart-shaped, the lower leaves rounder, crenate; stipules leaflike, narrowly ovate, with narrow, pointed teeth.

HABITAT In rich, often calcareous woods.

FLOWERING April to June.

NOTES This may be distinguished from our other blue violets by the somewhat flattened appearance of the flower, its more lilac color, and the very noticeable long spurs.

MONOCOTYLEDONS

ACORACEAE *Calamus Family*

This genus was once placed within the family Araceae, but it is now placed in its own family. A single genus *Acorus; A. americanus* considered native, while *A. calamus* considered introduced from Asia. Both with leaves lemon-scented when crushed, traditionally used medicinally, and are sometimes grown in water gardens.

The two species can be distinguished as follows:

1 Leaves lacking a single prominent raised midvein, but with 2–several clearly separate veins of ± equal strength along with numerous fainter veins, the leaf thus not obviously ridged to the naked eye (best observed at about 1/2 to 2/3 of the length of the leaf, the veins become reduced to a single vein towards the leaf apex); fruit maturing ...... ***A. americanus***

1 Leaves with a single prominent raised midvein that is much more conspicuous than any other veins, appearing as a ridge to the naked eye; fruits not maturing. ...... ***A. calamus***

Sweetflag

Acorus americanus (Raf.) Raf., *Acorus calamus* L.

A. americanus

A. calamus

Aromatic perennials up to 1 m tall from a thick, creeping, aromatic rootstock. **Flowers** tiny, yellowish-green, borne on a spikelike spadix which extends obliquely outward from a long leaflike spathe; sepals 6; stamens 6, with kidney-shaped anthers; ovary solitary; **fruit** dry but gelatinous inside, with 1 or more seeds. Flowering stalk 3-angled, having a sweet odor when broken. **Leaves** flattish, sword-like, thicker at the middle, the edges sharp, up to 1 m or more long and 0.6–2 cm wide.

SWEETFLAG

HABITAT In clumps and large masses in wet places along the borders of quiet water and in small woodland pools.

FLOWERING May to August. Most Michigan plants are apparently the native *Acorus americanus* (synonym *Acorus calamus var. americanus* Raf.).

NOTES When not in flower these plants could be mistaken for Iris, but are readily distinguished by their aromatic odor when crushed or broken, and by the flat leaves, which are a yellowish-green and are not folded at the base, in contrast to the blue-green folded leaves of the Iris.

USES In the past, the roots were boiled in water to remove the bitterness, then sliced thin and boiled in a heavy syrup to make a confection.

ALISMATACEAE *Water-Plantain Family*

This family includes 18 genera and about 119 species; 3 genera occur in Michigan, 2 being common. The species of the family are widely distributed in freshwater swamps and marshes in the northern hemisphere. The arrowheads (*Sagittaria*) are primarily American plants; 7 of the 43 species grow in Michigan. Water-plantain (*Alisma*) is a small genus comprising about 10 species, 2 of which may be found in Michigan. The family is of limited economic importance, but some species of *Sagittaria* are cultivated for food by the Chinese, and several species are grown as ornamentals.

AMERICAN WATER-PLANTAIN

American Water-Plantain

Alisma subcordatum Raf.

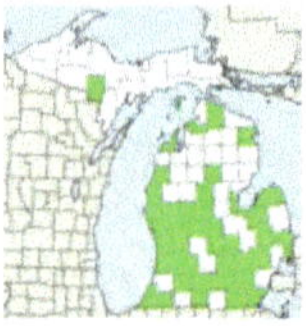

Aquatic perennial with scape-like stems and long-petioled leaves rising above the water. **Flowers** white or rarely rose, small (up to 1.3 cm wide), numerous, borne on threadlike pedicels in many-branched panicles, which rise in whorls from the stiff scape; petals 3, soon falling; sepals 3, green, persistent; stamens 6; pistils numerous, borne in a circle on the flattened receptacle; **achenes** leathery. **Leaves** basal, ascending or erect on long petioles, the blades ovate-elliptic, rounded to subcordate at base, up to 2.5 dm long and 1.5 dm broad, the principal veins few, the cross veins strong, parallel.

HABITAT In shallow water in low fields, mud flats, and ditches, and along stream banks.

FLOWERING late June to September.

NOTES This plant can be grown in a wet-soil garden, where it gives an attractive lacy effect.

BROADLEAF ARROWHEAD

Broadleaf Arrowhead, Duck-Potato

Sagittaria latifolia Willd.

Plants up to 1 m tall from a tuber embedded in mud. **Flowers** white, 1.5–3 cm broad, in whorls of 3 on erect or ascending pedicels, each whorl subtended by a pair of somewhat papery bracts, the flowers usually unisexual (rarely bisexual), the lowest whorls pistillate and the upper staminate (but some plants are unisexual). Petals 3, separate; sepals 3, thin, pale green, separate; stamens numerous, filaments slender, usually longer than the anthers; pistils numerous, distinct, borne on a globular receptacle; achenes borne in a globose head, flat, with membranous wing and incurved or horizontal beak. Flowering stem leafless, hollow, ridged, smooth or nearly so.

Leaves basal, extremely variable but usually arrow-shaped (sagittate), the upper part of the blade ovate to linear and usually more than half as long as the body, but sometimes the blade merely ovate or linear-lanceolate and tapering to the base; veins few, strong, parallel, the cross veins conspicuous; petiole long, strongly ribbed, hollow, sheathing the stem at base.

HABITAT In ditches, along creeks and lakes, in marshy fields and forest openings, on mud flats and beaches.

FLOWERING July to September.

NOTES This extremely variable species has several well-defined forms and varieties which are distinguished chiefly by leaf characters.

USES The tuberous roots of this and other species of *Sagittaria* were called **Wapato** by American Indians, for whom they formed an important food staple, as well as a medicine for indigestion and a poultice for wounds. The tubers have a bitter milky juice when fresh, but are rendered sweet and palatable by cooking or drying, and Lewis and Clark virtually subsisted on them during a winter spent at the mouth of the Columbia River. The plants were harvested by the Indian women, who waded waist-deep into the water, using their toes to loosen the tubers, which then rose to the surface of the water and could be gathered into canoes.

These attractive plants can be easily cultivated in pools, but they spread rapidly by underground stems and hence are not suitable for small pools.

ARACEAE *Arum Family*

This predominantly tropical family, with 143 genera and more than 4,000 species, is represented in Michigan by 9 genera. The chief value of the family in North America is for ornamentals, about three dozen species being cultivated. The florist's calla (*Zantedeschia*) is perhaps the best-known example. *Philodendron, Caladium* (Elephant's-Ear), and *Monstera* (which has edible fruits) are common house plants. In other parts of the world, arums are used for food, taro being quite important. Most Michigan representatives of this family are readily recognized by the minute flowers borne on a cylindric or globose spadix which is subtended, and often enveloped, by a single spathe.

The **duckweeds** (*Lemna*) and 3 other aquatic genera (*Spirodela, Wolfia, Wolfiella*) were previously placed in a separate plant family, the Lemnaceae, but are now placed in Araceae. These are small, mostly floating plants of ponds and quiet waters.

NOTE The Duckweeds (*Lemna*) and other tiny aquatic plants are not keyed here.

KEY TO ARACEAE (ARUM FAMILY) SPECIES

1 Leaves compound .... *Arisaema triphyllum, A. dracontium*
1 Leaves simple 2
  2 Spathe white and petal-like .... *Calla palustris*
  2 Spathe not white and petal-like 3
3 Spadix short and thick to nearly globose .... *Symplocarpus foetidus*
3 Spadix long and slender .... *Peltandra virginica*

Green Dragon

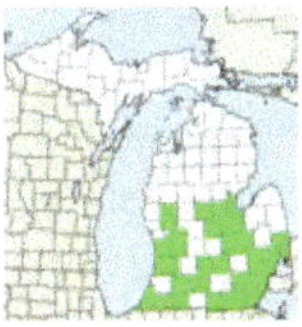

Arisaema dracontium (L.) Schott

Perennial, 6–10 dm tall. **Flowers** tiny, similar to those of Jack-in-the-pulpit, but the spadix slender and tapering to a long slender point that extends far beyond the spathe; the mature fruiting head slenderly conical, the **berries** reddish-orange. **Leaf** solitary, compound, the leaflets 5–15, unequal in size.

GREEN DRAGON

JACK-IN-THE-PULPIT

HABITAT In rich woods, in thickets, and in swales.

FLOWERING May and June. This plant can be readily grown in the woodland garden, but is neither so attractive nor so common as Jack-in-the-pulpit.

Jack-in-the-Pulpit

Arisaema triphyllum (L.) Schott

SYNONYM *Arisaema atrorubens* (Ait.) Blume

Erect, unbranching, smooth perennial up to nearly 1 m tall. **Flowers** tiny, unisexual, borne on the base of the pale-green or purple, mottled, slenderly club-shaped spadix (the "jack"); spathe (the "pulpit and canopy") tubular below, the tube shallowly corrugated and with a narrow flange above, becoming expanded and arching over the spadix, all green or purple to bronze with pale greenish stripes (quite variable in coloration and markings); the staminate flowers a cluster of lavender anthers, the pistillate consisting of a bright-green, 1-celled, ovoid pistil; fruiting heads ovoid to globose, the berries shiny and bright scarlet. Stems stout, sheathed at base, whitish below, green above, often mottled with purple. **Leaves** 1–3 (usually 2), expanding at flowering time, trifoliate, the lateral leaflets strongly unsymmetrical; venation pinnate. Tuber brown, globose, rooting and producing basal offsets.

HABITAT In rich woods and thickets, usually in rather moist locations.

FLOWERING late April and May.

USES This highly variable species is one of our most interesting wildflowers. Children often call it Indian Turnip and entice others to eat the tuber, which has a very sharp taste due to crystals of calcium oxalate. It is said that even after being boiled in several changes of water these tubers are too pungent to eat, but if thoroughly dried for several weeks they become edible. Even though the fresh plants are poisonous, they have medicinal qualities and were often collected for drug use. American Indians used them for treating sore eyes and ulcers. The pungent taste makes them unacceptable to most wildlife.

Jack-in-the-pulpit is very easy to grow in the wildflower garden; it flourishes without special attention.

Water-Dragon

Calla palustris L.

Low-growing, smooth perennial 1.5–4 dm tall, the solitary flowering stalk and long-petioled leaves rising from thick, often green, creeping rootstocks. Inflorescence showy, the spathe large, white, petal-like but thick, ovate, abruptly pointed, up to 3.5 cm wide and 1.5–7.5 cm long; **flowers** small, borne on the knob-like spadix, perfect or the upper ones staminate, the perianth lacking; stamens 6; ovary 1-celled, forming in fruit a head of large red berries. **Leaves** numerous, thick, broadly ovate, heart-shaped or rounded at the base and tapering abruptly to a pointed apex, bright green, glossy, 5–10 cm in width and length, the parallel veins numerous.

HABITAT In low wet places, particularly swamps and bogs, pond margins, quagmires.

FLOWERING May to July.

USES American Indians used the underground parts of Wild Calla for poultices and also for food. The plant may be grown in the cool bog garden. The florist's calla is somewhat similar in appearance to this plant but belongs to the genus *Zantedeschia*.

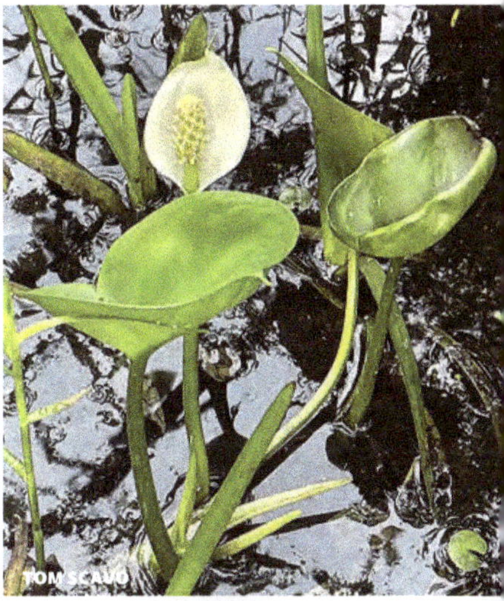

WATER-DRAGON

Tuckahoe, Arrow-Arum

Peltandra virginica (L.) Schott

Perennial 2–3.5 dm tall from thick subtuberous roots. **Flowers** minute, borne on a slender, tapering, whitish spadix, the staminate flowers above, covering a much larger portion than the pistillate; anther masses 4–6-celled, embedded in the margin of a shield-shaped connective; ovaries 1-celled, surrounded by 4–5 distinct, scalelike staminodes; **berries** green or amber, in an ovoid, fleshy head enveloped by the base of the spathe. **Leaves** basal, arrow-shaped, the basal lobes usually well developed and spreading, the margin entire; blade 1–3.5 dm long, the principal veins 3, one to each of the basal lobes, the secondary veins pinnate, fine, very numerous, arching within the margin and forming a fine network; petioles long, sheathing.

HABITAT In marly lake margins, muddy shores, bogs, mud flats, and shallow water.

FLOWERING June and July.

TUCKAHOE, ARROW-ARUM

USES This was the "breadroot" of the eastern American Indians. The large bitter roots were roasted in pits for a day or so, then dried and ground into meal. The boiled spadix and berries were considered luxuries. The berries are eaten by wood ducks and, infrequently, by other birds.

Skunk-Cabbage

Symplocarpus foetidus (L.) Salisb. ex Nutt.

Stemless perennial with a skunk-like odor. **Flowers** tiny, inconspicuous, covering a fleshy, globose, yellowish, stalked spadix which is partially enclosed by the fleshy, nearly sessile, ovoid spathe; spathe protruding only a little way

SKUNK-CABBAGE

from the ground, green, spotted and streaked with purple and yellowish-green, the margins inrolled; at least some flowers perfect; stamens 4, sepals 4; **fruit** a globular to ovoid mass consisting of the enlarged spongy spadix with the spherical seeds just under the surface. **Leaves** appearing after the spathes, all basal and becoming large (3–6 dm long), smooth, entire, the midrib fleshy, the veins pinnate, the petiole short and fleshy.

HABITAT In low wet woods, flood plains, open or shady stream banks, swampy forests, or wet thickets. Most common in southern Michigan.

FLOWERING mid-March to mid-May.

USES The tender leaves while still folded and just pushing through the ground are said to be edible if cooked in 2 or 3 changes of water. A number of game birds, including the ring-necked pheasant, grouse, and quail, eat the seeds. This species may be grown in a moist shady garden, but requires considerable space.

BUTOMACEAE *Flowering-Rush Family*

This small aquatic or marsh-inhabiting family includes a single genus, *Butomus,* and 1 or 2 species. It is relatively uncommon in Michigan. The plants are sometimes cultivated in water gardens but may be invasive if introduced in the wild.

FLOWERING-RUSH

Flowering-Rush

Butomus umbellatus L.

Marsh or aquatic perennial, up to 1 m or more tall from a thick, fleshy rootstock in the mud, which late in the season produces many easily detached grain-like tubers. **Flowers** showy, roseate tinged with green, up to 2.5 cm wide, numerous, in terminal umbels which are at first enclosed by 3 large, purple-tinged, papery bracts. Petals and sepals 3 each, the sepals petal-like but slightly tinged with green; stamens 9, the anthers red; pistils 6; barely united at base; **follicle** inflated, long-beaked, 5–10 mm long. **Leaves** basal, sword-shaped, 3-angled at base.

HABITAT In shallow water, marshes, river margins. Introduced from Europe.

FLOWERING June to September.

COMMELINACEAE *Spiderwort Family*

A large tropical family of about 36 genera and about 740 species; 5 species in 2 genera occur in Michigan. Members of both genera, Spiderwort (*Tradescantia*) and Dayflower (*Commelina*), are grown in gardens.

Bluejacket, Spiderwort

Tradescantia ohiensis Raf.

Erect glaucous and glabrous perennial with mucilaginous juice, the stem fleshy, often branching, up to 7 dm tall. **Flowers** lasting a very short time, violet, blue, rose, or rarely white, up to 3.5 cm wide, borne on purplish to pale-yellow pedicels which droop in the bud and after flowering, the cymes solitary and terminal and having 2 leaflike reflexed bracts at the base which are nearly as long as the leaves. Petals 3, nearly round, narrowed to a short claw; sepals 3, green or tinged with rose-purple, smooth or with short hairs at the tip; stamens 6, all alike, the filaments colored like the petals and densely bearded with long jointed hairs; ovary superior, the **fruit** a capsule. **Leaves** linear-lanceolate, sheathing at their base, up to 5 dm long.

BLUEJACKET, SPIDERWORT

HABITAT On banks, in thickets, meadows, and woodlands.

FLOWERING April to June.

NOTES Spiderworts were formerly very common garden plants. They are easily grown.

RELATED SPECIES An uncommon species, ***Tradescantia virginiana*** L., has pubescent pedicels and sepals and thin, dull-green leaves (Michigan **Threatened**).

IRIDACEAE *Iris Family*

This family of perennial herbs includes 69 genera and about 2,500 species distributed over much of the earth except for the colder regions. Two genera are native to Michigan. The family is of considerable economic importance. Ornamentals include Crocus, Iris, Gladiolus, Belamcanda, Sisyrinchium, Freesia, and Tigridia. Orris-root powder, to which many people are allergic, is from a species of Iris; it was formerly much used in flavoring dentrifices and scenting face powder. The stigmas of a species of Crocus are used in making saffron dyes.

The members of this family are low-growing. The leaves are parallel-veined, mostly basal and folded forward and clasping below. The flowers are usually showy and subtended by a spathe of 2 or more bracts. The flower parts are in 3's, and the ovary is inferior.

The flowers of Iris are large and conspicuous; the sepals and petals are unlike; and the stamens are hidden by the petal-like styles. *Sisyrinchium* has much smaller flowers; the sepals and petals are similar; and the stamens are clearly visible.

DWARF LAKE IRIS

NORTHERN BLUE FLAG

Dwarf Lake Iris

Iris lacustris Nutt.

Simple dwarf plants, the flowering stems up to 1.5 dm tall. **Flowers** blue, showy, with typical iris structure; perianth tube very slender, 1.3–1.8 cm long, nearly equaling or exceeding the wedge-shaped, petal-like sepals in length; petals smaller than the sepals; **capsules** 1–1.7 cm long. **Leaves** arching, broadly linear, 4–6 cm long at flowering time, later up to 18 cm long, 5–15 mm wide.

HABITAT On beaches and cliffs, in sandy woods and bogs.

FLOWERING late May to early July.

STATUS Threatened in Michigan.

NOTES This attractive little species is found only in the upper Great Lakes region. It forms large patches and is abundant in certain localities on the shores of Lakes Huron and Michigan at the tip of the Lower Peninsula.

Northern Blue Flag

Iris versicolor L.

Stiff, erect perennial up to 1 m tall. **Flowers** several, bluish-violet, rarely white, very showy, 10–12 cm wide, each rising from a 2-bracted spathe, the bracts sub-herbaceous or papery, Perianth tubular at base, constricted above the ovary and fairly long; petals 3, erect, 1/2 to 2/3 as long as the sepals, the blade round to obovate, the claw long; sepals 3, petal-like, spreading, the blade nearly round and deep blue-violet, the claw long, narrow, variegated at base, and greenish-yellow or yellow and white with deep-purple veins at center; stamens arching, attached to the perianth tube opposite the sepals; ovary inferior, the style with 3 long, petal-like branches which arch over to the sepals and completely hide the stamens, the stigmas (just below the apex of the branches) white, plate-like, covered with fine white hairs and a sticky secretion; **capsule** ellipsoid to thick-cylindric, obtusely 3-angled, 3-celled, beaked, opening slowly, often persisting over winter, the inner surface lustrous as though varnished. Stems round, simple or with 1 or 2 branches above. **Leaves** mostly from the lower part of the stem and shorter than the stem, folded and clasping at the base, ascending, firm, the fresh tufts purplish at base. **Rootstock** irregularly branched, thick and fleshy.

HABITAT In sunny or sometimes partially shady wet places such as meadows, marshes, and edges of streams or swamps.

FLOWERING late May to July.

USES American Indians used the rootstocks of Wild Iris medicinally, some tribes cultivating it in ponds.

SIMILAR SPECIES Southern Blue Flag, *Iris virginica* L., resembles *I. versicolor* in general appearance and habitat but differs from it in several respects: petals nearly as long as the sepals, the blades oblong to ovate; a downy bright yellow spot at the base of the sepal; the **capsule** long and narrow, often unsymmetrical, soon disintegrating, the inner surface dull; leaves quite broad, and the leaf tufts buff or pale brown at the base when young. Both of these Blue Flags are quite variable through hybridization, and intermediate forms are common in Michigan, especially near the Straits of Mackinac, where the northern *I. versicolor* and southern *I. virginica* meet.

Yellow Iris, *Iris pseudacorus* L., a European species which is common along streams in the east, is now found in parts of Michigan. This is a somewhat larger plant with very showy yellow flowers. Other cultivated species, including **Bearded Iris**, sometimes appear to grow wild.

White Blue-Eyed-Grass

Sisyrinchium albidum Raf.

Tufted, stiffly erect, pale-green perennial 1.5–4.5 dm tall. **Flowers** pale violet or white, about 1 cm wide, borne in few-flowered umbels in a spathe of 2 leaflike bracts; spathes generally 2, terminal on the simple stems, subtended by an erect leaflike bract and each enclosing several smaller, dry, often purple-tinged inner bracts, the margin of the outer bract free to the base. Perianth nearly wheel-shaped; sepals and petals 3 each, alike, yellow inside at the base, giving the flowers a yellow eye-spot, separate, obovate, attached above the ovary; stamens 3, attached to the top of the ovary, the white filaments united into a tube to the summit and enclosing the style, the anthers yellow, forming a cluster just below the stigmas; pistil 1, the ovary inferior, the style branches threadlike, the stigmas 3; **capsule** pale straw-color, obscurely 3-angled, about 4 mm in diameter, the seeds numerous, globular. Stems stiff, flattened, and distinctly 2-winged, less than 4 mm, wide. **Leaves** mostly basal, flat, linear, clasping and distinctly folded at base, 1–3 mm wide, pointed at tip, usually less than half as long as the stems.

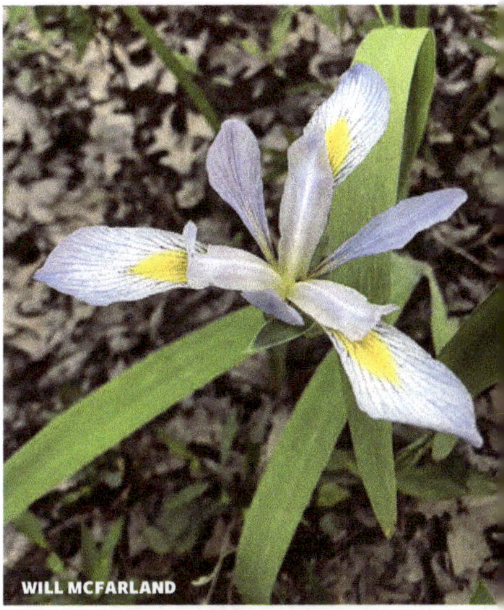

SOUTHERN BLUE FLAG

YELLOW IRIS

WHITE BLUE-EYED-GRASS

HABITAT On prairies, in thin woodlands and open grassy places around lakes and ponds, often in sandy soil.

FLOWERING May and June.

NARROW-LEAF BLUE-EYED-GRASS

Narrow-Leaf Blue-Eyed-Grass

Sisyrinchium angustifolium P. Mill.

Deep green plants up to 4.5 dm tall. **Flowers** pale blue to dark violet. Spathes borne on peduncles from the axils of the leaflike bracts; peduncles 2–5, ascending, winged, 2–15 cm long, the bracts of the spathe subequal, the outer bract with its margins united above the base; pedicels 1–5, threadlike, long and slender in fruit, arching or recurving, much longer than the spathe. Stems usually forked, broadly winged, flexuous or abruptly bent. **Leaves** 1.5–6 mm wide.

HABITAT On low ground in meadows, along damp shores and ridges, and in thickets.

FLOWERING June and July.

JUNCAGINACEAE *Arrow-Grass Family*

This small family of 3 genera and 34 species is widely distributed; a single genus and 2 species occur in Michigan, and both inhabit marshes or bogs.

SEASIDE ARROW-GRASS

Seaside Arrow-Grass

Triglochin maritima L.

Rush-like perennial with several flowering stems up to 8 dm tall. **Flowers** small, greenish, in a very slender, spikelike terminal raceme up to 1.3 dm long; perianth of 3 concave greenish sepals and 3 similar petals; stamens 6; pistils usually 6, sometimes 3; **fruit** ovoid, about 5 mm long, beaked at apex, rounded at the base. **Leaves** all basal, fleshy, roundish in cross section, 4–15 cm long and 1–4 mm thick above the sheathing base.

HABITAT In mud flats, near sand dunes, in beach pools, and on wet shores; in swamps, bogs, and marshes.

FLOWERING May to August.

USES American Indians parched and ground the seeds for food and also roasted them in the manner of coffee. The leaves, fresh or dried, of both our species of Arrow-grass are poisonous, producing hydrocyanic acid, and the plants should not be allowed in pastures or included in hay.

Marsh Arrow-Grass

Triglochin palustris L.

Similar to the preceding species but smaller, more slender, and usually having only 1–2 flowering stems 2–4 dm tall; the **fruit** 3-celled, thinner and longer (7 mm), linear or clavate, tapering to a slender, pointed base, blunt at the apex, separating from below upward when ripe.

Habitat and flowering time similar to those of the preceding.

MARSH ARROW-GRASS

LILIACEAE *Lily Family*

(and related families)

As currently defined, this large, widely distributed family includes about 15 genera and about 610 species. It is especially abundant in warm-temperate and tropical regions. Seven genera occur in Michigan. Many members (and former members) of the family are listed for horticultural purposes, including a large number of our most attractive ornamentals. Tulips, lilies, fritillarias, yuccas, lily-of-the-valley, hyacinths, and many others are extensively cultivated. Such foods as asparagus, onion, garlic, and leek are former members of this family.

Recent changes in our understanding of this formerly very large and diverse family have resulted in assigning most former members of the family to new or other families, but not all authorities agree on the new family placements. For convenience, we have included these often similar appearing plants under a **"lily-like" group**, arranged in the following families:

- **AMARYLLIDACEAE, Daffodil Family**, p. 238
- **ASPARAGACEAE, Asparagus Family**, p. 239
- **ASPHODELACEAE, Onionweed Family**, p. 242
- **COLCHICACEAE, Autumn-Crocus Family**, p. 243
- **HYPOXIDACEAE, Yellow Star-Grass Family**, p. 244
- **LILIACEAE, Lily Family**, p. 245
- **MELANTHIACEAE, False Hellebore Family**, p. 248
- **NARTHECIACEAE, Asphodel Family**, p. 250
- **SMILACACEAE, Greenbrier Family**, p. 251
- **TOFIELDIACEAE, Featherling Family**, p. 252

NOTE The species are arranged in alphabetical order under their new family name.

KEY TO LILIACEAE (AND FORMER LILIACEAE) SPECIES

1 Flowers or flower clusters rising from axils of stem leaves **2**

1 Flowers or flower clusters at end of stem or scape **5**

  2 Sepals and petals united into a tube nearly to tip of corolla .... ***Polygonatum pubescens*** **or *P. biflorum*, pp. 241, 242**

  2 Sepals and petals separate to base **3**

3 Flowers borne in globose clusters .... ***Smilax lasioneuron*, p. 251**
3 Flowers solitary or in pairs, hanging from axils 4
4 Flowers yellow; fruit a capsule ..... ***Uvularia grandiflora*** or ***U. sessilifolia*, pp. 243, 244**
4 Flowers not yellow; fruit a berry .... ***Streptopus lanceolatus***
5 Leaves borne in 2 distinct whorls, one at summit, the other about middle of stem .... ***Medeola virginiana*, p. 247**
5 Not as in alternate choice 6
6 Flowers solitary.. 7
6 Flowers borne in clusters.. 10
7 Plants having only 2 or 3 leaves 8
7 Plants normally with numerous leaves.. 9
8 Leaves 2, basal .... ***Erythronium americanum* or *E. albidum*, pp. 245, 246**
8 Leaves 3, at top of stem .... ***Trillium*, p. 249**
9 Leaves narrow, elongate; stem unbranched below; flowers large and showy............ .... ***Lilium philadelphicum*** or ***L. michiganense*, p. 246**
9 Leaves oblong or ovate; stem usually forked; flowers not very showy .... ***Uvularia grandiflora*** or ***U. sessilifolia*, pp. 243, 244**
10 Flowers large and showy, funnelform and lily-like 11
10 Flowers typically smaller (usually less than 3 cm)................................ 12
11 Stem leafless or nearly so; leaves long and slender; sepals and petals united at base .... ***Hemerocallis fulva*, p. 242**
11 Stem bearing whorled or scattered leaves; sepals and petals entirely separate .... ***Lilium philadelphicum*** or ***L. michiganense*, p. 246**
12 Flowers borne in umbels (all rising from same place)........................... 13
12 Flowers in racemes or panicles.. 14
13 Flowers white; plants strongly onion-scented .... ***Allium tricoccum*, p. 239**
13 Flowers yellow, plant not strongly scented .... ***Clintonia borealis*, p. 245**
14 Leaves chiefly at base of plant 15
14 Leaves alternate on stem .... ***Maianthemum*, pp. 239–241**
15 Sepals and petals separate.. 16
15 Sepals and petals united into a tube at base .... ***Aletris farinosa*, p. 250**
16 Stem sticky to the touch .... ***Triantha glutinosa*, p. 252**
16 Stem not sticky to the touch .... ***Anticlea elegans*, p. 248**

LILY-LIKE GROUP

■ AMARYLLIDACEAE *Daffodil Family*

Herbaceous, mainly perennial and bulbous flowering plants in 71 genera and about 1,600 species; 3 genera in Michigan, the most important being the onions (*Allium*).

Leaves are usually linear, and the flowers are usually bisexual and symmetrical, arranged in umbels on the stem. The petals and sepals are undifferentiated (tepals), which may be fused at the base into a floral tube.

The family is found in tropical to subtropical areas of the world and includes many ornamental garden plants and vegetables.

■ Wild Leek, Ramp

Allium tricoccum Ait.

Onion-scented perennial, leafless at flowering time, the erect flowering stems 1.5–3 dm tall. **Flowers** white, small, borne on stout pedicels in a hemispherical terminal umbel on the glabrous flowering stem; umbel bracts usually 2, enclosing the flowers, soon falling. Sepals and petals each 3, similar, separate, white, rather dry, 1-nerved; stamens 6, about the same length as the perianth, the filaments about as broad as the sepals and petals at the base, tapering above; ovary superior, deeply 3-lobed; style 1, rather slender; **seeds** globose, black, 1 in each cell. **Leaves** present in early spring, flat, fleshy, elliptic-lanceolate, tapering to a long slender petiole, 1–3 dm long, 2–5.5 cm wide, the veins fine. Bulbs slender, long-ovoid, white, very strongly onion-flavored, clustered.

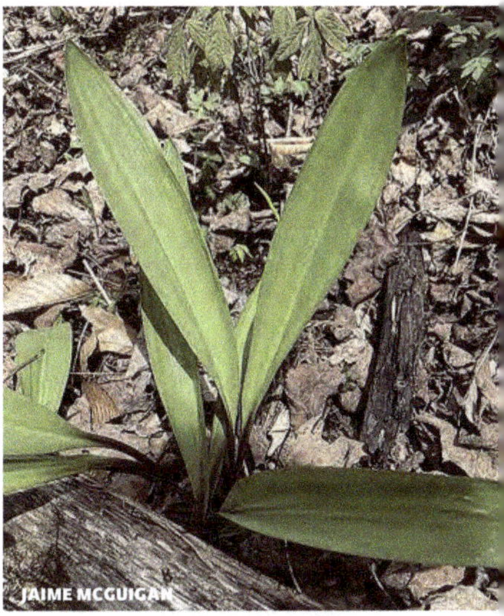

WILD LEEK, RAMP

HABITAT In deep rich woods and bottom land, often forming large beds.

FLOWERING in June and July.

NOTES Very common in the deciduous woods in the northern part of the Lower Peninsula. The clumps of leaves form the most conspicuous ground-cover in some woods in late April, but not a vestige of the leaves can be found by flowering time.

USES The bulbs taste strongly of onion-almost burning. However, the bulbs, both raw and dried, were used by American Indians for food, and an emetic was also derived from them. These plants will grow in a wild garden, but are not especially attractive.

LILY-LIKE GROUP

■ ASPARAGACEAE *Asparagus Family*

122 genera, 2,900 species; 13 genera in Michigan.

KEY TO MAIANTHEMUM SPECIES

1 Leaves heart-shaped at base; stamens 4 .... *M. canadense*
1 Leaves becoming narrower toward base; stamens 6 2
  2 Flowers borne in branched racemes (panicles) .... *M. racemosum*
  2 Flowers borne in unbranched racemes 3
3 Leaves 1–4, glabrous beneath .... *M. trifolium*
3 Leaves 6–14, finely pubescent beneath .... *M. stellatum*

■ False Lily-of-the-Valley

Maianthemum canadense Desf.

Perennial with erect zigzag stems up to 2.5 dm tall from a horizontal, thread-like rootstock which bears stalked tuberous enlargements. **Flowers** white, often fragrant, about 4 mm wide, on short pedicels, usually rising in pairs in a terminal raceme which is 1–4 cm long. Perianth of 4 similar, separate, lanceolate segments; stamens 4; pistil 1, 2-celled, the stigma 2-lobed; **berries** pale red, speckled, about 5 mm in diameter, with 1 or 2 seeds. **Leaves** usually 2 or 3 (one on stemless plants, in which the petiole comes directly from the rootstock), ovate or lanceolate, heart-shaped at base, 2.5–10 cm long, sessile or nearly so.

HABITAT In moist woods, thickets, and recent clearings; also in rather dry sandy soil under aspens and in cedar bogs.

FLOWERING May to June.

NOTES This species is not like the cultivated Lily-of-the-valley, which has a bell-shaped 6-lobed corolla and belongs to the genus *Convallaria.*

USES American Indians used the rootstock of Wild Lily-of-the-valley for medicinal purposes and the berries for food. Grouse and small mammals, such as chipmunks and mice, also eat the berries.

FALSE LILY-OF-THE-VALLEY

FALSE SOLOMON'S-SEAL

False Solomon's-Seal

Maianthemum racemosum (L.) Link

SYNONYM *Smilacina racemosa* (L.) Desf.

Coarse, unbranched, arched or ascending perennial up to 1 m tall from a rootstock that is knotty and jointed in appearance, fleshy, brown, 9–12 mm thick. **Flowers** white, up to 6 mm in diameter, in terminal stalked or nearly sessile panicles 5–14 cm long, the branches of panicle and the pedicels white and hairy. Perianth segments 6, alike, separate, narrower and shorter than the filaments, spreading and often drooping slightly; stamens 6, the filaments white; pistil white, ovary 3-celled, stigma obscurely 3-lobed; **berries** red, often speckled with purple, about 6 mm in diameter, aromatic. **Leaves** alternate, 5–13, elliptic, tapering to a blunt apex and rounded to the short petiole, 10–18 cm or more long, 3–8 cm wide, lighter beneath, the margin entire; veins parallel, numerous, 3 quite distinct, the others smaller.

HABITAT In dry or moist open woods or thickets; often growing on hillsides or shady river banks.

FLOWERING May to July. Berries ripe July to September.

USES This is a satisfactory garden plant and easy to cultivate. The berries have a strong, pleasant odor and seem to be liked by birds. They are also sometimes eaten by people.

Starry False Solomon's-Seal

Maianthemum stellatum (L.) Link

SYNONYM *Smilacina stellata* (L.) Desf.

Arching or inclined unbranched perennial up to 1 m tall from a slender, freely forking rootstock. **Flowers** white, small, 6–8 mm in diameter, borne on short pedicels in a few-flowered terminal raceme; raceme sessile or nearly so, sometimes zigzag, 2–6 cm long. Perianth parts 6, white, alike, slightly longer than the stamens; stamens 6; **berry** green, with black stripes or sometimes all black, usually turning bronzy at maturity and usually 6-seeded. **Leaves** alternate, 7–14, lanceolate or narrowly elliptic, sessile or slightly clasping at base, 4–14 cm long, 1.5–2.3 cm wide, downy beneath.

HABITAT In rich sandy or gravelly soil in open woods or thickets, on sandy lake shores, between dune ridges, and along margins of bogs.

FLOWERING mid-May to mid-July. Berries ripe July to September.

USES Easily grown in the garden, this species spreads rapidly by means of the rootstock.

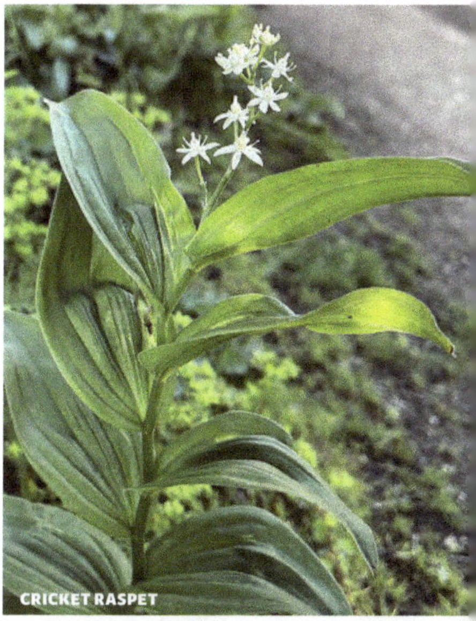

STARRY FALSE SOLOMON'S-SEAL

Three-Leaf False Solomon's-Seal

Maianthemum trifolium (L.) Sloboda
SYNONYM *Smilacina trifolia* (L.) Desf.

Smaller and weaker than the preceding two species, usually erect, up to 2.4 dm tall from a slender, whitish, extensively creeping rootstock. **Flowers** white, in a stalked, few-flowered terminal raceme, the pedicels longer and more nearly erect than in the preceding species. Perianth segments longer than the stamens, remaining attached and withering and drooping as the fruit matures; **berry** dark red. **Leaves** 2–4, usually 3, alternate, elliptic or ovate, somewhat clasping at the base, 3–10 cm long, glabrous.

HABITAT In bogs, wet places in the forest, clearings, and swamps. More common in the northern part of the state.

FLOWERING mid-May to July (occasionally September). Berries ripe late July to September.

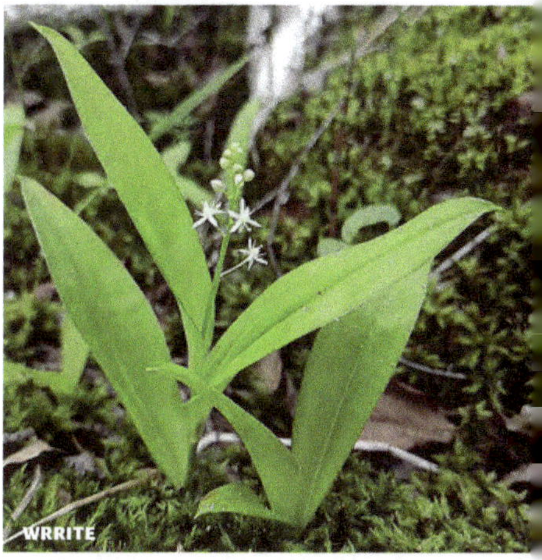

THREE-LEAF FALSE SOLOMON'S-SEAL

NOTES Difficult to grow in the garden because it requires cool, acidic soil.

Hairy Solomon's-Seal

Polygonatum pubescens (Willd.) Pursh

Unbranched and erect, inclined or arched perennial 3–6 dm tall from a knotty, jointed rootstock near the surface. **Flowers** greenish or greenish-white, 8–12 mm long, 1–2 drooping on smooth, slender peduncles from the leaf axils. Sepals and petals united into a slender tube which has 6 rounded lobes; stamens 6, shorter than the perianth tube and inserted on it; pistil 1, the style long, the stigma of 3 tufts of fine hairs; **berry** globose, bluish, about 5 mm in diameter. **Leaves** narrowly elliptic to broadly oval, 5–10 cm long, 1–5 cm wide, sessile

HAIRY SOLOMON'S-SEAL

HAIRY SOLOMON'S-SEAL

SMOOTH SOLOMON'S-SEAL

or on very short petioles, green and glabrous above, grayish-green and stiff-hairy along the veins beneath.

HABITAT In rich woods and thickets.

FLOWERING May and June.

NOTES This and other species of Solomon's-seal are readily grown in the woodland garden. The common name refers to the large seal-like scars on the rootstock.

Smooth Solomon's-Seal

Polygonatum biflorum (Walter) Elliott

SYNONYM *Polygonatum canaliculatum* (Muhl.) Pursh

Generally larger and coarser than the preceding species, the stem stout and up to 2 m tall from a thick, deeply buried, scarcely constricted rootstock. **Flowers** 2–10 from an axil 1.7–2 cm long. **Leaves** not quite flat, somewhat corrugated and puckered at the margins, glabrous on both sides, narrowed to a slightly sheathing base.

HABITAT In rich woods and alluvial thickets, along roadsides. This large plant is quite conspicuous along roadsides in southeastern Michigan.

FLOWERING in May and June.

LILY-LIKE GROUP

ASPHODELACEAE *Onionweed Family*

Orange Day-Lily

Hemerocallis fulva L.

Showy perennial 1–2 m tall above clustered basal leaves. **Flowers** large, showy, tawny-orange, deeper-colored toward the center, up to 10 cm long, 3–15 borne at the top of the naked flowering stem, produced in succession through a long period, but each flower lasting but a single day. Perianth funnel-shaped, the tube narrow below, the 6 segments abruptly flaring above, the 3 inner segments wider than the outer ones, obtuse, with wavy margins; stamens inserted in the throat of the tube, the filaments colored like the corolla and directed slightly downward, the tips curving upward and on casual inspection giving the impression that the flower is unsymmetrical; ovary superior, the style slender, curving, longer than the perianth. **Leaves** basal, very long and narrow (3–6 dm long, 1–1.5 cm wide), tapering to both ends and sheathing at the base, usually curved downward.

HABITAT Along roadsides, borders of thickets, in fields, meadows, along streams. Introduced from Europe.

FLOWERING May to July.

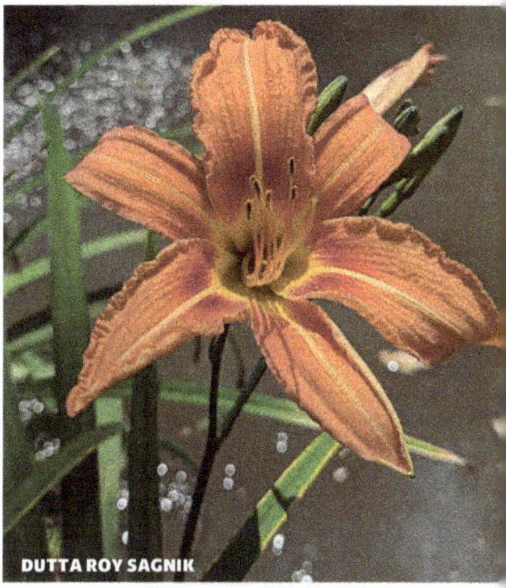

DUTTA ROY SAGNIK

ORANGE DAY-LILY

SIMILAR SPECIES The **Yellow Day-Lily**, *Hemerocallis lilioasphodelus* L., is often found in the vicinity of abandoned houses, around cemeteries, and sometimes along roads. It has similar but smaller and yellow flowers, with a pleasing fragrance. Day-Lilies are very popular garden flowers, and many beautiful hybrids have been developed in recent years. Orange Day-Lily has lost the ability to produce fertile seeds.

LILY-LIKE GROUP

■ COLCHICACEAE *Autumn-Crocus Family*

■ Large-Flower Bellwort

Uvularia grandiflora Sm.

Slender perennial, simple below, forking above, up to 8 dm tall, the stem zigzag and leafy above, sheathed below, from a short rootstock with fleshy root fibers. **Flowers** straw-colored to orange-yellow, narrowly bell-shaped, 2–5 cm long, drooping, solitary on peduncles, terminal at first but appearing axillary in fruit. Perianth parts separate, 6, similar, long and narrow, often twisted longitudinally, green at the swollen base, glabrous inside; stamens 6, the filaments short, closely appressed to the ovary, the anthers 1–1.5 cm long; pistil solitary, superior, the ovary ovoid, green, the style 3-cleft; **capsule** truncate, 3-lobed. **Leaves** alternate, perfoliate, 1 or 2 below the fork of the stem, several above, lance-oblong, bright green on upper surface, pubescent and grayish beneath, the veins few.

HABITAT In rich woods and thickets; often abundant.

KEN-ICHI UEDA

LARGE-FLOWER BELLWORT

SESSILE-LEAF BELLWORT

FLOWERING April to early June.

USES American Indians made an infusion from the roots to treat backache and also used this species with lard in massaging sore muscles. This plant grows readily in a woodland garden and will sometimes grow in borders.

Sessile-Leaf Bellwort

Uvularia sessilifolia L.

Similar to *U. grandifolia*, but shorter (up to 4.5 dm tall); rootstock elongate; **leaves** sessile, not perfoliate, whitish and glabrous beneath; **capsule** distinctly stalked, ellipsoid.

HABITAT In woods, thickets, clearings.

FLOWERING April to mid-June.

LILY-LIKE GROUP

HYPOXIDACEAE *Yellow Star-Grass Family*

YELLOW STAR-GRASS

Yellow Star-Grass

Hypoxis hirsuta (L.) Coville

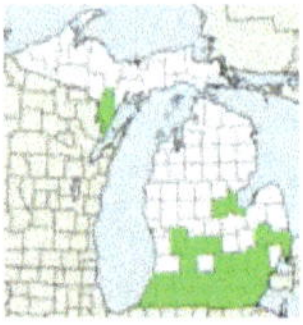

Stiff, tufted, grasslike perennial 1–6 dm tall. **Flowers** yellow, 7–18 mm wide, 2–7 in a small umbel on a hairy, threadlike, ascending or somewhat reclining scape which is usually shorter than the leaves. Perianth 6-parted, the divisions alike, spreading, narrowly oblong, coherent with the ovary at the base; stamens 6, attached to the perianth segments, the filaments yellow, the anthers slender, arrow-shaped, yellow, becoming brownish; pistil 1, the ovary inferior, 3-celled; **capsule** not opening, the seeds black, lustrous. **Leaves** all basal, linear, 1–8 dm long, 3–6 mm wide, long-hairy or nearly smooth.

HABITAT In open woods and meadows or wet marshy places around lakes.

FLOWERING late April to June, but some flowers may be produced in August and September.

LILY-LIKE GROUP

LILIACEAE *Lily Family*

The flowers are usually regular, with a perianth of 6 petal-like parts, 6 stamens, and a 3-celled ovary with 3 stigmas or one 3-lobed stigma. The fruit is a capsule or berry. The leaves are basal or on the stem, and the blade may be expanded or linear.

Yellow Bluebead-Lily

Clintonia borealis (Ait.) Raf.

Perennial, 1.5–4 dm tall from a slender creeping rootstock. **Flowers** yellow or greenish-yellow, 2–2.5 cm long, bell-shaped, usually drooping, 2–8 in a loose terminal umbel on a pubescent, leafless or bracted flowering stem, sometimes with a small secondary umbel of 2–3 flowers below the terminal one. Perianth of 6 separate, spreading, similar segments in 2 rows, the segments linear, hairy outside; stamens 6, the filaments long, pale greenish-yellow, the anthers hanging below the corolla; pistil ovoid, style long; **berry** shining blue (white in one form), 5–6 mm long. **Leaves** basal, 3–5, elliptic, oblong, or oval, 1–3 dm long, very thick, finely hairy on margins, the veins fine, the petioles sheathing the flowering stem at base.

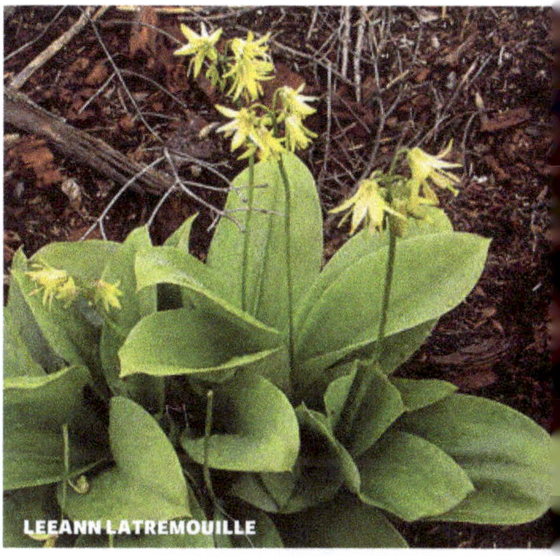

YELLOW BLUEBEAD-LILY

HABITAT In woods and thickets, usually in moist rich soil.

FLOWERING late May to late June. Berries ripe August, September.

USES The beautiful, shiny blue berries are both more striking and more attractive than the pale flowers. The Chippewa American Indians believed that dogs could use the roots of Beadlily to poison their teeth in order more easily to overcome their prey. If a person was bitten by a dog with teeth so poisoned, it was thought necessary to get a root of the same species and put it on the wound to draw the poison out.

Beadlily is a difficult species to grow. It requires a cool, strongly acid, peaty soil.

Yellow Trout-Lily

Erythronium americanum Ker -Gawl.

Nearly stemless perennial up to 3 dm tall from a deep-seated scaly bulb which sends out numerous elongate underground shoots. Flower lily-like, yellow, 1.8–4 cm long, solitary and nodding at the end of the flowering stem. Perianth of 6 spreading separate divisions in 2 rows, the 3 inner divisions with small projections at the base, pale yellow within and often spotted near the base; stamens 6, the filaments tapering to apex, the anthers linear, yellow or red; ovary 1, the style elongate, the stigmas 3, erect. **Leaves** of mature plant 2, opposite, fleshy, mottled or plain, elon-

YELLOW TROUT-LILY

gate-elliptic, tapering to the petiole, which sheathes the flowering stem; leaves of young plants solitary, elliptic, usually mottled.

HABITAT In rich woods, thickets, bottomlands, and meadows; usually in extensive colonies.

FLOWERING late March to June.

NOTES This attractive woodland flower often carpets open woods in early spring. Because the underground shoots produce numerous new bulbs, extensive colonies are formed. The flowers are short-lived. Several years, usually 6 or more, are required to produce flowering plants from seed.

USES American Indians steeped the leaves to make an infusion used for stomach distress. The species is popular in the garden but tends to die out. Many beautiful kinds are offered for sale by nurseries.

WHITE TROUT-LILY

White Trout-Lily

Erythronium albidum Nutt.

Quite similar to the preceding species, but the flowers pinkish- or bluish-white, the stigmas spreading or drooping, the leaves typically less mottled.

HABITAT In woods and thickets.

FLOWERING April to June.

NOTES In southern Michigan this species is usually through blooming at about the time the yellow species begins to flower.

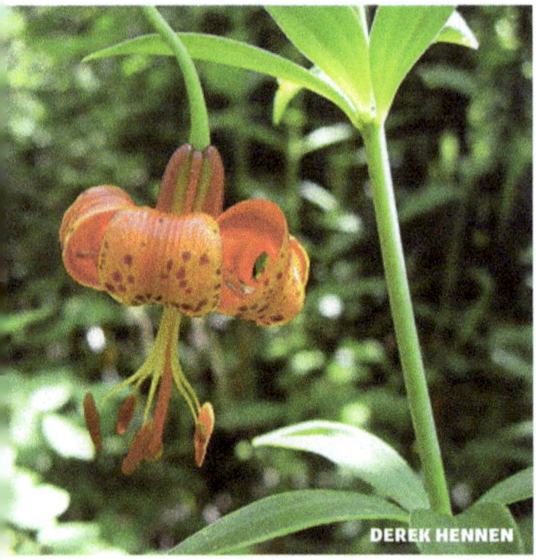

MICHIGAN LILY

Michigan Lily

Lilium michiganense Farw.

Erect perennial, up to 2 m or more tall, from yellowish bulbs. **Flowers** 1 to several, chiefly in 1–3 umbels, nodding, orange to orange-red, the sepals and petals alike and strongly recurved, their tips extending back to or beyond the base of the perianth tube; stamens 6, the filaments curving strongly outward; stigma broadly 3-lobed. **Leaves** whorled, lanceolate, tapering toward both base and apex, the margins and veins beneath usually with minute spicules (points).

HABITAT In meadows, low woods, thickets, or bogs.

FLOWERING late June to August.

USES This attractive plant can be grown in the garden if the bulbs are protected from rodents.

Wood Lily

Lilium philadelphicum L.

Stiffly erect perennial up to 1 m tall, branching only at the top. **Flowers** erect, showy, faded-orange to deep orange-red, spotted, open bell-shaped, 1–5 at the top of the plant, 5.5–7.5 cm long and 10–12 cm wide. Perianth segments alike, lanceolate, obtuse at apex, tapering to a long claw at base, spotted with conspicuous red or purple oblong dots near the base, somewhat spreading; stamens 6, the filaments the same color as the perianth, the anthers dark purple, about 1 cm long, attached near the middle, free-swinging; ovary green, 3-celled, superior; style long, orange to orange-red, the stigma blackish, 3-lobed; **capsule** rounded at summit, tapering at base, 3–5 cm long. **Leaves** in whorls or scattered along the stem, lanceolate or linear-lanceolate, 5–10 cm long, sessile.

HABITAT In open woods and clearings, along roadsides, and in bogs. This species is very common along highways at the northern tip of the Lower Peninsula.

FLOWERING mid-June to mid-August.

USES Wood Lily can be cultivated in the garden if rodents are controlled; it prefers an acidic soil. American Indians used the bulbs like potatoes. They made a medicinal charm from the plants.

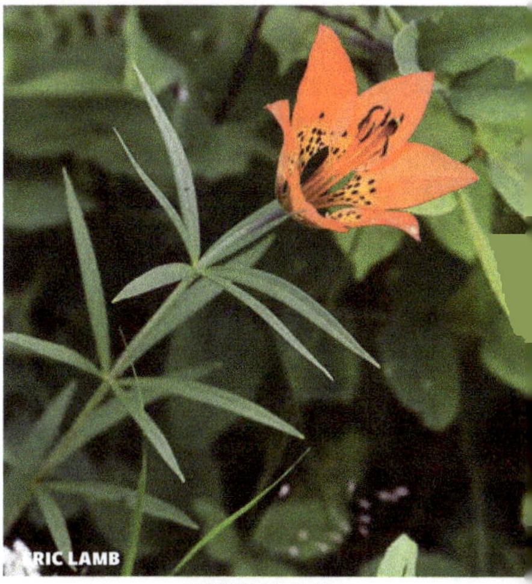

WOOD LILY

Indian Cucumber-Root

Medeola virginiana L.

Simple, erect perennial, 2–9 dm tall from a thick white rootstock which somewhat resembles the cucumber in taste and smell. **Flowers** pale yellow to greenish-yellow, usually drooping, borne in a loose, few-flowered terminal umbel above a whorl of leaves. Perianth parts 6, alike, oblong, recurved; stamens 6; styles 3, long and threadlike, spreading and recurved; **berry** dark purple, globose, 3-celled. **Leaves** in 2 whorls, those of lower whorl 5–9, obovate to lanceolate; those of upper whorl 3 (rarely 4 or 5), smaller.

HABITAT In rich woods, sandy soil under hardwoods, and in cedar bogs.

FLOWERING May and June.

USES The rootstock of this species is edible and tastes like cucumber.

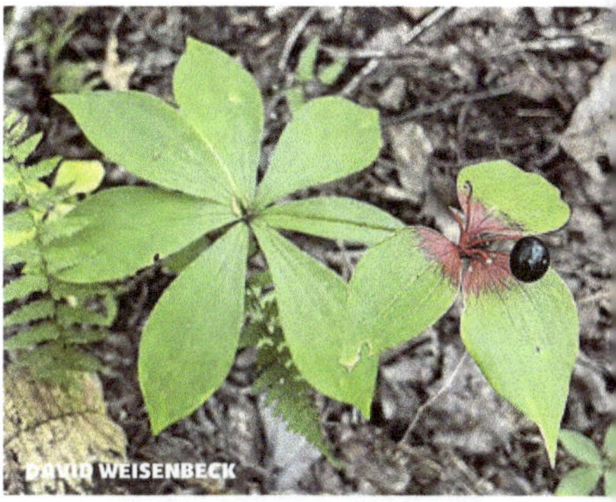

INDIAN CUCUMBER-ROOT

Lance-Leaf Twisted-Stalk

Streptopus lanceolatus (Aiton) Reveal

SYNONYMS *Streptopus amplexifolius* (L.) DC., *Streptopus roseus* Michx.

Rather stout perennial, 2.5–9 dm tall, with forking and diverging branches usually covered above with stiff, many-celled hairs. **Flowers** greenish-white to deep purple,

LANCE-LEAF TWISTED-STALK

spreading bell-shaped, drooping singly or in pairs on bent or twisted threadlike peduncles from the leaf axils. Perianth parts 6, alike, 6–12 mm long, the tips curving back in age; stamens 6, the anthers arrow-shaped; pistil 1, the stigma entire to 3-cleft; **fruit** a nearly globose or obscurely 3-lobed cherry-red berry 5–10 mm in diameter. **Leaves** alternate, sessile and sometimes clasping, the margin finely hairy, the veins parallel.

HABITAT In moist woods and thickets.

FLOWERING April to July.

LILY-LIKE GROUP

MELANTHIACEAE *False Hellebore Family*

MOUNTAIN DEATHCAMAS

Mountain Deathcamas

Anticlea elegans (Pursh) Rydb.

SYNONYM *Zigadenus glaucus* Nutt.

Stiffly erect, slender, glaucous, rarely branched perennial 2.5 dm–1 m tall. **Flowers** greenish-white, 1.2–2 cm in diameter, borne in a terminal raceme or panicle, each on a long pedicel which has a linear bract at the base. Perianth parts 6, pale green to whitish, sometimes bronze or purple on the back, each with a conspicuous, inverted heart-shaped, greenish gland at the base; stamens 6; ovary superior, forming a 3-lobed **capsule** with 3 recurving beaks. **Leaves** mostly crowded at the base of the plant, linear, leathery, up to 5 dm long, the stem leaves similar but smaller.

HABITAT In calcareous gravel and sand along shores, on cliffs, and in bogs. Common in the northern part of the Lower Peninsula, especially along the shores of Lakes Huron and Michigan.

FLOWERING mid-July to September.

These plants are poisonous; all parts contain poison. They retain their toxic properties even after drying. Milk from cows that have eaten this species can cause a mild to fatal sickness.

KEY TO TRILLIUM SPECIES

1 Flowers normally white when fresh 2

1 Flowers normally brown-purple to purple; strongly ill-scented .... *T. erectum*

2 Flowering stem usually straight and holding flower well above leaves; petals more than 3.5 cm long .... *T. grandiflorum*

2 Flowering stem usually curving so that flower hangs below leaves; petals less than 3 cm long .... *T. cernuum*

Nodding Trillium

Trillium cernuum L.

Erect, unbranched perennial with 1 to several stems. **Flowers** white, solitary, sweet-scented, nodding and nearly hidden under the leaves, 2.3–3 cm broad, with peduncles 0.5–4 cm long, coming from the whorl of leaves and recurved beneath them. Petals 3, oblong-lanceolate with recurving tips, 1.5–2.5 cm long and 0.5–1.7 cm broad, longer than the sepals; sepals 3, green, 1.7–2.5 cm long; stamens 6, the anthers slightly longer than the filaments, pink or pale pinkish-purple, over-topping the styles; styles 3, stout, recurved; ovary white or tinged with pink or pinkish-purple, 6-angled, **berry** broadly ovoid, reddish-purple. **Leaves** 3, whorled at top of stem, broadly rhombic, narrowed at base and sessile or nearly so, pale green, the veins netted. Rootstock thick (often over 2 cm), brown or grayish-brown, ascending.

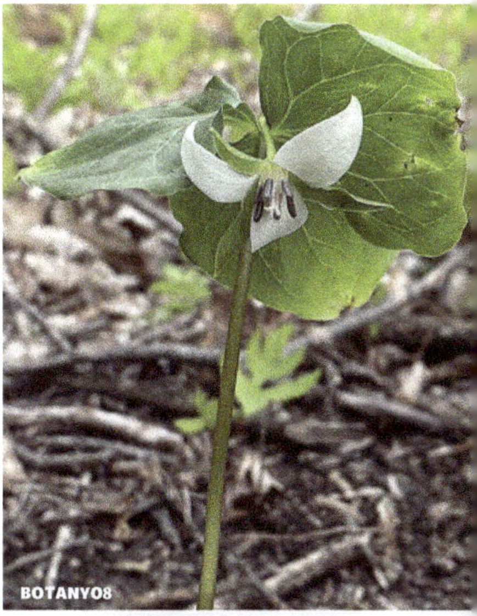

NODDING TRILLIUM

HABITAT In damp or peaty woods or thickets and in low woodlands, usually in acid soil.

FLOWERING May to late June.

SIMILAR SPECIES Larger-flowering **Nodding Trillium**, *Trillium flexipes* Raf., is coarser, the flowers about twice as large; petals white, maroon, or purple, 2–5 cm long, not recurved; ovary usually pale or white, but, with the filaments, sometimes purple; anthers creamy white, at least twice as long as the filaments. Peduncles straight, divergent or reflexed. Leaves sessile.

Stinking-Benjamin, Red Trillium

Trillium erectum L.

Stout, erect plants 1.5–6 dm tall. **Flowers** ill-scented, variable in color, crimson, purple, or purple-brown at base fading into whitish above, or greenish, clear yellow, or white, 3–7 cm wide; peduncle straight, erect, divergent or rarely bending down from the whorl of leaves. Petals 2.5–5.5 cm long and 1–3 cm broad, spreading from the base; sepals about the same length as the petals but narrower; stamens protruding beyond the spreading stigmas; ovary usually purple but pale in the pale flowers; **berry** dark red, 6-angled. Stems 1 to several. **Leaves** 3, whorled, broadly rhombic-ovate, sessile, the veins netted, Rootstock stout, up to 3 cm thick, brown.

STINKING-BENJAMIN, RED TRILLIUM

HABITAT In rich woods.

FLOWERING April to early June.

LARGE WHITE TRILLIUM, WAKEROBIN

WHITE COLICROOT

Large White Trillium, Wakerobin

Trillium grandiflorum (Michx.) Salisb.

Erect, unbranched perennial 1.5–4.5 dm tall. **Flowers** white, often fading to pink or rose, showy, 5–7.5 cm broad, solitary and erect, or nearly so, at the end of the 5–15-cm. long peduncle. Petals oblong, 4–7.5 cm long, 1–3.5 cm wide, erect at base, spreading toward tip; sepals 3, green, lanceolate, spreading, shorter than the petals; stamens 6, the anthers pale yellow, the filaments stout, shorter than the anthers; pistil solitary, the ovary 6-angled, pale, the stigmas 3, not tapering to the tip, spreading or erect; **berry** black, globose but slightly lobed, 2–2.5 cm in diameter. **Leaves** a single whorl of 3 at top of stem, broadly rhombic-ovate, pointed at ends, 5–15 cm, long, sessile, the veins very prominent, netted. Rootstock coarse, brown, scaly, 1–3 cm thick.

HABITAT In rich woods and thickets, especially in ravines and on upland slopes.

FLOWERING April to June.

USES This is one of our best wildflowers for the garden. It can be obtained from several nurseries. All parts of the plant are subject to considerable modification, and one can find specimens with a whorl of many leaves or with many petals.

American Indians used the roots of this and other species of *Trillium* in decoctions for treating stomach disorders, rheumatism, and sore ears.

LILY-LIKE GROUP

NARTHECIACEAE
Asphodel Family

White Colicroot

Aletris farinosa L.

Very bitter, rosette-forming plants with flowering stems 3–9 dm tall from a short thick rootstock. **Flowers** white, 5–8 mm long, borne in a terminal spiral spike 0.8–2.5 dm long on an unbranched stem which is clothed with a few small bracts. Sepals and petals united into a tube, the tube glandular outside, 6-lobed at the end, adherent to the base of the ovary; stamens 6, attached at the base of the lobes and enclosed in the perianth; style 3-lobed; **capsule** ovoid, beaked, partially enclosed by the dry, roughened perianth. **Leaves** in a basal rosette, linear to narrowly lanceolate or oblanceolate, up to 2 dm long.

HABITAT In dry or moist peat, sand, and gravels.

FLOWERING late May to August.

LILY-LIKE GROUP

SMILACACEAE *Greenbrier Family*

Midwestern Carrion-Flower

Smilax lasioneuron Hook.

Stems elongate, climbing by tendrils which arise from the axils of the middle and upper leaves. **Flowers** unisexual, small, greenish, strongly carrion-scented, borne in globose 20–100-flowered umbels rising from the leaf axils, the staminate flowers often the larger; peduncles up to twice as long as the petioles which subtend them; pedicels 5–20 mm long. Perianth segments 6, soon falling, those of the staminate flowers 3.5–6 mm long, the stamens 6, inserted on the base of perianth; stamens of pistillate flowers greatly reduced and threadlike, or lacking, the ovary 3-celled, the stigmas thick and spreading; **berry** black, with a bloom. **Leaves** oblong-ovate to rounded, truncate or heart-shaped at base, glaucous or pale and hairy beneath, strongly parallel-ribbed, net-veined, the petioles 2.5–9 cm long, the leaves at base of stem reduced to bracts.

HABITAT In rich low woods, thickets, meadows, and along stream banks.

FLOWERING May and June.

SIMILAR SPECIES Another herbaceous species, **Upright Carrion-Flower**, *Smilax ecirrhata* (Engelm. ex Kunth) S. Wats., has erect or leaning stems up to 1 m tall, which lack tendrils or have only a few weak terminal ones. The umbels are few-flowered, and the peduncles arise from the bracts below the foliage leaves.

NOTES Several woody species of *Smilax* also grow in Michigan. The medicinal qualities of the several kinds were known to American Indians, who made decoctions of the stalks for various ailments and chewed the berries to relieve hoarseness. The berries are eaten extensively by birds, and the plants provide cover for rabbits and other small mammals.

MIDWESTERN CARRION-FLOWER

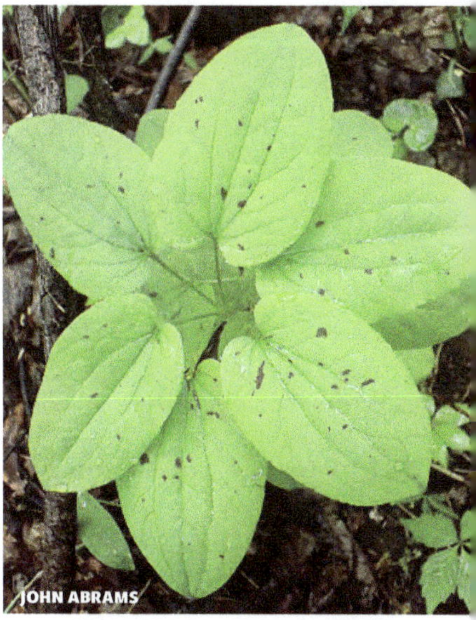

UPRIGHT CARRION-FLOWER

LILY-LIKE GROUP

TOFIELDIACEAE *Featherling Family*

STICKY FALSE ASPHODEL

Sticky False Asphodel

Triantha glutinosa (Michx.) Baker

SYNONYM *Tofieldia glutinosa* (Michx.) Pers.

Simple, erect, sticky perennial up to 5 dm tall, the stem somewhat bulbous at base, covered (as are also the pedicels) with red to black glands. **Flowers** white, 5–15 mm wide, borne in a terminal spikelike compound raceme, the short pedicels mostly in groups of 3, the flowers having a calyx-like, rather close fitting involucre, the raceme compact at first, becoming 1–9 cm long and 2.5 cm thick. Perianth segments 6, petal-like, spreading, the outer ones broader; stamens 6, the anthers pink; styles 3, short, persistent; **capsule** oblong, the seeds numerous and bearing small tail-like appendages at each end. **Leaves** few, sheathing, mostly at or near the base, in 2 rows, grasslike, 5–17 cm long.

HABITAT In sphagnum or marly bogs, calcareous marshes, and on damp ledges and shores. Often abundant along the shores of the Great Lakes.

FLOWERING June to August.

SIMILAR SPECIES The related **False Asphodel**, *Tofieldia pusilla* (Michx.) Pers., is reported for the Keweenaw Peninsula. It has smaller flowers, the seeds are without appendages, and the stem is not sticky (Michigan **Threatened**).

ORCHIDACEAE *Orchid Family*

This family, which includes more than 700 genera and approximately 28,000 species and varieties, is second in number of species only to the family Asteraceae (with an estimated 32,000 species). The family is cosmopolitan but is most highly developed in the tropics and subtropics. About 50 species grow in Michigan.

Orchids are of value chiefly for ornamental purposes and are the basis of a multimillion-dollar floral business in the United States alone. They are prized both for their exotic beauty and for the long life of most of their flowers. Vanilla is extracted from the unripened pods of various species of tropical orchids.

The irregular, 3-petaled flowers are frequently showy and strikingly beautiful. They are solitary or borne in spikes or racemes and are always subtended by leafy bracts. The lower petal, known as the lip, is usually different from the lateral ones and larger; in the Lady's-slippers it is greatly inflated. The 3 sepals may be green or colored and petal-like, alike or the middle one different, separate or united to each other or to the petals. The stamens and pistil are united in a complex structure known as the column. The Lady's-slippers have a stamen on each side of the column and bear a thick fleshy petal-like staminode (sterile stamen). In all other genera there is a single 2-celled anther borne at or near the top of the column. The pollen is granular in some species, but in most species the grains are waxy and stick together in masses (pollinia). The stigma is below the anther on the column and is sticky or rough, depending on the species. The ovary is inferior, long and cylindric, often so slightly swollen at flowering time that it is mistaken for the pedicel. The 1-celled, 3-valved capsule contains an incredible number of minute seeds, but these have no stored food, and the great majority do not produce new plants. The leaves of orchids are alternate or basal, simple, entire, and parallel-veined; in a few species no leaf is present at flowering time. The plants are perennial from fibrous or tuberously thickened corms or bulbs.

These flowers are completely dependent on insects for pollination. Many species can be pollinated only by a specific kind of insect. The flower parts are arranged in such a manner that the visiting insect while trying to reach the nectary touches the stigma and deposits pollen upon it. In leaving, the insect touches the anther, thus picking up more pollen, which is deposited on the next flower. The flowers have many interesting devices that insure cross pollination, but they cannot be dealt with here.

Since all native species of orchids are protected in Michigan, they should not be picked or transplanted. The more striking ones are offered for sale by nurseries, and people interested in growing orchids can obtain them from that source. However, these plants are exacting in their requirements, and few people have the patience to successfully grow them.

KEY TO ORCHIDACEAE (ORCHID FAMILY) SPECIES

1 Flowers usually solitary (sometimes 2 or 3), rather large and showy, lip large and conspicuous 2
1 Flowers typically numerous, lip not greatly enlarged 5
  2 Lip having a prominent beard; leaf lacking or solitary at flowering time 3
  2 Lip lacking a beard, inflated, margin smooth and inrolled; several leaves present . . . . . ***Cypripedium*, p. 258**
3 Lip inflated, slipper-like or pouch-like; leaf round, definitely petioled . . . ***Calypso bulbosa***
3 Lip broad but not inflated; leaf longer than broad 4
  4 Leaf flat, well developed at flowering time . . . . . ***Pogonia ophioglossoides***
  4 Leaf somewhat folded lengthwise, immature or absent at flowering time . . . . . ***Arethusa bulbosa***
5 Flowers having a spur at base 6
5 Flowers lacking a spur at base 7
  6 Lip white, sepals and petals pink to purplish . . . . . ***Galearis spectabilis***
  6 Lip, sepals, and petals colored alike . . . . . ***Platanthera*, *Dactylorhiza*, p. 266**
7 Flowers bright rose pink, lip at top of flower; leaf grasslike . . . . . ***Calopogon tuberosus***
7 Not as in alternate choice 8
  8 Flowering stalk spirally twisted 9
  8 Flowering stalk not spirally twisted 10
9 Leaves mottled or variegated with white; leaves broad and flat . . . . . ***Goodyera***
9 Leaves neither mottled nor variegated with white; often grasslike . . . . ***Spiranthes*, p. 272**
  10 Leaves borne on stem 11
  10 Leaves basal or none at flowering time 12

11 Stem leaves 2 .... ***Neottia convallarioides*** or ***N. cordata***
11 Stem leaf solitary .... ***Malaxis monophyllos***
  12 Leaves several in a rosette, usually mottled or variegated with white ..... ***Goodyera***
  12 Leaves 2 or fewer 13
13 Leaves 2 .... ***Liparis loeselii***
13 Leaf solitary or none 14
  14 Plant typically green, one leaf produced (may be lacking at flowering time); stem rising from a tuber .... ***Aplectrum hyemale***
  14 Plant lacking chlorophyll, not green, no true leaf produced; stem rising from a coral-like mass .... ***Corallorhiza, p. 256***

PUTTY-ROOT, ADAM-AND-EVE

Putty-Root, Adam-and-Eve

Aplectrum hyemale (Muhl. ex Willd.) Torr.

Smooth perennial up to 3.5 dm tall, leafless at flowering time, the flowering stalk rising from a horizontal rootstock which typically has 2 (sometimes more) subglobose, glutinous corms connected by slender stolons. **Flowers** greenish, yellowish, or whitish, marked with madder-purple or unspotted; raceme 4–14 cm long, the flowers ascending, fruit drooping. Petals long (about 1.2 cm) and narrow; lip 3-lobed, white, spotted with magenta. **Leaf** solitary, basal, dark green, the veins whitish or tinged with purple, elliptic, short-petioled, appearing in autumn, decaying before the appearance of the flowering stalk in the spring.

HABITAT In rich woods, wet mucky soil, low moist hardwoods, peat bogs, and tamarack swamps.

FLOWERING May to June.

NOTES Because two joined corms are common, this plant is frequently called "Adam and Eve."

USES A sticky paste can be made by adding water to the ground corms of this species. A paste so made was formerly used to mend broken pottery.

Dragon's-Mouth, Swamp Pink

Arethusa bulbosa L.

Slender perennial, usually leafless at flowering time, the sheathed flowering stalk 5–30 cm tall. **Flowers** usually magenta-pink, sometimes bluish-lilac or white, 2–5.5 cm long, solitary and slightly nodding at the tip of the unbranched stem. Sepals and petals quite similar, arching over the column; lip partly erect, the apical portion drooping and having 3–6 fringed, yellow or white crests, the margin fringed, often spotted and striped with purplish-red; column arching upward, petal-like at apex; **capsule** erect, long-beaked. **Leaf** solitary, linear, usually hidden in the sheaths until after the flower opens.

HABITAT In open sphagnum bogs, in tamarack, cedar, and other swamps and in peaty meadows.

FLOWERING late May to mid-July.

NOTES This strange little flower has somewhat the appearance of a startled animal. The only other species in this genus occurs in Japan. Arethusa was formerly used as a remedy for toothache.

DRAGON'S-MOUTH, SWAMP PINK

Tuberous Grass-Pink

Calopogon tuberosus (L.) B.S.P.
SYNONYM *Calopogon pulchellus* (Salisb.) R. Br.

Slender perennial with solitary flowering stem up to 5 dm tall. **Flowers** showy, rose-purple, magenta, lilac, rose-pink or white, 3–3.7 cm wide, borne in a loose, zigzag, terminal raceme of 2–12 flowers. Sepals and petals nearly alike, separate, spreading, the flower inverted so that the lip is uppermost; lip broadened at the apex, flanged on each side toward the base, copiously bearded with 3 longitudinal rows of short, fleshy, white, yellow-tipped hairs; column colored like the perianth, bending down, then arching upward and outward, but position varying with age and weather conditions; ovary much shorter than the perianth segments. **Leaf** solitary, narrow, arising from the upper sheath and remaining close to the stem, 10–20 cm long and 4–20 mm wide.

HABITAT In wet, open grassy meadows, sphagnum bogs, tamarack and cedar swamps.

FLOWERING mid-June to mid-July.

NOTES This flower blooms at the same time as Arethusa, Rose Pogonia and White Fringed Orchid, and often grows in association with them. Frequently very abundant.

TUBEROUS GRASS-PINK

Calypso, Fairy-Slipper

Calypso bulbosa (L.) Oakes

Small, smooth perennial 6–22 cm tall. **Flowers** purplish to rose-purplish, solitary, showy, nodding, about 3 cm long. Sepals and petals separate, alike, purplish, magenta, crimson, or rarely white, linear-lanceolate, spreading; lip larger than the rest of the flower, sac-like, expanded in front, forming a spreading, whitish, purple-spotted apron over the 2-horned

CALYPSO, FAIRY-SLIPPER

apex, bearded at the middle of the base with 3 rows of golden-yellow or brown-spotted (rarely white) hairs; column winged, petal-like, nearly round. **Leaf** solitary, oval to nearly round, 2–6.5 cm long, 1.5–5 cm wide, plaited, petioled, the margin undulate, the veins prominent, produced from the summit of the corm in autumn, overwintering and shriveling soon after the flowering season.

HABITAT In cool mossy forests, chiefly calcareous, and in tamarack and cedar swamps.

FLOWERING mid-May to early July.

STATUS Threatened in Michigan.

NOTES This beautiful little species is rare in Michigan. It usually grows singly, but often in association with Twinflower and Fringed Polygala. Michigan plants are treated as var. *americana* (R. Br.) Luer.

KEY TO CORALLORHIZA (CORALROOT) SPECIES

1 Lip not lobed, tongue-shaped with inrolled edges; flowers with reddish-purple stripes .... *C. striata*
1 Lip with a small lobe on each side; flowers not striped 2
 2 Lip spotted with crimson or magenta; sepals usually having 3 veins ...... *C. maculata*
 2 Lip seldom spotted; sepals usually with 1 vein .... *C. trifida*

SPOTTED CORALROOT

Spotted Coralroot

Corallorhiza maculata (Raf.) Raf.

Erect, unbranching fleshy perennial, up to 6 dm tall, the leafless stem yellow to brown, purple-brown or reddish-brown. **Flowers** pale yellowish to whitish with red or purple spots, few to many, in racemes 4–20 cm, long, ascending at flowering time but drooping after pollination. Petals separate; lip white, spotted, 3-lobed, the middle lobe the largest; spur small, sepals 3-veined.

HABITAT In dry woods, common on slopes and frequent at the base of wooded dunes.

FLOWERING late June to August.

NOTES This, our most common coralroot, is quite variable in color. In addition to the typical form described above, three other forms may be distinguished. In one the lip is unspotted; in another the perianth is yellowish-brown; in the third, the flower stalk, sheaths, and perianth are reddish-purple and the lip is spotted.

SIMILAR SPECIES Autumn Coralroot, *Corallorhiza odontorhiza* (Willd.) Poir., is a frail, slender plant, light brown to madder-purple. It has a purple-spotted white lip, often wider than long, with an eroded, wavy margin, and the sepals have one vein. This is our only coralroot with a thickened, bulbous base. It flowers in late summer and fall.

Striped Coralroot

Corallorhiza striata Lindl.

Unbranched, leafless, reddish to purplish perennial up to 4 dm tall, the stoutish, succulent stem bearing a few large, striped, sheathing bracts and rising from a cluster of many-branched, coral-like roots. **Flowers** madder-purple or nearly ruby-red, somewhat drooping, borne in a loose terminal raceme 8–16 cm long. Perianth 1–1.5 cm long, the petals similar to the sepals but usually 5-veined; lip sharply turned down, unlobed, boat-shaped with an inrolled margin, deep red at apex, somewhat striped at the base; sepals sometimes yellowish or whitish, with 3 conspicuous madder-purple veins.

HABITAT In rich woods, either hardwoods or mixed forest; said to do best in calcareous areas.

FLOWERING late May to July.

NOTES Coralroots, like other saprophytic plants, cannot be transplanted successfully. They may be grown from seeds if soil conditions are suitable. It is said to take from 5 to 10 years for them to attain blooming size. Most species in this genus have been used for medicinal purposes at one time or another.

Yellow Coralroot

Corallorhiza trifida Chatelain

Slender, unbranched, pale yellow or yellowish-green perennial up to 4.5 dm tall, the solitary or clustered sheathed stems rising from a whitish, intricately branched root. **Flowers** pale greenish-yellow, sometimes (particularly in age) tipped with brown, borne in loose terminal, 2–20-flowered racemes. Sepals and petals alike, unspotted; lip white, unspotted or sometimes with a few red spots at the base, notched

AUTUMN CORALROOT

STRIPED CORALROOT

YELLOW CORALROOT

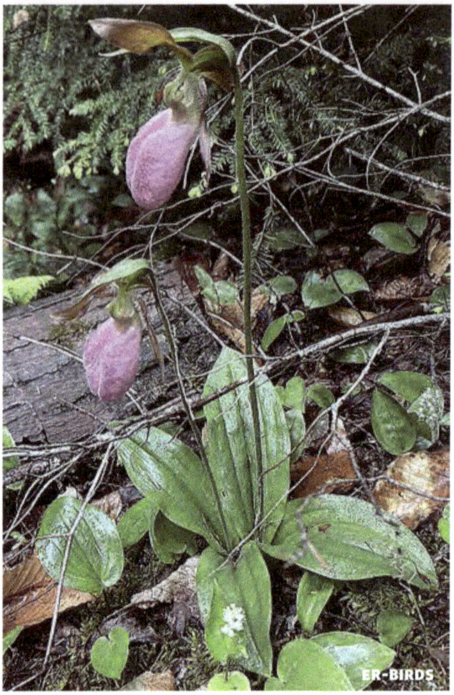

PINK LADY'S-SLIPPER

on each side toward the base and blunt or bluntly rounded and notched at the apex.
HABITAT In damp woods and thickets, cedar or tamarack swamps, and in bogs.
FLOWERING May to July.

KEY TO CYPRIPEDIUM (LADY'S-SLIPPER) SPECIES

1 Lip (pouch) yellow .... *C. parviflorum*
1 Lip not yellow 2
 2 Leaves only 2, borne at base of stem .... *C. acaule*
 2 Leaves more than 2, borne on stem........ 3
3 Lip a moccasin-like inflated pouch, 2 sepals united to, or almost to, the tip 4
3 Lip a cone-shaped pouch; sepals separate nearly to base .... *C. arietinum*
 4 Lip magenta, rose, or white, 2.5–5 cm long; petals wide-spreading .... *C. reginae*
 4 Lip waxy-white, veined with purple inside only at base, up to 2.5 cm long; petals twisted .... *C. candidum*

Pink Lady's-Slipper

Cypripedium acaule Ait.

Downy perennial with 2 basal leaves and flowering stem 2–4.5 dm tall. **Flowers** showy, pink to deep rose, solitary, nodding, 5–6 cm long. Sepals and petals yellowish-green to greenish-brown, often striped with purple or brown; lower sepals entirely united; petals somewhat spirally twisted, up to 5.5 cm long; lip a greatly inflated pouch, pink to deep rose, occasionally pure white, veined on the outer surface with rose, having a long fissure with infolded edges extending down the front. **Leaves** 2 (rarely 3), opposite or nearly so, broadly elliptic, strongly ribbed, 1–2.3 dm long and 4–14 cm wide.

HABITAT In moist or dry woods, on wooded dune ridges, in and near tamarack, spruce, or cedar swamps, or in bogs.

FLOWERING May to mid-July.

NOTES One of the most widely distributed orchids in Michigan, this species has been reported from most counties. It frequently grows in large patches but is one of the most difficult of our native plants to cultivate.

■ Ram's-Head Lady's-Slipper

Cypripedium arietinum R. Br.

Slender, somewhat glandular-pubescent, erect, unbranched perennial 0.7–3.4 dm tall. **Flowers** white and purplish-brown or madder-purple, about 2.5 cm long, solitary and nodding at the top of the stem. Petals resembling the lower sepals, linear, sometimes twisted; lip enlarged, prolonged downward into a blunt conical pouch, white or pinkish, strongly netted with crimson or purple, 1.2–2.5 cm long; sepals entirely separate, the upper one the largest, dark purplish-brown to madder-purple. Stems often somewhat twisted, covered with brown tubular sheaths below, becoming leafy at about the middle. **Leaves** 3–4, narrowly elliptic to elliptic-lanceolate, 5–10 cm long and 2–3 cm wide.

HABITAT In sphagnum, cedar, or tamarack swamps and bogs, in coniferous forests, and on wooded rocky slopes or wooded dunes.

FLOWERING mid-May to early June.

NOTES This little species is rare, but it is sometimes locally abundant in the northern and eastern part of the state. In contrast to most orchids, this species has short-lived fragrant flowers—they remain in their prime but a single day. An even more rare, all white, form is known.

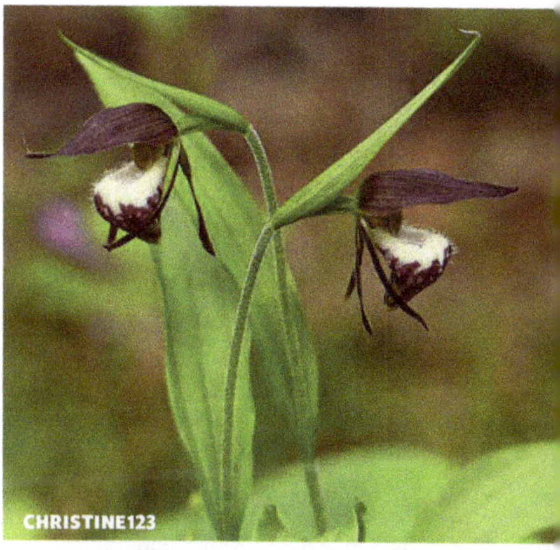

CHRISTINE123

RAM'S-HEAD LADY'S-SLIPPER

■ Small White Lady's-Slipper

Cypripedium candidum Muhl. ex Willd.

Generally similar to **Showy Lady's-Slipper** (*C. reginae*) but smaller throughout, up to 4 dm tall; the **flowers** white, usually solitary, slightly fragrant, the lip 2–3 cm long, white, with rose-purple veins inside, the petals greenish-yellow, veined with purple or maroon, the two lower sepals united to the tip; leaves 3–6, nearly erect, narrowly elliptic to lanceolate.

NORMA MALINOWSKI

SMALL WHITE LADY'S-SLIPPER

HABITAT In marly bogs, open marshes, mossy glades, sphagnum bogs, and sometimes on dry rocky hills and in woods.

FLOWERING May and June.

STATUS Threatened in Michigan.

NOTES This is the only species of Lady's-slipper that grows in open prairies in Michigan. It is often associated with the **Prairie White Fringed Orchid**, ***Platanthera leucophaea*** (Nutt.) Lindl., in the southwestern part of the state.

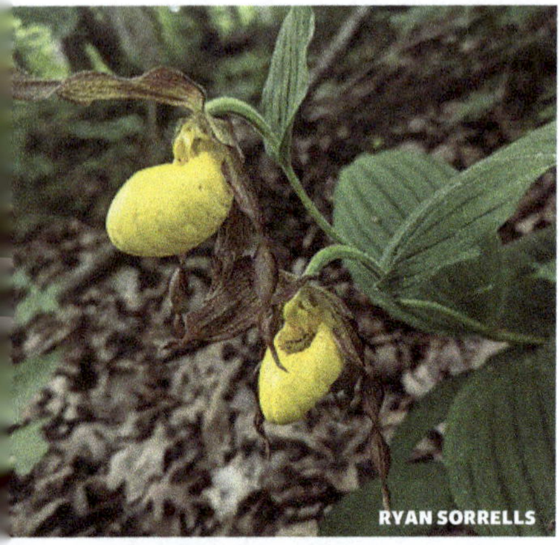

LARGE YELLOW LADY'S-SLIPPER

SMALL YELLOW LADY'S-SLIPPER

Large Yellow Lady's-Slipper

Cypripedium parviflorum var. *pubescens* (Willd.) Knight

SYNONYM *Cypripedium calceolus var. pubescens* (Willd.) Correll

Downy, erect perennial with unbranched, often clustered, stems, 1.5– 7 dm tall. **Flowers** solitary or sometimes 2, nodding at the top of the stem, golden-yellow and greenish, or brownish-yellow to madder-purple, sometimes fragrant. Petals striped or mottled with purple or brown, linear, spirally twisted, spreading and drooping, 5–9 cm long and about 8 mm wide; lip sac-like, yellow with dots and stripes of maroon, about 3 cm long; sepals colored like the petals, the upper one erect, ovate to lanceolate, 5–6 cm long and up to 2.5 cm wide, the lower ones partially united and appearing as one except for the forked tip; staminode large, thick and petal-like, yellow, often spotted, the stigma moist, roughened, obscurely 3-lobed. **Leaves** 3–6, alternate, broadly ovate or elliptic, narrowed and clasping at base, strongly ribbed, parallel-veined, downy, 7–20 cm long, and about half as wide, the upper floral one erect behind the flower, "framing" it, the margin entire or somewhat undulate, the blade somewhat plaited.

HABITAT In swampy, open or shaded ground, in bogs, on wet wooded beaches, and in rich woods.

FLOWERING mid-April to late June.

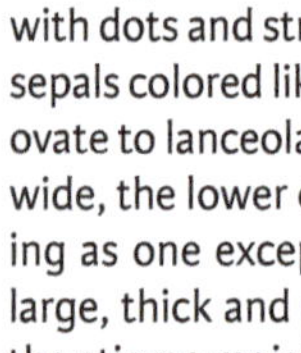

SIMILAR SPECIES Another variety of Yellow Lady's-slipper grows in Michigan, but the two so intergrade that it is frequently difficult or impossible to distinguish them: **Small Yellow Lady's-slipper**, ***Cypripedium parviflorum*** **var.** ***makasin*** (Farw.) Sheviak, is a generally smaller plant, 1.5–5.5 dm tall, with 3–4 leaves, the largest being 2–9 cm broad; the sepals and petals are usually madder-purple, the upper sepal 2.5–5 cm long; the flowers are strongly fragrant. This variety tends to bloom just as the other fades. Generally it grows in wetter places, but the two may grow together.

Showy Lady's-Slipper

Cypripedium reginae Walt.

Glandular-hairy, usually erect, unbranched perennial 3–9 dm, tall. **Flowers** showy, white and magenta or rose, usually 1 to 2 but sometimes 3 at the top of the plant, the larger ones about 9 cm wide (somewhat wider than high). Petals wide-spreading, white, narrower than the sepals; lip white, striped and dotted with magenta or rose, greatly inflated, 2.5–5 cm long, the opening at the top nearly circular and covered with long, fine, upward-pointing hairs; sepals one. white, the upper one large and erect, the lower pair appearing as **Leaves** alternate, broadly elliptic, strongly ridged.

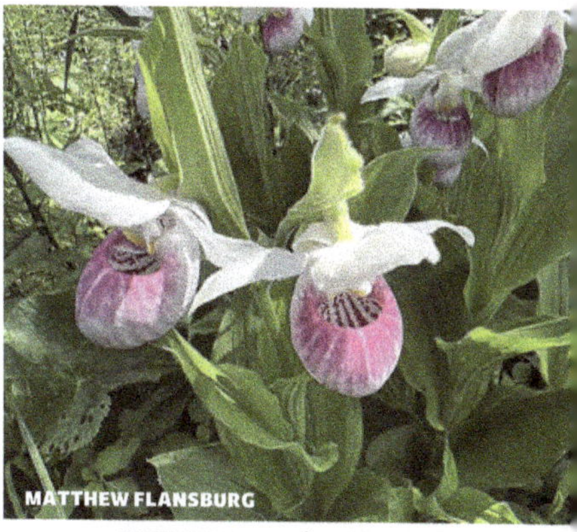

SHOWY LADY'S-SLIPPER

HABITAT In mossy swamps, bogs, cedar swamps, or in woodland glades, usually in neutral or slightly alkaline soil.

FLOWERING June and early July.

NOTES This beautiful species is worthy of its name and has been adopted as the state flower of Minnesota. It may form colonies of dozens, hundreds, or—according to some authorities—even thousands (but many populations have been lost over the years). In northern Michigan it is often found in cedar bogs, flowering at the same time as the Wood Lily. Some people get a severe dermatitis similar to that of poison ivy from touching this plant. Many of the cells contain a large number of needlelike crystals which make it unattractive to grazing animals. The species is susceptible to parasitic fungi and is difficult to maintain in cultivation.

Showy Orchis

Galearis spectabilis (L.) Raf.

SYNONYM *Orchis spectabilis* L.

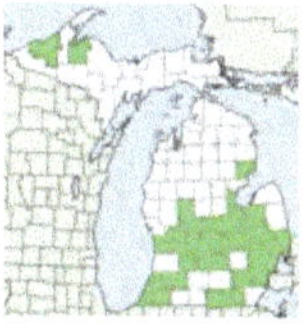

Low-growing, smooth, succulent perennial 1–4.5 dm tall, with basal leaves. **Flowers** showy, pink to mauve and white, about 2 cm long, borne in lax 2–5-flowered racemes up to 10 cm long, the leaflike bracts equaling or slightly exceeding the flowers in length. Petals and sepals all free, erect or nearly so and forming a hood over the column, pink to mauve, rarely white; lip white, rarely pink, spurred below; column stout, the anther cells parallel, the 2 pollen masses borne on slender stalks attached to the viscid stigmatic disks, which are contained in a pouch (bursicle) just above the opening of the spur; capsule ellipsoid, up to 2.5 cm long. Flowering stalk stout, 4–5-angled. **Leaves** 2, basal, sub-orbicular to oblong-obovate or broadly elliptic, narrowed to an indistinct sheathing petiole, 7–18 cm long, 5.5–7 cm wide.

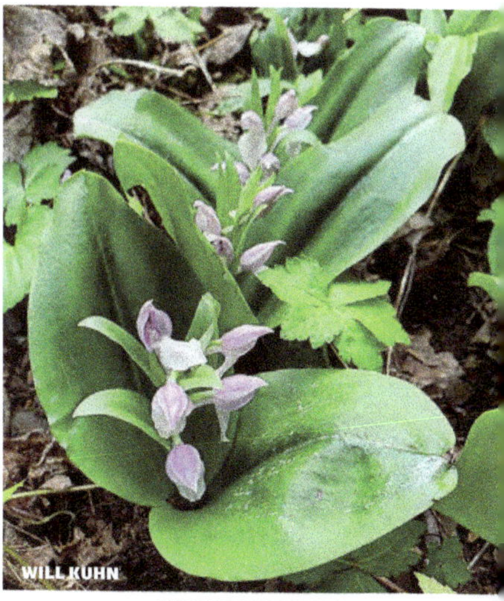

SHOWY ORCHIS

HABITAT In rich hardwood or, less frequently, coniferous forests, in wooded ravines, and on slopes.

FLOWERING May to early July.

STATUS Threatened in Michigan.

NOTES This is one of our earliest-flowering woodland orchids. It may form extensive colonies.

KEY TO GOODYERA (RATTLESNAKE-ORCHID) SPECIES

1 Flowers 6–9 mm long; leaves green over all or mottled with white .... ***G. oblongifolia***

1 Flowers 2.5–5 mm long; some veins and/or margin of leaves bordered with white or pale green 2

2 Leaves 1–3 cm long, veins bordered with white .... ***G. repens***

2 Leaves 2–7 cm long, veins bordered with white to pale green .... ***G. tesselata***

WESTERN RATTLESNAKE-ORCHID

Western Rattlesnake-Orchid

Goodyera oblongifolia Raf.

Perennial up to 4.5 dm tall, the racemes loosely spiral or 1-sided. Perianth 8–9 mm long, the lip elongated, only slightly sac-like, the margin incurved. **Leaves** 4–10 cm long, uniformly green or mottled or with the midrib white, without evident netted venation, often reddish at least in dried specimens.

HABITAT In deep cedar woods, other coniferous forest, and in rather dry, deciduous or mixed woods.

FLOWERING July and August.

USES Species of Rattlesnake-Orchid are sometimes used in enclosed “dish gardens” or terraria, where they may thrive for a considerable length of time, sometimes even blooming.

DWARF RATTLESNAKE-ORCHID

Dwarf Rattlesnake-Orchid

Goodyera repens (L.) R. Br.

Creeping evergreen perennial, the unbranched flowering stalk up to 27 cm tall from an evergreen rosette; often spreading by slender runners. **Flowers** white or greenish-white, 10–20 in a loose 1-sided raceme, the flowering stalk, bracts, ovaries and perianth glandular-hairy. Upper sepal and petals united into a hood over the lip, the lateral sepals free; lip sac-like, inflated, with a recurved tip, spurless. **Leaves** all basal, thick, smooth, ovate with a blunt apex, tapering abruptly to a winged, sheathing petiole, the blade 1–3 cm long, dark green, with about 5 parallel veins, these and the horizontal to slightly oblique cross veins bordered with white on the upper surface (sometimes the blade blotched with white), the fine veins netted.

HABITAT In damp or dry cold woods, usually under conifers, sometimes in bogs or swamps.

FLOWERING July to early September.

NOTES The name Rattlesnake-Orchid is preferable to the more commonly used name, **Rattlesnake-Plantain**, because it expresses the true affinity of these plants and because a well-known group of dicotyledons are called plantains. Since several species of this genus hybridize freely, particularly in the Great Lakes region, it is often very difficult to identify them.

SIMILAR SPECIES None of the Michigan species of this genus is easy to cultivate, but the **Downy Rattlesnake-Orchid**, *Goodyera pubescens* (Willd.) R. Br., is probably the easiest. This grows up to 4.5 dm tall and has a dense cylindric raceme of globose flowers, the perianth is 4–5 mm long, and the lip is beaked; the 5–10 leaves are dark green above, with 5–7 white nerves and numerous white cross veins.

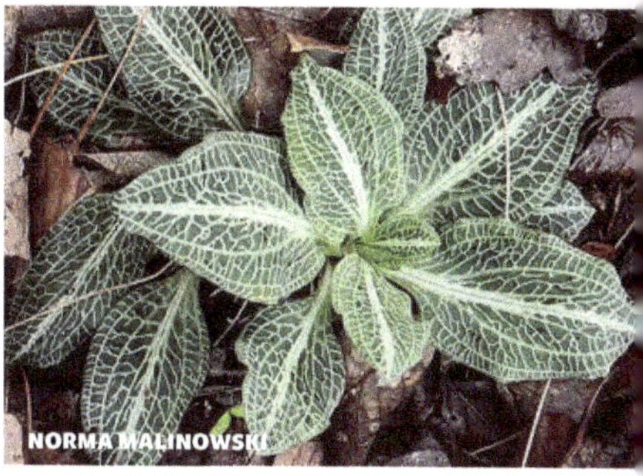

DOWNY RATTLESNAKE ORCHID

Checkered Rattlesnake-Orchid

Goodyera tesselata Lodd.

Plants 1–3.5 dm tall, the raceme loosely spiral or somewhat 1-sided. Perianth about 5 mm long, the lip only slightly inflated, its tip slightly recurved. **Leaves** 2–7 cm long, 5–6-nerved, the bordering of the veins usually a light green.

HABITAT In dry to moist deciduous, coniferous, or mixed woods and cedar swamps.

FLOWERING mid-July to Sept.

Loesel's Twayblade, Fen Orchid

Liparis loeselii (L.) L.C. Rich.

Small, erect, pale-green or yellowish-green perennial 6–26 cm tall. **Flowers** yellowish-green or whitish, 2–25 on short ascending pedicels in a slender terminal raceme. Petals thread-like, tubular; sepals oblong-lanceolate or sub-orbicular, with wavy margin; column short and

CHECKERED RATTLESNAKE ORCHID

LOESEL'S TWAYBLADE, FEN ORCHID

stout; capsule ellipsoid, erect. **Leaves** 2, basal, greasy-looking, sheathing the stem at base, oblong-elliptic to elliptic-lanceolate, keeled beneath, 3–19 cm, long and 1–6 cm wide.

HABITAT In bogs, in sedge mats around bogs, in peaty meadows, damp thickets, and woods, cedar or tamarack swamps, on marshy shores, or around beach pools.

FLOWERING June to mid-July.

White Adder's-Mouth

Malaxis monophyllos (L.) Sw.

SYNONYM *Malaxis brachypoda* (Gray) Fern.

Slender perennial 5–15 cm tall. **Flowers** small, yellowish-green, borne in a very slender elongate terminal raceme 1.5–2 cm long. Pedicels and ovaries about the same length at flowering time; perianth parts strongly divergent; lip drooping, entire, ovate, with a slender pointed tip. **Leaf** usually solitary, less than half as tall as the flowering stalk, basal or sub-basal on the stem, oblong to broadly elliptic-oval, 1–9 cm long; petioled (in one form a second, smaller, leaf may also be present).

HABITAT In bogs, peaty places, swales, swamps, damp calcareous gravels, in crevices in wet shady places, and on ledges,

FLOWERING June to mid-August.

SIMILAR SPECIES Green Adder's-Mouth, ***Malaxis unifolia*** Michx., is somewhat less common. It is quite similar in general aspect, but the raceme is oblong-cylindric and more dense at the developing tip; the leaf is a darker green and usually about midway on the scape; the lip (drooping at first) becomes erect, is 2-lobed at the summit, and may have a small central tooth.

WHITE ADDER'S-MOUTH

GREEN ADDER'S-MOUTH

Broad-Lipped Twayblade

Neottia convallarioides (Sw.) Rich.

SYNONYM *Listera convallarioides* (Sw.) Nutt.

Stout, erect, glandular-pubescent perennial with solitary stem 6–30 cm tall. **Flowers** pale green or yellowish green, about 1 cm long exclusive of the ovary, borne on short, slender, glandular-hairy pedicels in rather loose terminal racemes 2–12 cm long. Petals and sepals separate, reflexed and closely appressed against the ovary, the petals almost linear; lip long and narrow, forward pointing, dilated at apex and with 2 rounded lobes, bearing a tooth at each side of the base; column arching upward, green, thought by some to resemble the open mouth of a snake. **Leaves** 2, opposite, sessile, about the middle of the stem, broadly ovate to nearly round, 2.5–5 cm long and 2–4 cm broad, 3–9-veined.

HABITAT In leaf mold in peaty or mossy glades, mixed woods, thickets, and swamps, and along shores.

FLOWERING June and July.

SIMILAR SPECIES Another member of this genus, the rare **Auricled Twayblade**, ***Neottia auriculata*** (Wiegand) Szlach., is known from the Upper Peninsula. It has pale-green or greenish-white flowers; the lip is dilated and deeply cleft above (the sinus narrow), and has 2 short incurved teeth or auricles at the base.

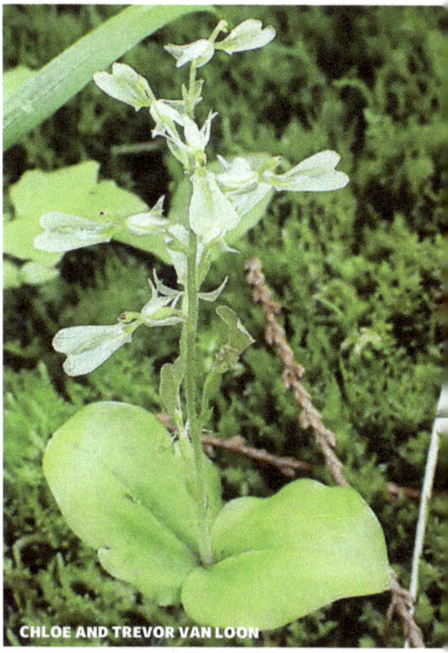

BROAD-LIPPED TWAYBLADE

Heartleaf Twayblade

Neottia cordata (L.) Rich.

SYNONYM *Listera cordata* (L.) R. Br.

Delicate, erect perennial 5–24 cm tall, glabrous except near the leaves. **Flowers** small, dark purplish or greenish to straw-colored, borne in a slender, terminal, long-peduncled raceme 2.5–10 cm long on pedicels that are much longer than the subtending bracts. Petals ovate-lanceolate; lip linear-oblong, deeply cleft into 2 narrow, spreading lobes, with a pair of hornlike teeth at the base; sepals ovate. **Leaves** 2, opposite, sessile, about the middle of the stem, heart-shaped or rounded triangular, much shorter than the peduncle of the raceme.

HEARTLEAF TWAYBLADE

HABITAT In damp mossy coniferous forests, mixed woods, bogs, and evergreen swamps.

FLOWERING June to mid-July.

KEY TO PLATANTHERA AND DACTYLORHIZA SPECIES

1 Lip fringed along margin ... 2
1 Lip not fringed along margin ... 4
  2 Lip deeply 3-parted ... 3
  2 Lip not parted, tongue-shaped ... *Platanthera blephariglottis*
3 Flowers lilac to rose-purple ... *Platanthera psycodes*
3 Flowers yellowish green, dirty yellow, or tinged with bronze or rose ... *Platanthera lacera*
  4 Leaf only 1 ... 5
  4 Leaves more than 1 per plant ... 6
5 Leaf basal ... *Platanthera obtusata*
5 Leaf borne about middle of stem ... *Platanthera clavellata*
  6 Leaves roundish, nearly flat on ground ... 7
  6 Leaves typically several, borne along stem ... 8
7 Flowers greenish white, spur thickened, curved; stem having 1–5 bracts ... *Platanthera orbiculata*
7 Flowers greenish to yellowish green, spur tapering to a point; stem bractless ... *Platanthera hookeri*
  8 Lower bracts in raceme much longer than flowers; lip 2- or 3-toothed at apex ... *Dactylorhiza viridis*
  8 Lower bracts in raceme no longer, or only slightly longer, than flowers; lip obtuse or tapered at apex, but not toothed ... 9
9 Flowers white ... *Platanthera dilatata*
9 Flowers greenish to yellowish green ... 10
  10 Lip margin not toothed ... *Platanthera huronensis*
  10 Lip margin having a single tooth on each side near base ... *Platanthera flava*

LONG-BRACT FROG ORCHID

Long-Bract Frog Orchid

Dactylorhiza viridis (L.) R.M. Bateman, Pridgeon & M.W. Chase

SYNONYM *Habenaria viridis* var. *bracteata* (Muhl.) Gray

Stout, erect, leafy-stemmed perennial, up to 5 or 6 dm tall. **Flowers** small, green, borne on stout pedicels in a raceme up to 20 cm long, with very conspicuous, linear or lanceolate, spreading floral bracts, the lower bracts 2–6 times as long as the flowers, upper ones shorter. Flower structure typical, the lip often tinged with bronze, 2–3-toothed at the apex, 2–4 times as long as the short, sac-like spur; capsule ellipsoid, up to 6.5 cm long. **Leaves** 2–7, the lower ones oblong, the upper ones lanceolate, all sheathing at the base.

HABITAT In rich woods and thickets, meadows, bogs, open grassy swamps, and beach meadows.

FLOWERING mid-May to early August.

White Fringed Orchid

Platanthera blephariglottis (Willd.) Lindl.

SYNONYM *Habenaria blephariglottis* (Willd.) Hook.

Slender, erect perennial up to 6 dm tall. **Flowers** white, crowded in flat-topped, terminal raceme 2.5–15 cm long and 4.5–12 cm wide. Upper concave sepal and the 2 smaller petals form an upstanding hood, the 2 lateral sepals spreading, the lip fringed around the margin, oblong or tongue-shaped, not cleft, the spur slender, 2–3 cm long.

HABITAT In bogs, peaty soil, in spruce, tamarack, or other swamps, often in sphagnum around bog lakes.

FLOWERING July and August.

NOTES This species is often locally abundant in sphagnum mats around bog lakes, where it may be found blooming with Rose Pogonia, Grass Pink, and *Arethusa*. This is not as common as our other *Platanthera* species.

Green Club-Spur Orchid

Platanthera clavellata (Michx.) Luer

SYNONYM *Habenaria clavellata* (Michx.) Spreng.

Slender perennial, up to 4.5 dm tall with a single well-developed leaf; **flowers** few, greenish, greenish-yellow, or greenish-white, borne in an oblong-cylindric spike, often turned so that the lip and spur project from the side, the lip with 3 short, rounded teeth at apex, the spur slenderly club-shaped, upward-curved, equaling or exceeding the ovary in length. Well-developed **leaf** (rarely 2) at about the middle of the stem, narrow, oblong, tapering to the base, 5 or 10 times as long as broad; other leaves reduced to bracts.

HABITAT In water and at the edge of water in swamps, woods, meadows, and bogs.

FLOWERING June to August.

NOTES This is one of our most common and easily recognized rein-orchids. This species is unlike most orchids in being self-pollinated and able to produce more seed-filled capsules than is typical.

ER-BIRDS

WHITE FRINGED ORCHID

TOM SCAVO

GREEN CLUB-SPUR ORCHID

WHITE BOG ORCHID

PALE-GREEN ORCHID

White Bog Orchid

Platanthera dilatata (Pursh) Lindl. ex L.C. Beck
SYNONYM *Habenaria dilatata* (Pursh) Hook.

Leafy-stemmed perennial, up to 1 m tall. Floral bracts usually incurved against the stem, giving the raceme a slender, wandlike appearance. **Flowers** usually milk-white but sometimes yellowish or greenish-white, the fragrance strong and spicy, somewhat like that of cloves or vanilla. Petals and sepals soft, the lip dilated at the base and projecting outward; spur about the same length as the lip. **Leaves** narrow, up to 30 cm long and 5.5 cm wide.

HABITAT In moist meadows, swamps, bogs, marshes, and wet forests; on beaches and along streams.

FLOWERING late May to early September.

Pale-Green Orchid

Platanthera flava (L.) Lindl.
SYNONYM *Habenaria flava* var. *herbiola* (R. Br.) Ames & Correll

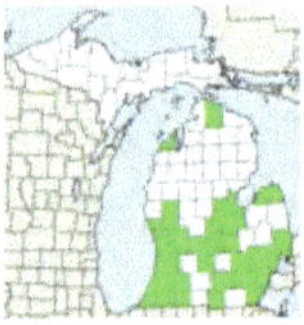

Rather stout perennial with 2–5 stem leaves, distinguished from the preceding species by the yellowish-green flowers, which have a tongue- shaped lip with a small tooth on each side at the base and a central tubercle near the base.

HABITAT In similar habitats and flowering at about the same time.

Lake Huron Green Orchid

Platanthera huronensis (Nutt.) Lindl.
SYNONYM *Habenaria hyperborea* var. *huronensis* (Nutt.) Farw.

Somewhat fleshy and coarse leafy-stemmed perennial, up to 1 m tall. **Flowers** greenish, often faintly fragrant, about 6 mm long, borne in a dense (rarely loose) raceme, the lower floral bracts longer than the flowers. Perianth herbaceous, the upper sepal united to the petals to form a hood; lip not dilated at the base, upturned or projecting forward, lanceolate, blunt and entire at apex, the spur slender, usually shorter than or about equaling the lip and shorter than the plump ovary, curving outward. **Leaves** alternate, decreasing in size upward, the lower leaves linear-lanceolate, tapering to a point at the apex and to the sheathing base, up to 20 cm long and 5 cm broad, veins obscure.

HABITAT In wet meadows, shady places, margins of beach pools, marshes, swamps, cedar bogs, and roadsides.

FLOWERING late June to September.

Blunt-Leaf Orchid

Platanthera obtusata (Banks ex Pursh) Lindl.

SYNONYM *Habenaria obtusata* (Pursh) Richards.

Slender, unbranching perennial with 4-angled leafless stem, up to 3.8 dm tall. **Flowers** greenish-white, about 1 cm long, 3–15 borne on very short pedicels in a loose, slender terminal raceme. Lateral petals curving upward, the upper sepal large, forming a hood; lip entire, tapering, the spur long, curving downward, the lower sepals turned back. **Leaf** typically solitary, basal, obovate to oblanceolate, rounded or blunt at apex, tapering below to the sheathing base, up to 13 cm long; occasionally one stem leaf present.

HABITAT In deep shade in cold boggy or mossy forests; in cedar, tamarack, or balsam-spruce swamps.

FLOWERING late June to mid-August.

Round-Leaf Orchid

Platanthera orbiculata (Pursh) Lindl.

SYNONYM *Habenaria orbiculata* (Pursh) Torr.

Unbranched perennial with bracted flowering stem up to 5 dm tall above pair of basal leaves. **Flowers** greenish-white, borne on short, slender pedicels in a loose terminal raceme 2.5–7.5 cm in diameter and up to 30 cm, long. Lateral petals lanceolate, the lip narrowly lanceolate with a blunt tip, bent abruptly downward, the spur about twice as long as the lip, slender or somewhat club-shaped, incurved toward the apex. **Leaves** growing flat against the ground, nearly round, 7–20 cm in diameter, smooth and shining green above, silvery beneath, rather thick and fleshy, the veins numerous and quite conspicuous.

HABITAT Under hardwoods, in mixed woods or coniferous forests, on dunes and in rich, sandy or swampy ground.

FLOWERING July to mid-September.

LAKE HURON GREEN ORCHID

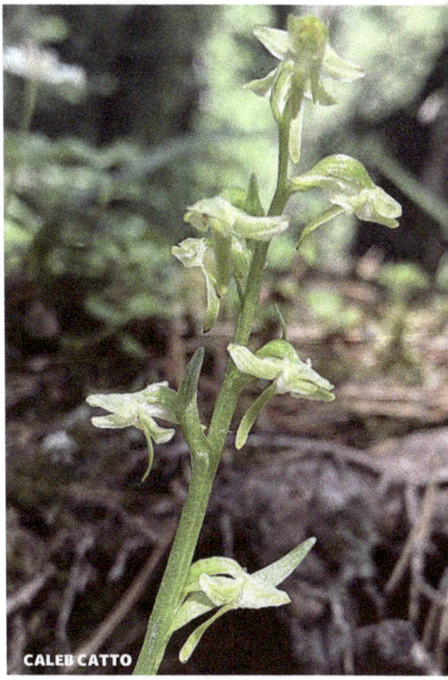

BLUNT-LEAF ORCHID

ROUND-LEAF ORCHID

NOTES A much larger plant with whiter flowers having a very long, drooping spur is sometimes treated as ***Platanthera macrophylla*** (Goldie) P.M. Brown; by other authors it is considered merely a large form of *P. orbiculata* [***Platanthera orbiculata*** **var.** ***macrophylla*** (Goldie) Luer].

Hooker's Orchid

Platanthera hookeri (Torr.) Lindl.

SYNONYM *Habenaria hookeri* Torr.

Greatly resembling **Round-Leaf Orchid**, this species may usually be distinguished by the absence of bracts on the stem, by the more yellow cast to the flowers, and by the tapering of the spur.

HABITAT In dryish woods or sometimes in swampy places.

FLOWERING June and July.

Lesser Purple Fringed Orchid

Platanthera psycodes (L.) Lindl.

SYNONYM *Habenaria psycodes* (L.) Spreng.

Slender, unbranched, leafy-stemmed perennial, up to 9 dm tall. **Flowers** numerous (up to 80), fragrant, lilac-purple to deep rose, pink, or white, crowded in cylindrical, terminal racemes 5–20 cm long and 2.5–5 cm thick. Petals and upper sepal entire or finely toothed on the margin, the lip broad, usually cleft nearly to the

HOOKER'S ORCHID

LESSER PURPLE FRINGED ORCHID

base into 3 finely fringed, fan-shaped divisions; spur about equaling or exceeding the ovary in length; capsule long and slender. **Leaves** alternate, oblong-lanceolate or elliptic, 7–25 cm long, 1.5–7.5 cm wide (the upper ones smaller); margins entire, base sheathing.

HABITAT In wet meadows, swamps, marshes, the margins of beach pools, along wet roadsides, at the edge of woods and thickets, and in open hardwood forests.

FLOWERING mid-July through August.

RAGGED FRINGED ORCHID

Ragged Fringed Orchid

Platanthera lacera (Michx.) G.Don

SYNONYM *Habenaria lacera* (Michx.) Lodd.

As in the preceding species, the lip 3-cleft nearly to the base; the 3 divisions even more deeply fringed with long, threadlike segments, the **flowers** fragrant, yellowish-green, sordid, or bronzy. The plants grow 2–8 dm tall and have 4–9 leaves.

HABITAT In open swamps and marshes, dry to wet meadows, wet open woods, sometimes in dryish woods and clearings.

FLOWERING in July and early August.

ROSE POGONIA

Rose Pogonia

Pogonia ophioglossoides (L.) Ker-Gawl.

Slender, erect perennial up to 6 dm tall, usually with a single leaflike bract just below the flower, and a leaf about the middle of the stem. **Flowers** usually solitary but occasionally 2, showy, fragrant, pale to deep pink or rose, 1–2 cm long (exclusive of the ovary). Petals arching over the column; lip broadly wedge-shaped or spatulate, having 3 longitudinal rows of short, fleshy, yellow- or brown-tipped hairs, the margin lacerate; sepals spreading, column club-shaped, arching forward. **Stem leaf** one (occasionally 2-3), narrow, lanceolate or obovate, up to 6 cm long, ascending close to the stem, sessile and clasping, 3–5-ribbed; **basal leaf** (not often seen) lanceolate, oblong or narrowly obovate, long-petioled.

HABITAT In open sunny bogs and peaty swales, on wet mossy shores, grassy flats, low wet open woods and old beach pools along the Great Lakes.

FLOWERING mid-June to mid-August.

NOTES This beautiful orchid is frequently found growing abundantly with Grass Pink, Pitcher-plant (*Sarracenia*), and cranberries in open sphagnum bogs.

KEY TO SPIRANTHES (LADIES'-TRESSES) SPECIES

1 Leaves linear to lanceolate, sheathing; flowers in several longitudinal rows 2

1 Leaves ovate to elliptic; flowers in a single, twisted longitudinal row .... *S. lacera*

2 Flowers in spiral or nearly straight longitudinal rows; lip tongue-shaped or only slightly narrowed about middle; flowers nodding .... *S. arcisepala*

2 Flowers in compact spiral rows; lip decidedly narrowed near the middle; flowers ascending .... *S. romanzoffiana*

SPHINX LADIES'-TRESSES

SHINING LADIES'-TRESSES

Sphinx Ladies'-Tresses

Spiranthes incurva (Jenn.) M.C. Pace

SYNONYM *Spiranthes cernua* auct. non (L.) L.C. Rich.

Perennial 1–6 dm tall. **Flowers** small, creamy or white, usually with a vanilla-like odor, in 2–4 spiral or nearly vertical rows forming a rather compact spike. Perianth downy, the upper sepal and petals somewhat united at the base to form the hood, the lateral sepals free, spreading; lip tongue-shaped, with wrinkled or slightly eroded margin at the apex and 2 small basal callouses. **Leaves** few, mostly basal and upstanding, linear to linear-lanceolate, petioled, sheathing at the base, pale green.

HABITAT In bogs, swamps, marshes, boggy edges of lakes and streams, wet fields, meadows, sand dunes and low woods; Michigan's most common *Spiranthes*.

FLOWERING August to early October.

SIMILAR SPECIES **Shining Ladies'-Tresses**, *Spiranthes lucida* (H.H. Eat.) Ames, has a yellow lip and dark-green, oblong-elliptic leaves.

Slender Ladies'-Tresses

Spiranthes lacera (Raf.) Raf.

Stiff, erect perennial 1–5 dm tall. **Flowers** white, small, borne in a single twisted row on one-sided spikes 2–5 cm long. Sepals and petals united into a forward-pointing tube, the lip trough-like and flaring at end, green except at tip. **Leaves** basal, thin, ovate to elliptic, usually withered at flowering time.

HABITAT In sand under jack-pines or aspen, in dry or moist peaty meadows, dune hollows, barren fields, or thickets.

FLOWERING mid-July to early September.

Hooded Ladies'-Tresses

Spiranthes romanzoffiana Cham.

Stiff, erect perennial, the stem up to 5 dm tall, sometimes leafy below, bracted above. **Flowers** white, creamy, or straw-colored, fragrant, about 7 mm long, ascending, borne in 3 spiral rows in a compact spike, each flower subtended by a leafy bract which may be longer than the flower. Sepals and petals joined to form an upward arching hood; lip, when spread open, constricted below the apex and appearing fiddle-shaped. **Leaves** mostly basal, linear-lanceolate, pointed at apex, the lowest usually petioled, clasping at the base.

HABITAT On wet marly lake shores and slopes, in rich damp meadows, sandy edges of bogs and marshes, old beach pools, swamps in dune areas, and thickets,

FLOWERING mid-July through September.

SLENDER LADIES'-TRESSES

HOODED LADIES'-TRESSES

PONTEDERIACEAE *Pickerelweed Family*

A small family of aquatic or marsh perennials. There are only 2 genera (both found in Michigan) and about 40 species. The members of this family are of some importance as ornamentals. Like the introduced **Water Hyacinth**, *Pontederia crassipes* Mart., Pickerelweed is often grown in pools. Unlike Pickerelweed, however, the Water Hyacinth becomes a pest in the southeast (and in many warm regions of the world), where it grows so profusely that it often clogs waterways, and should never be planted.

PICKERELWEED

Pickerelweed

Pontederia cordata L.

Creeping perennial with stout, erect flowering stems having a single well-developed leaf and several sheathing bractlike basal leaves; the rootstock thick. **Flowers** lasting a very short time, intensely blue, rarely white, about 5 mm long, borne in an erect, dense, terminal spike which has a sheathing bract at the base; perianth irregular, funnel-shaped, 2-lipped, the 3-lobed upper lip with a pair of yellow spots, the 3 divisions of the lower lip spreading, their claws forming the lower part of the curving tube and more or less separate to the base, the tube withering above and becoming coiled, the base hardening around the fruit; stamens 6, some of them often sterile or imperfect, blue to purple; ovary superior, 3-celled, the mature **fruit** 1-seeded, with 6 toothed ridges. **Leaves** heart-shaped-ovate, the veins parallel, the sheath longer than the petiole and completely enclosing the stem, the margin wavy or with small rounded teeth.

HABITAT In shallow water or on muddy shores.

FLOWERING June to September.

TYPHACEAE *Cat-Tail Family*

This cosmopolitan family contains 2 genera, *Typha* and *Sparganium* (the latter a recent addition to the family), with some 50 species. Cattails are of little economic importance but are sometimes used in mattings and chair bottoms. The rootstocks, young shoots, and young inflorescences are edible. Cattail marshes are among those most favored by muskrats and also provide shelter and nesting cover for birds such as blackbirds and marsh wrens. Geese eat the rootstocks.

Cattails are characterized by erect, sword-like leaves with parallel veins; brown sausage-like terminal spikes of very densely compacted flowers; and by their habitat in water or wet places.

Sparganium are marsh or aquatic perennials with unisexual flowers in compact heads scattered along the upper part of the leafy stems. The hard, bur-like pistillate heads decay slowly and are sometimes found as fossils.

Broad-Fruit Bur-Reed

Sparganium eurycarpum Englm. ex Gray

Stout, erect, simple or branched perennial up to 1.5 m tall. **Flowers** greenish or yellowish, densely crowded into globose heads. Staminate heads numerous, uppermost, the flowers sessile, consisting of 3 stamens and numerous scales; pistillate heads fewer, larger, the perianth of 3–6 scales, the ovary 1-celled and topped by 2 threadlike stigmas; **fruits** sessile, crowded into bur-like heads 1.5–2.5 cm in diameter with prominent pointed beaks. Stems stout, erect or floating, simple or branched. **Leaves** alternate, nearly erect, linear, almost flat, sheathing at their base.

HABITAT In shallow water in sloughs and ponds, in marshes and other very wet ground, chiefly in alkaline or clay soils.

FLOWERING May to August.

USES The seeds are eaten by waterfowl and marsh-birds. Muskrats eat the entire plant.

BROAD-FRUIT BUR-REED

Broad-Leaf Cattail

Typha latifolia L.

Stout, stiff, erect perennial 1–3 m tall. **Flowers** tiny, unisexual, borne in a tall, very dense, cylindrical terminal spike, the staminate flowers on the upper 6–11 cm, the pistillate flowers on the lower 3–20 cm (where the spike is considerably thicker), the spike in fruit 1–3 cm thick, dark brown, with a pebbly appearance under a lens. Perianth of bristles in flowers of both sexes, the staminate flowers with 2–7 stamens, the pistillate with a single, small, short-stalked ovary which develops into a nutlet with copious soft down. Stem round, very stout. **Leaves** stiffly ascending, linear, fleshy, grayish-green or pale green, sheathing at the base, 6–23 mm wide, the lower leaves reduced to sheaths.

BROAD-LEAF CATTAIL

HABITAT In shallow water in marshes, swamps, and edges of rivers, streams, or lakes, and in roadside ditches.

FLOWERING May to August.

USES American Indians used the leaves in basketry and for making waterproof mattings to cover their wigwams (overlapping them like clapboards and stitching them together with fibers of Swamp Milkweed or Indian Hemp). They used the down for dressing burns and other wounds, for filling pillows, and in making comforts for cradles. Both American Indians and early settlers ate the young -shoots and young inflorescences, ground the roots into meal, and used the rootstocks medicinally.

NARROW-LEAF CATTAIL

Narrow-Leaf Cattail

Typha angustifolia L.

Distinguished from the preceding species by the typically longer, more slender spikes, in which the staminate portion is usually distinctly separated from the pistillate; and by the narrower (usually 5–7 mm) leaves. Believed to be introduced in North America, it is less common than *Typha latifolia* but grows in similar habitats.

FLOWERING late May through July.

XYRIDACEAE *Yellow-Eyed-Grass Family*

A small family of 5 genera and nearly 400 species, most of which grow in the tropics; 3 species of *Xyris* occur in Michigan.

NORTHERN YELLOW-EYED-GRASS

Northern Yellow-Eyed-Grass

Xyris montana Ries

Tufted or densely matted, rush-like plants up to 4.5 dm, tall from forking rootstocks. **Flowers** yellow, solitary in the axils of leathery, scalelike, tightly overlapping bracts which form brown, conelike, terminal spikes or heads on naked wiry flowering stems, the spikes 4–10 mm long and 2–7 mm in diameter; petals 3, yellow, soon falling; sepals 3, the lateral ones with brown or purplish tips extending beyond the bracts, the third sepal keeled, enclosing the corolla in bud and falling off with it; fertile stamens 3, inserted on the claws of the petal and alternating with the 3 sterile, bearded filaments; **fruit** a many-seeded capsule. **Leaves** sheathing the base of the stem, grasslike, less than half as long as the stem and up to 2.5 mm broad.

HABITAT In wet peat and sand.

FLOWERING July to early September.

FAMILY KEYS

A few plant families are easily recognized on sight. For example, the **Brassicaceae,** or **Mustard Family,** with its cross-shaped flowers and 6 stamens of unequal length, is readily recognized. However, many plants offer more difficulty, and help is needed in learning to identify them. Even if one knows the family or genus to which a specimen belongs, there may be many similar species, and some quicker method of identification is needed than that of reading through many descriptions. A botanical "key" offers such a method.

The process of "running down" a specimen in a key consists, first, of selecting from a pair of identically numbered statements the one applicable to the specimen at hand. When this choice has been made, the leaders to the right of the statement indicate either a name—in which instance identification is made—or another number. If it is a number, reference is made to the pair of statements bearing that number. Of this pair, the one best fitted to the specimen is again selected. This process is continued until selection leads to a name. Reference is then made to the description and illustration of the species, and the specimen is checked to see if the identification is correct.

If further attempts do not succeed, either the species is not included in this book or an error in judgment has been made. Since plants vary considerably, it is often difficult to evaluate a character correctly. Nevertheless, though not infallible, a key provides both help and a short-cut in identifying plants.

The keys below include the families and species described in *Michigan Wildflowers,* Remember, though, that the plants described in this book represent less than a fourth of the flowering plants known to occur in Michigan. For identification of the many species not included, the books and websites cited in the **Selected References** (p. 296) may be consulted. See the **Glossary** (p. 287) for definitions of botanical terms.

SECTIONS OF THE KEY

Below are the primary divisions of the keys to the families, genera, and species covered in this book. Keys to the plants within each section begin on page 278.

■ **Monocots: Section A.** Flower parts mostly in 3's or multiples of 3; leaves simple, margins entire, veins usually parallel, extending from base to apex; veins in stem scattered, not appearing in a ring in cross section.

■ **Dicots:** Flower parts mostly in 4's or 5's or multiples thereof; leaves simple or compound, margins entire, toothed, or lobed; veins netted, principal veins either pinnate or palmate in arrangement; veins in stem appearing in a ring in cross section.

Section B. Perianth lacking or consisting of only one (or only one apparent) circle of parts, often colored and/or petallike.

Section C. Calyx and corolla both present, petals distinct and separate, often falling separately. ***Part 1.*** Flowers regular. ***Part 2.*** Flowers irregular.

Section D. Calyx and corolla both present, petals united at least at base, corolla falling as a unit. ***Part 1.*** Flowers regular. ***Part 2.*** Flowers irregular.

■ **Some Monocots and Dicots: Section E.** Flowers several to many, usually small, sessile or nearly so, borne in a structure resembling a single flower: 1, on a spadix subtended by a spathe; 2, inside a cuplike structure; 3, in heads or dense clusters subtended by at least 4 bracts.

SECTION A: MONOCOTS

1 Flowers small, several to many, in a dense, compact, globose or cylindrical inflorescence. 2

1 Flowers often large, individual ones distinct, or, if borne in clusters, not densely compact 5

2 Flower cluster with a distince leaflike or fleshy and colored spathe at base (if flowers blue, see *Pontederia*) .... **ARACEAE, p. 229**

2 Flower cluster lacking a spathe 3

3 Flowers borne in dense, terminal, brown, cylindrical spikes .... ***Typha latifolia*** or ***T. angustifolia*, pp. 275, 276**

3 Not as in alternate choice 4

4 Flowers greenish white, borne in dense, globose heads along stem .... ***Sparganium eurycarpum*, p. 275**

4 Flowers yellow, borne in solitary, terminal, conelike heads ...... ***Xyris montana*, p. 276**

5 Plants rushlike, leaves round in cross section; flowers greenish white .... ***Triglochin maritima*** or ***T. palustris*, pp. 236, 237**

5 Not as in alternate choice 6

6 Sepals green, petals white or colored 7

6 Sepals and petals colored alike 10

7 Leaves 3, in a whorl at top of stem .... ***Trillium*, p. 249**

7 Leaves not whorled 8

8 Flowers blue .... ***Tradescantia ohiensis*, p. 233**

8 Flowers white or sometimes pinkish 9

9 Flowers 1.5-3 cm. wide, of 2 kinds (staminate and pistillate separate), usually 3 in a whorl; leaves usually arrow-shaped .... ***Sagittaria latifolia*, p. 228**

9 Flowers smaller, perfect, leaves never arrow-shaped .... ***Alisma subcordatum*, p. 228**

10 Flowers rose, in an umbel, leaves long and narrow .... ***Butomus umbellatus*, p. 232**

10 Not as in alternate choice 11

11 Flowers regular 12

11 Flowers irregular 15

12 Ovary superior .... **LILIACEAE (and *Lily-like group*), p. 237**

12 Ovary inferior 13

13 Petals and sepals similar in size, shape, and position; flowers not over 2.5 cm broad ... 14

13 Petals and sepals quite different in size and/or shape, petals erect, sepals hanging down; flowers 3 cm or more broad .... ***Iris versicolor*** or ***I. lacustris*, p. 234**

14 Flowers yellow .... ***Hypoxis hirsuta*, p. 244**

14 Flowers blue to whitish .... ***Sisyrinchium albidum*** or ***S. angustifolium*, pp. 235, 236**

15 Flowers bright blue, leaves elongate heart-shaped .... ***Pontederia cordata*, p. 274**

15 Not as in alternate choice .... **ORCHIDACEAE, p. 252**

SECTION B: DICOTS

Perianth lacking or, actually or apparently, consisting of only one circle of parts.

1 Leaves simple and not deeply cut 2
1 Leaves compound or cut nearly to base or midrib 13
  2 Nodes of stem covered by tubular sheaths . . . . **POLYGONACEAE, p. 182**
  2 Nodes not covered by tubular sheaths 3
3 Plants stemless or creeping 4
3 Plants having erect leafy stems . . . .
  4 Stems creeping; flowers brown-purple, bell-shaped, 3-lobed . ***Asarum canadense*, p. 41**
  4 Plants stemless; flowers of separate white, pink, or blue sepals; leaves 3-lobed . . . . ***Hepatica*, p. 196**
5 Juice of plant milky . . . . ***Euphorbia*, p. 120**
5 Juice of plant watery 6
  6 Leaves mostly in whorls . . . . ***Galium*, p. 211**
  6 Leaves not in whorls 7
7 Leaves opposite; stem swollen at nodes 8
7 Leaves alternate or scattered 9
  8 Flowers pink or magenta; leaves with definite blade and petiole . . . . ***Mirabilis nyctaginea*, p. 158**
  8 Flowers usually not pink; if pink, leaves apparently sessile **CARYOPHYLLACEAE, p. 100**
9 Flowers white or pink 10
9 Flowers greenish, red, or yellow 12
  10 Flower clusters opposite the leaves . . . . ***Phytolacca americana*, p. 174**
  10 Flower clusters at end of stem 11
11 Leaves usually 2-3, heart-shaped . . . . ***Maianthemum canadense*, p. 239**
11 Leaves numerous, elliptic . . . . ***Comandra umbellata*, p. 214**
  12 Flowers bright yellow, showy . . . . ***Caltha palustris*, p. 194**
  12 Flowers greenish or red, tiny, in dense clusters . . . . ***Blitum capitatum* or *Chenopodium album*, p. 25**
13 Plant a vine . . . . ***Clematis virginiana* or *C. occidentalis*, p. 195**
13 Plant an erect herb 14
  14 Flowers in a many-branched terminal corymb; stamens 3; leaves opposite . . . . ***Valeriana uliginosa*, p. 99**
  14 Flowers in umbels; stamens 5; leaves alternate . . . . **APIACEAE, p. 26**

SECTION C: DICOTS

Sepals present; petals present, distinct from each other and usually falling separately.

Part 1. Flowers Regular

1 Leaves, flowers, or both, floating; or the plants insectivorous, with hollow, pitcherlike leaves or leaves covered with sticky glands 2
1 Not as in alternate choice 4
  2 Leaves or flowers floating . . . . **NYMPHAEACEAE, p. 159**
  2 Not as in alternate choice 3
3 Flowers large, showy, solitary . . . . ***Sarracenia purpurea*, p. 214**
3 Flowers small, several on a coiled stalk . . . . ***Drosera*, p. 110**
  4 Flowers small, numerous, borne in an umbel 5
  4 Not as in alternate choice 7

5 Plants climbing by tendrils .... *Smilax lasioneuraon*, p. 251
5 Plants not climbing, lacking tendrils........ 6
  6 Fruit dry, splitting into 2 seedlike bodies; petioles enlarged and sheathing at base .... APIACEAE, p. 26
6 Fruit a berry; petioles not sheathing .... ARALIACEAE, p. 38
7 Stem, leaves, or both, thick and fleshy; leaves lacking, scalelike, or thick and fleshy; plants green or not........ 8
7 Not as in alternate choice 10
  8 Plants not green; stem having small scales in place of leaves .... ERICACEAE, p. 111
  8 Plants green........ 9
9 Stem enlarged into fleshy, padlike sections, usually leafless, bearing prickles or spines; flowers large and showy .... *Opuntia cespitosa*, p. 93
9 Stem bearing fleshy leaves and numerous small flowers up to 1 cm wide .... *Sedum acre* or *Hylotelephium telephium*, pp. 108, 109
  10 Erect shrub, flowers white; lower surface of leaves covered with rusty wool........ *Rhododendron groenlandicum*, p. 118
  10 Not as in alternate choice 11
11 Sepals or calyx lobes 2, often soon falling, sometimes very small........ 12
11 Sepals or calyx lobes more than 2 15
  12 Juice of plant colored .... PAPAVERACEAE, p. 170
  12 Juice of plant not colored........ 13
13 Leaves 2, opposite; petals 5 .... *Claytonia virginica* or *C. caroliniana*, p. 157
13 Leaves usually more than 2; petals 2 14
  14 Leaves opposite .... *Circaea canadensis* or *C. alpina*, p. 161
  14 Leaves alternate .... *Maianthemum canadense*, p. 239
15 Sepals 3 16
15 Sepals or calyx lobes more than 3 17
  16 Leaves 3, borne in a whorl at top of stem; flowers large and showy.... *Trillium*, p. 249
  16 Leaves more than 3, mostly in a basal rosette; flowers tiny... *Rumex acetosella*, p. 185
17 Sepals or calyx lobes and petals 4 18
17 Sepals or calyx lobes and/or petals more than 4 20
  18 Stamens 6, 2 shorter than the other 4; leaves alternate or basal BRASSICACEAE, p. 86
  18 Stamens same number or twice as many as petals; leaves often opposite........ 19
19 Stems swollen at nodes, ovary superior........ CARYOPHYLLACEAE, p. 100
19 Stems not swollen at nodes, ovary inferior .... ONAGRACEAE, p. 161
  20 Leaves having translucent or black dots; stamens clustered in 3-5 groups; petals 5, oblique at ends, rolled lengthwise in bud .... *Hypericum*, p. 139
  20 Not as in alternate choice 21
21 Calyx bearing a band of hooked bristles .. *Agrimonia gryposepala* or *A. parviflora*, p. 203
21 Not as in alternate choice 22
  22 Stamens same number as petals........ 23
  22 Stamens more numerous than petals........ 31
23 Leaves compound or deeply lobed 24
23 Leaves simple 26
  24 Flowers rose to rose-purple........ *Erodium cicutarium*, p. 137
  24 Flowers white or yellowish........ 25
25 Flowers solitary, white .... *Jeffersonia diphylla*, p. 81
25 Flowers in clusters, yellowish to greenish or bronze .... *Caulophyllum thalictroides*, p. 81

26 Leaves opposite 27

26 Leaves alternate, scattered, whorled, or basal 29

27 Stems swollen at nodes .... CARYOPHYLLACEAE, p. 100

27 Stems not swollen at nodes 28

28 Flowers borne in axillary clusters; stems recurved and rooting at tip .... *Decodon verticillatus*, p. 153

28 Flowers in slender, elongate terminal spikes; stems erect ... *Lythrum salicaria*, p. 154

29 Leaves in a basal rosette .... SAXIFRAGACEAE, p. 215

29 Not as in alternate choice 30

30 Flowers blue .... *Linum usitatissimum*, p. 153

30 Not as in alternate choice .... PRIMULACEAE, p. 186

31 Stamens twice as many as petals 32

31 Stamens more than twice as many as petals 39

32 Flowers white, solitary, drooping from fork between two leaves .... *Podophyllum peltatum*, p. 82

32 Not as in alternate choice 33

33 Leaves cloverlike, having 3 distinct leaflets .... *Oxalis*, p. 169

33 Not as in alternate choice 34

34 Leaves opposite or whorled 35

34 Leaves alternate or basal, at most a single pair opposite 37

35 Stem swollen at nodes .... CARYOPHYLLACEAE, p. 100

35 Stem not swollen at nodes 36

36 Flowers borne in axillary clusters; stems recurved and rooting at tip .... *Decodon verticillatus*, p. 153

36 Flowers borne in slender, elongate terminal spikes, stems erect .... *Lythrum salicaria*, p. 154

37 Flowers rose-purple; fruits with a long, pointed beak; leaves deeply lobed or compound .... *Geranium maculatum* or *G. robertianum*, pp. 137, 138

37 Not as in alternate choice 38

38 Leaves often evergreen, rather thick or lustrous; anthers inverted and opening by basal pores .... ERICACEAE, p. 111

38 Leaves thin; anthers opening by slits .... SAXIFRAGACEAE, p. 215

39 Filaments united into a tube at base, stamens numerous, forming a bottle-brushlike structure in center of flower .... MALVACEAE, p. 154

39 Not as in alternate choice 40

40 Sepals unequal in size; pistil only 1 .... *Crocanthemum canadense*, p. 106

40 Not as in alternate choice 41

41 Sepals separate; stamens borne on surface of receptacle; stipules absent .... RANUNCULACEAE, p. 189

41 Sepals united at least at base, calyx often having a row of 5 sepallike bractlets; stamens borne on edge of a cuplike structure or on calyx; stipules often present .... ROSACEAE, p. 202

Part 2. Flowers Irregular

1 Leaves simple 2

1 Leaves compound 5

2 Flowers borne singly 3

2 Flowers borne in clusters 4

3 Flowers hanging horizontally on slender, drooping pedicels; leaves ovate to elliptic .... *Impatiens capensis*, p. 80
3 Flowers on erect peduncles; leaves ovate, cordate, lanceolate, or palmately dissected.......... *Viola*, p. 223
4 Leaves in a basal rosette .... *Goodyera*, p. 262
4 Leaves borne along stem.......... *Polygala*, p. 179
5 Flowers tiny, borne in umbels; petioles enlarged and sheathing at base . **APIACEAE, p. 26**
5 Not as in alternate choice 6
6 Sepals 2, separate.......... **PAPAVERACEAE, p. 170**
6 Sepals united into a 5-lobed calyx .... **FABACEAE, p. 121**

SECTION D: DICOTS

Calyx and corolla both present; petals united at least at base, and falling together.

Part 1. Corolla regular or nearly so

1 Flowers few to many on a common receptacle or disk subtended by involucre of bracts .... **ASTERACEAE, p. 42**
1 Not as in alternate choice 2
2 Juice of plant milky.......... 3
2 Juice of plant not milky.......... 6
3 Plants twining or trailing .... *Convolvulus*, p. 106
3 Plants principally erect, not twining 4
4 Flowers blue .... *Campanula*, p. 94
4 Flowers not blue.......... 5
5 Flowers bell-shaped or tubular, white to pinkish .... *Apocynum androsaemifolium* or *A. cannabinum*, p. 35
5 Not as in alternate choice.......... *Asclepias*, pp. 36–38
6 Plants not green; leaves reduced to scales .... **ERICACEAE, p. 111**
6 Plants having normal green leaves 7
7 Plants shrubby, prostrate, creeping, or vinelike.......... 8
7 Not as in alternate choice.......... 13
8 Flowers white, borne in spherical heads.......... *Cephalanthus occidentalis*, p. 211
8 Not as in alternate choice 9
9 Plants climbing by tendrils.......... *Echinocystis lobata*, p. 109
9 Not as in alternate choice 10
10 Leaves opposite; flowers borne in pairs 11
10 Not as in alternate choice1 2
11 Flowers joined together at base .... *Mitchella repens*, p. 213
11 Flowers separate, drooping from tips of slender stalks .... *Linnaea borealis*, p. 98
12 Flowers violet to purple.......... *Solanum dulcamara*, p. 221
12 Not as in alternate choice .... **ERICACEAE, p. 111**
13 Leaves simple 14
13 Leaves compound or cut nearly to base or midrib.......... 36
14 Flowers orange, having a crownlike circle formed of 5 erect hoods .... *Asclepias tuberosa*, p. 37
14 Not as in alternate choice 15
15 Flowers funnelform 16
15 Not as in alternate choice.......... 17

16 Rank, erect plants .... *Datura stramonium*, p. 219
16 Low, weak plant, often twining at tip .... *Calystegia spithamaea*, p. 107
17 Leaves alternate or scattered 18
17 Leaves opposite, whorled, or in basal rosettes................................ 24
18 Stamens numerous, filaments united at base, resembling a bottle-brush in center of flower .... MALVACEAE, p. 154
18 Not as in alternate choice 19
19 Flower cluster coiled when young, flowers borne mostly on one side 20
19 Not as in alternate choice 21
20 Leaves palmately lobed and coarsely toothed....... *Hydrophyllum canadense*, p. 138
20 Leaves pinnately veined, entire................................ BORAGINACEAE, p. 83
21 Flowers nearly flat or saucer-shaped; fruit a berry 22
21 Not as in alternate choice 23
22 Flower white with yellow center .... *Leucophysalis grandiflora*, p. 220
22 Flower yellowish with dark center................... *Physalis heterophylla*, p. 220
23 Flowers blue, bell-shaped or wheel-shaped ...... *Campanula*, p. 94, *Palustricodon*, p. 97
23 Not as in alternate choice .... see keys for OROBANCHACEAE, p. 165 .... PLANTAGINACEAE, p. 174 and SCROPHULARIACEAE, p. 218
24 Leaves usually joined together at base; fruit resembling a small, hard tomato .... *Triosteum perfoliatum*, p. 98
24 Not as in alternate choice 25
25 Flowers small, lavender, borne in fairly large, compact, conelike, spiny heads................................ *Dipsacus fullonum*, p. 97
25 Not as in alternate choice 26
26 Leaves having several conspicuous parallel veins extending from base to apex; corolla tubular, its lobes twisted in bud........................... GENTIANACEAE, p. 133
26 Not as in alternate choice 27
27 Flowers pink or magenta, 1-5 in a cuplike involucre; leaf blades heart-shaped at base .... *Mirabilis nyctaginea*, p. 158
27 Not as in alternate choice 28
28 Corolla blue to rose pink, having a slender tube and a 5-lobed limb............... 29
28 Not as in alternate choice................................ 30
29 Leaves borne along stem .... *Phlox divaricata* or *P. pilosa*, p. 178
29 Leaves borne in a basal rosette .... *Primula mistassinica*, p. 189
30 Leaves in whorls along stem .... *Galium*, p. 211
30 Leaves not in whorls along stem 31
31 Ovary inferior; lobes of calyx and corolla .... *Houstonia longifolia*, p. 213
31 Ovary superior; lobes of calyx and corolla generally................................ 32
32 Fruit consisting of 4 seedlike nutlets; ovary deeply 4-lobed or divided 33
32 Fruit a capsule 34
33 Foliage usually aromatic; flowers usually clustered in axils; style attached at base of ovary .... LAMIACEAE, p. 141
33 Foliage not aromatic; flowers blue to roseate, borne in slender, tapering terminal spikes; style attached at top of ovary .... *Verbena*, p. 221
34 Stamens same number as corolla lobes..................... PRIMULACEAE, p. 186
34 Stamens 4, fewer in number than corolla lobes .... *Agalinis purpurea* or *A. tenuifolia*, pp. 165, 166
35 Leaflets 3 36
35 Leaflets more than 3 37

36 Plants of bogs; leaflets narrow, flowers white .... *Menyanthes trifoliata*, p. 156
36 Plants of average soil; leaflets inverted heart-shaped; flowers yellow or pink or white. .... *Oxalis*, p. 169
37 Flowers white; leaves opposite .... *Valeriana uliginosa*, p. 99
37 Flowers blue, leaves alternate 38
38 Leaves pinnately compound .... *Polemonium reptans*, p. 179
38 Leaves palmately compound or lobed .... *Hydrophyllum canadense*, p. 138

Part 2. Corolla Irregular

1 Plants lacking chlorophyll, not green .... OROBANCHACEAE, p. 165
1 Plants having normal green color 2
2 Plants having minute or very finely dissected leaves; growing in water or wet places; flowers yellow .... *Utricularia*, p. 151
2 Not as in alternate choice 3
3 Flowers pealike, having wings and a keel 4
3 Not as in alternate choice 5
4 Leaves compound .... FABACEAE, p. 121
4 Leaves simple .... POLYGALACEAE, p. 179
5 Anthers united into a tube; juice of plant milky .... *Lobelia*, p. 95
5 Anthers separate; juice of plant watery 6
6 Leaves in a basal rosette; flowers borne on a scape 7
6 Leaves borne along stem 8
7 Flowers blue to violet .... *Pinguicula vulgaris*, p. 151
7 Flowers white .... *Goodyera*, p. 262
8 Fruit 1-4 seedlike nutlets 9
8 Fruit a capsule 12
9 Stamens 5; plants covered with bristly hairs; leaves alternate .... *Echium vulgare*, p. 84
9 Stamens 2-4; plants merely hairy to glabrous; leaves opposite 10
10 Flowers borne in distinct, opposite pairs; the elongate nutlet becoming bent down against stem .... *Phryma leptostachya*, p. 173
10 Not as in alternate choice 11
11 Flowers blue, only slightly irregular, borne in terminal spikes; style rising from apex of ovary; foliage not aromatic .... *Verbena*, p. 221
11 Not as in alternate choice; foliage usually aromatic .... LAMIACEAE, p. 141
12 Flowers hanging horizontally from slender pedicels rising from axils of upper leaves; capsule opening explosively .... *Impatiens capensis*, p. 80
12 Not as in alternate choice .... see keys for OROBANCHACEAE, p. 165 .... PLANTAGINACEAE, p. 174 and SCROPHULARIACEAE, p. 218

SECTION E: SOME MONOCOTS AND DICOTS

Flowers several to many, usually small, sessile or nearly so, borne in one of following ways: (1) on a spadix subtended by a spathe; (2) inside a cuplike structure; (3) in heads or clusters subtended by at least 4 bracts the whole structure resembling a single flower.

1 Plants having a milky juice 2
1 Plants not having a milky juice 3
2 Flowers greatly reduced, borne in a cup-shaped structure .... *Euphorbia*, p. 120
2 Flowers of normal structure, numerous in a head, with several green or greenish bracts at base .... ASTERACEAE, p. 42

3 Flowers borne on a spadix 4
3 Flowers not borne on a spadix 5
- 4 Flowers blue . . . . . ***Pontederia cordata*, p. 274**
- 4 Flowers not blue . . . . . **ARACEAE, p. 229**

5 Cluster of flowers surrounded by 4 (rarely 6) large, white, petal-like bracts . . . . . ***Cornus canadensis*, p. 108**
5 Not as in alternate choice 6
- 6 Involucre cuplike, 5-lobed, resembling a calyx; calyx petal-like . . . . . ***Mirabilis nyctaginea*, p. 158**
- 6 Involucre composed of separate bracts; petals present 7

7 Flowers distinctly 2-lipped; foliage aromatic; stems square . . . . . ***Monarda fistulosa*** or ***M. punctata*, p. 146**
7 Flowers not having 2 lips 8
- 8 Heads elongate, flowers interspersed with rigid, pointed bracts; stamens 4, separate . . . . . ***Dipsacus fullonum*, p. 97**
- 8 Not as in alternate choice; stamens united by their anthers . . . . . **ASTERACEAE, p. 42**

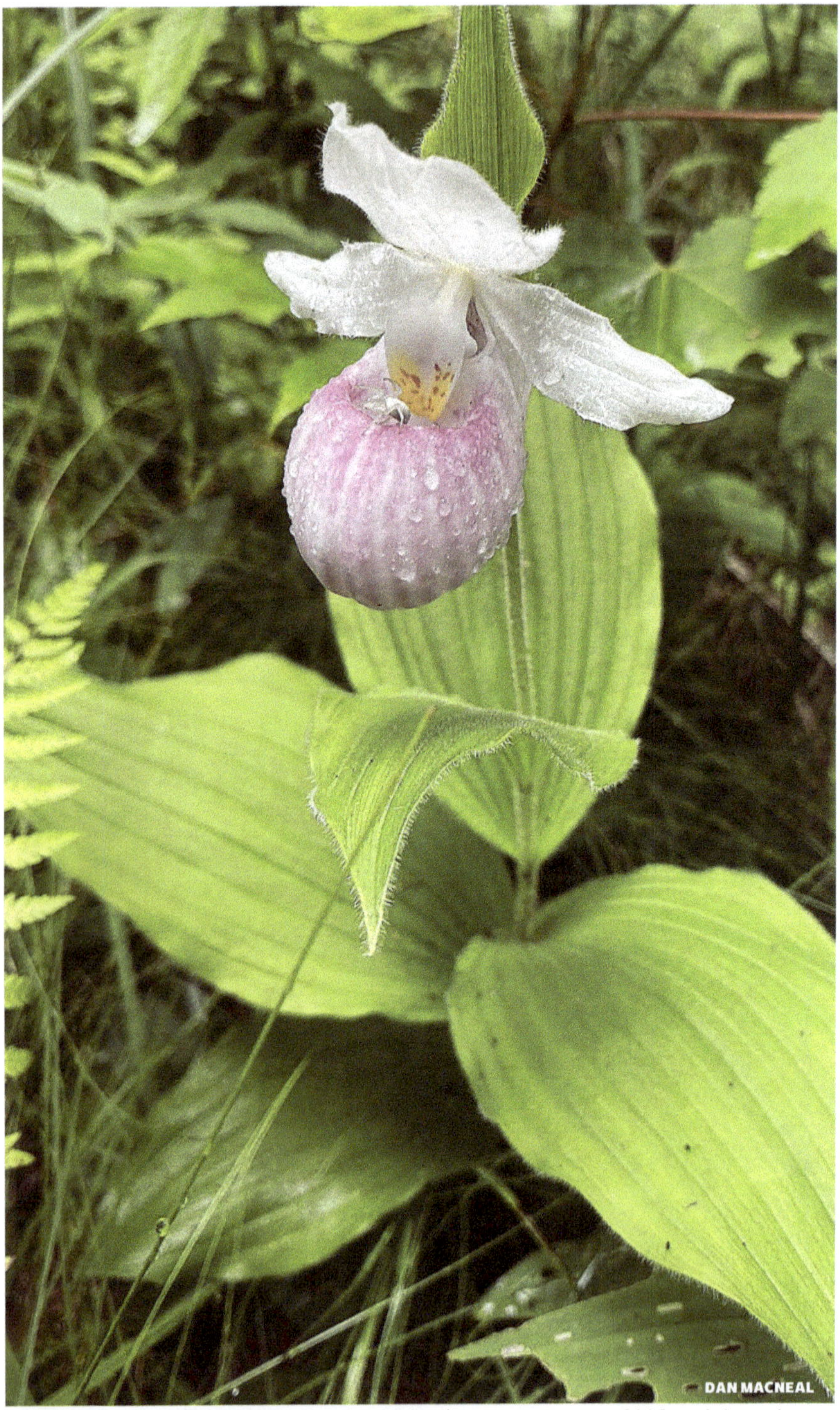

SHOWY LADY'S-SLIPPER (*Cypripedium reginae*), an uncommon orchid found throughout Michigan, usually where somewhat shaded and wet (see p. 261).

Achene. A small, dry, hard, 1-seeded fruit.

Alternate (leaves). Borne singly at a node, first on one side of the stem, then on the other.

Annual. Living a single season.

Anther. The pollen-producing part of the stamen.

Apex. Upper end or tip (of leaf, petal, etc.).

Appressed. Lying flat and close against.

Aril. A fleshy, often bright-colored, appendage to a seed.

Auricle. Ear-shaped appendage or lobe.

Awn. Slender, usually stiff, terminal bristle.

Axil. Angle formed between the leaf and stem, or branch and stem.

Axillary. In an axil.

Beak. A projection ending in an elongated tip.

Bearded. Bearing long or stiff hairs.

Berry. A fleshy fruit having a thin skin or outer covering, the seeds surrounded by the pulp.

Biennial. Living only two years and flowering the second year.

Bisexual. Having both stamens (male organs) and pistils (female organs).

Bract. A small leaf, often scalelike and usually subtending a flower or inflorescence.

Bracted. Having bracts.

Bulb. Underground leaf-bud with fleshy scales.

Calyx. The outer, usually green, portion of the flower.

Capitate. Headlike; arranged in a head, or dense cluster.

Capsule. A dry fruit composed of several cells, opening at maturity.

Carpel. A simple pistil, or one member of a compound pistil.

Chaff. The scales or bracts on the receptacle of many composite flowers.

Chlorophyll. The green coloring matter of plants.

Ciliate. Fringed with hairs.

Clasping. Partly surrounding another structure at the base, e.g., as a leaf surrounds a stem.

Clavate. Club-shaped; gradually thickened upward.

Compound. Composed of 2 or more similar parts united into one whole. **Compound leaf**, one divided into separate leaflets.

Connective. The part of a stamen that joins the two anther cavities.

Cordate. Heart-shaped, the broadest part at the base.

Corm. The enlarged, solid, bulblike base of some stems.

Corolla. The second or inner set of floral parts, composed of separate or united united petals; usually conspicuous by its size and/or color.

Corona. A crownlike structure on inner side of the corolla, as in narcissus, milkweed, etc.

Corymb. A flat-topped or convex flower cluster, the outer flowers opening first.

Crenate. Having much-rounded teeth.

Cuneate. Shaped like a wedge.

Cyathium. Cuplike involucre around the flowers in *Euphorbia*.

Cylindric. Shaped like a cylinder.

Cyme. A convex or flat flower cluster, the central flowers unfolding first.

Decumbent. Stems in a reclining position, but with the end ascending.

Decurrent. Extending downward (applied to leaves in which the blade is extended as wings along the petiole).

Dentate. With sharp teeth.

Diadelphous (referring to stamens). Combined in two, often unequal, sets.

Dicotyledon or Dicot. Plant with 2 seed-leaves or cotyledons. In practice usually recognized by the net-veined leaves, the parts of the flower in 4's or 5's, the fibers of the stem in a ring.

Disk. Central part of the head of many composite flowers.

Disk-flower. In composite flowers, the tubular flowers in the center of the head, or comprising the head.

Dissected. Cut or divided into narrow segments.

Drupe. A fleshy fruit with a single hard stone in the center, such as the peach, plum, or cherry.

Ellipsoid, Elliptic. Oval in outline, widest at the middle and narrowed about equally at the ends.

Entire. With a continuous, unbroken margin; without teeth, scallops, or lobes.

Fertile. Capable of normal reproductive functions. A fertile anther produces pollen; a fertile flower produces fruit.

Filament. The stalk of the stamen.

Filamentous. Threadlike.

Flexuous. Curved alternately in opposite directions.

Foliate. Having leaves, leaved, as 3-leaved.

Foliolate. Having leaflets.

Follicle. A dry fruit developed from a single ovary, opening by a slit along one side.

Fruit. The seed-bearing product of a plant; the ripened ovary with such other parts as may be attached to it.

Glabrous. Smooth, in the sense of lacking hairs (not necessarily smooth to the touch).

Glaucous. Covered with a fine bluish or whitish bloom.

Globose, globular. Spherical or nearly so.

Glutinous. Gluelike, sticky.

Hastate. Shaped like an arrowhead, but with the basal lobes pointed outward.

Herbaceous. Soft, not woody; leaflike in color or texture.

Hip. The fleshy, ripened fruit of the rose.

Hood. A cap-shaped or hood-shaped structure formed by the sepals and petals in orchids.

Horn. An incurved pointed projection.

Hypanthium. A cuplike receptacle that bears the flower parts on its upper margin.

Inferior. Lower or below (an inferior ovary is one that is seemingly below the calyx segments).

Inflorescence. A flower cluster.

Involucral. Belonging to an involucre.

Involucre. A whorl of small leaves or bracts below a flower or flower cluster.

Irregular (flower). Showing inequality in the size, form, or union of its similar parts.

Keel. The two fused lower petals of the flower of the Pea family.

Keeled. Having a central ridge.

Lanceolate. Shaped like a lance-head (much longer than wide, widest near the base and tapering to the apex).

Leaflet. One of the divisions of a compound leaf.

Legume. A fruit from a simple ovary, opening along 2 sides; e.g., a pea pod.

Ligulate. Furnished with a ligule (applied to the ray flowers in the Aster family).

Ligule. A straplike organ.

Limb. The expanded part of a corolla of united petals.

Lip. The upper or lower division of an irregular corolla or calyx, as in the mints. The characteristic (and apparently lower) petal of an orchid is called the lip.

Lobe. A partial division of a leaf or other structure.

Lyrate. Pinnately cut, with a large, rounded lobe at apex, but with the lower lobes small.

Membranous. Thin and somewhat transparent, like a membrane.

Midrib. The central rib or vein of a leaf.

Monocotyledon or Monocot. A plant with 1 seed-leaf (cotyledon). In practice usually recognized by the parallel veins of the leaves, the flower parts in 3's or 6's, the fibers of the stem scattered (not in a ring).

Naked receptacle. Without scales or bracts.

Nectar. A sweet, often fragrant liquid.

Nectariferous or Nectiferous. Producing or having nectar.

Nectary. Any place or organ where nectar is secreted.

Nerve. A principal, unbranched vein.

Net-veined. Having veins forming a network or reticulum.

Node. A point on a stem where leaves, branches, or buds are attached.

Nutlet. A small hard, 1-seeded fruit.

Ob-. Prefix meaning inverted, e.g., obovate, meaning ovate but with the broadest part toward the apex.

Oblong. Having the length two to three times the width, the sides nearly parallel.

Obtuse. Blunt or rounded at the end.

Opposite (leaves). A pair of leaves arising from the same node and on opposite sides of the stem; of stamens, inserted in front of the petals, and thus opposite or across from them.

Ovary. The part of the pistil that produces seeds.

Ovate. Egg-shaped, the broadest part basal.

Palmate. Having parts diverging from a common base, as the fingers from a hand.

Panicle. A loose, elongate, flower cluster with compound branching.

Papilionaceous. Having a standard, wings, and a keel and resembling a butterfly, as in a pea flower.

Pappus. The calyx in the Asteraceae, consisting of hairs, bristles, awns, or teeth at the top of the ovary or achene.

Parallel (venation), Parallel-veined; having veins rising at base of a leaf and continuing to apex in a nearly parallel manner.

Parasite. A plant that gets its nourishment from another living plant to which it is attached.

Parted. Cut or lobed more than halfway to the middle or base.

Pedicel. The stem of a single flower in a flower cluster.

Peduncle. The stem or stalk of a single flower, or of a cluster of flowers.

Perennial. A plant that is able to live year after year.

Perfect (flower). Having both stamens and pistils.

Perfoliate. Having the stem apparently passing through the leaf.

Perianth. The flower envelope, consisting of the calyx and the corolla (if present).

Petal. One of the parts of the corolla.

Petiole. The stem or stalk of a leaf.

Pinnate. Like a feather, having the parts arranged in two rows along a common axis.

Pistil. The seed-bearing part of the plant, produced in the center of the flower.

Pistillate. Provided with pistils.

Plumose. Featherlike, with fine, soft hairs along the sides.

Pod. A simple, dry fruit which opens when ripe; a legume.

Pollen. The spores (male) produced by the anthers.

Prostrate. Lying flat on the ground.

Pubescent. Hairy, the hairs soft.

Punctate. Covered with dots or pits.

Raceme. A flower cluster with pedicelled flowers borne along a somewhat elongated common axis, the lower flowers opening first.

Ray. One of the branches of an umbel; a strap-shaped flower in a head in which tubular disk flowers are also present.

Receptacle. The end of a flower stalk upon which the flower parts are borne, or the enlarged end of the peduncle upon which the flowers are borne in the Aster family.

Recurved. Curved downward or backward.

Reflexed. Bent downward or backward.

Regular (flower). Having all members of each kind similar in size and shape, e.g., the petals all alike.

Reticulate. Like a network.

Rhombic. Somewhat diamond-shaped.

Rib. A primary or prominent vein of a leaf.

Rootstock. (also called **rhizome**). A prostrate or underground stem, usually rooting at the nodes.

Rosette. A cluster of leaves in circular form, usually at the base of a plant.

Rotate (corolla). Wheel-shaped, having a short tube and widely spreading limb, as in the forget-me-not.

Runner. A horizontal, aboveground stem that may root and develop new plants.

Sagittate. Shaped like an arrowhead.

Salverform. Having a slender tube (relatively long) and widely spreading limb.

Scape. A leafless, flowering stalk arising from the ground or from a very short stem bearing basal leaves.

Seed. A mature ovule which can germinate and give rise to a new plant.

Sepal. One of the parts of the calyx or outer floral envelope.

Sessile. Lacking a stalk or stem.

Sheath. A tubelike part which surrounds another part, as a leaf may sheath a stem.

Shrub. A plant, shorter than a tree, with several to many woody stems which do not usually die at the end of the growing season.

Simple. Not compound; composed of a single piece or unit.

Sinus. The indentation between two lobes.

Solitary. Single.

Sordid. Dirty white, dingy.

Spadix. A spike of flowers on a fleshy axis.

Spathe. A large, often showy bract enclosing or subtending a flower cluster (usually a spadix).

Spatulate. Shaped like a spatula, oblong with a long, somewhat tapering base, the apex rounded.

Spike. An elongated flower cluster with sessile or nearly sessile flowers borne along a common axis.

Spur. A hollow, somewhat pointed projection.

Stamen. The part of the flower that bears the pollen.

Staminate. Having stamens.

Staminode. A sterile stamen; a stamen that does not bear pollen.

Standard. The upper, enlarged petal of a pea flower.

Stem. The major supporting system of a plant to which buds, leaves, and flowers are attached.

Sterile. Unproductive, as a flower without a pistil, or a stamen without an anther.

Stigma. The part of a pistil upon which pollen germinates.

Stolon. A basal branch rooting at the nodes to produce new plants; a runner.

Stoloniferous. Bearing or producing stolons.

Style. The portion of the pistil between the ovary and stigma.

Sub-. Prefix meaning somewhat or slightly, e.g., subcordate, meaning somewhat heart-shaped.

Subtend. To extend under, as a bract or involucre subtends a flower or flower cluster.

Succulent. Juicy or fleshy in texture, usually resistant to drying out.

Superior (ovary). Placed above the point of attachment of the other flower parts.

Tendril. A slender, almost threadlike, twining structure.

Throat. The part in a corolla or calyx where the tube and limb come together.

Tooth, teeth. Small, sharp-pointed projection or projections.

Tuber. A thickened, fleshy, modified, underground stem having numerous buds or eyes, e.g., the Irish potato.

Tubercle. A small, swollen, and usually hardened structure.

Tubular. Shaped like a tube.

Tufted. Forming clumps.

Twining. Winding spirally.

Umbel. A flower cluster with all the pedicels arising from the same point.

Undulate. Having a wavy margin or surface.

Unisexual. Of one sex, either staminate or pistillate only.

Urn-shaped (corolla). Having an enlarged, globular base, narrowed at the neck, and with a small limb.

Vein. A thread of visible fibrovascular (transporting) tissue.

Venation. Character or pattern of the veining.

Whorl. A group of three or more similar parts in a circle around a stem.

Wing. A thin expansion of tissue, e.g., along a petiole; a lateral petal, as in a pea flower.

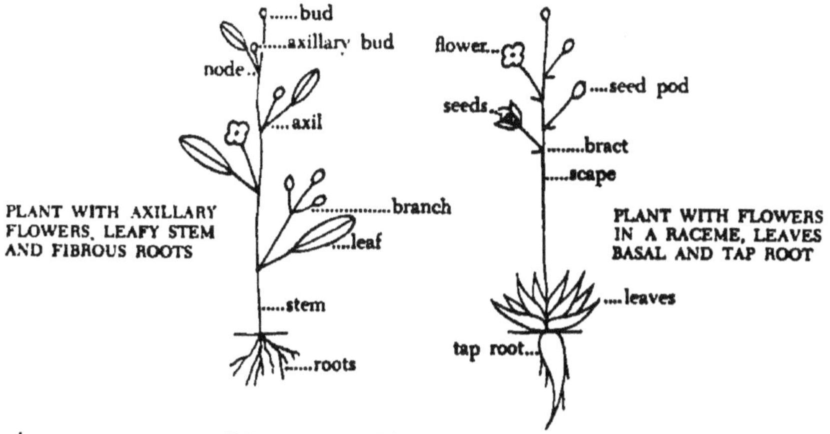

PLANTS AND THEIR PARTS

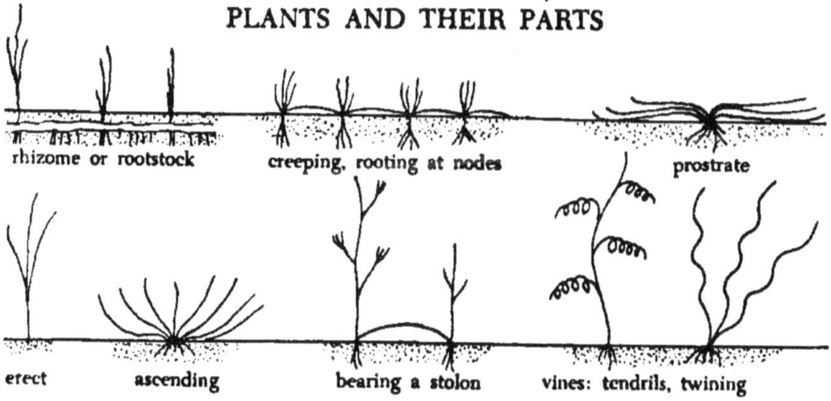

GROWTH PATTERNS OF STEMS

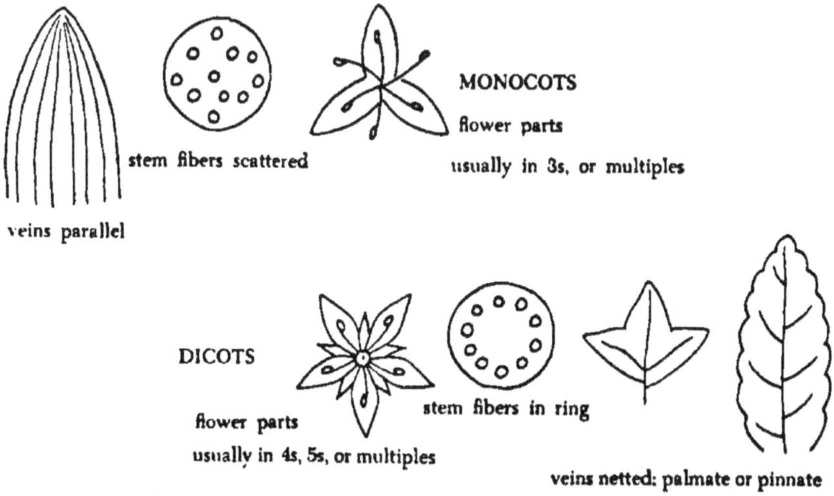

MONOCOT AND DICOT CONTRASTS

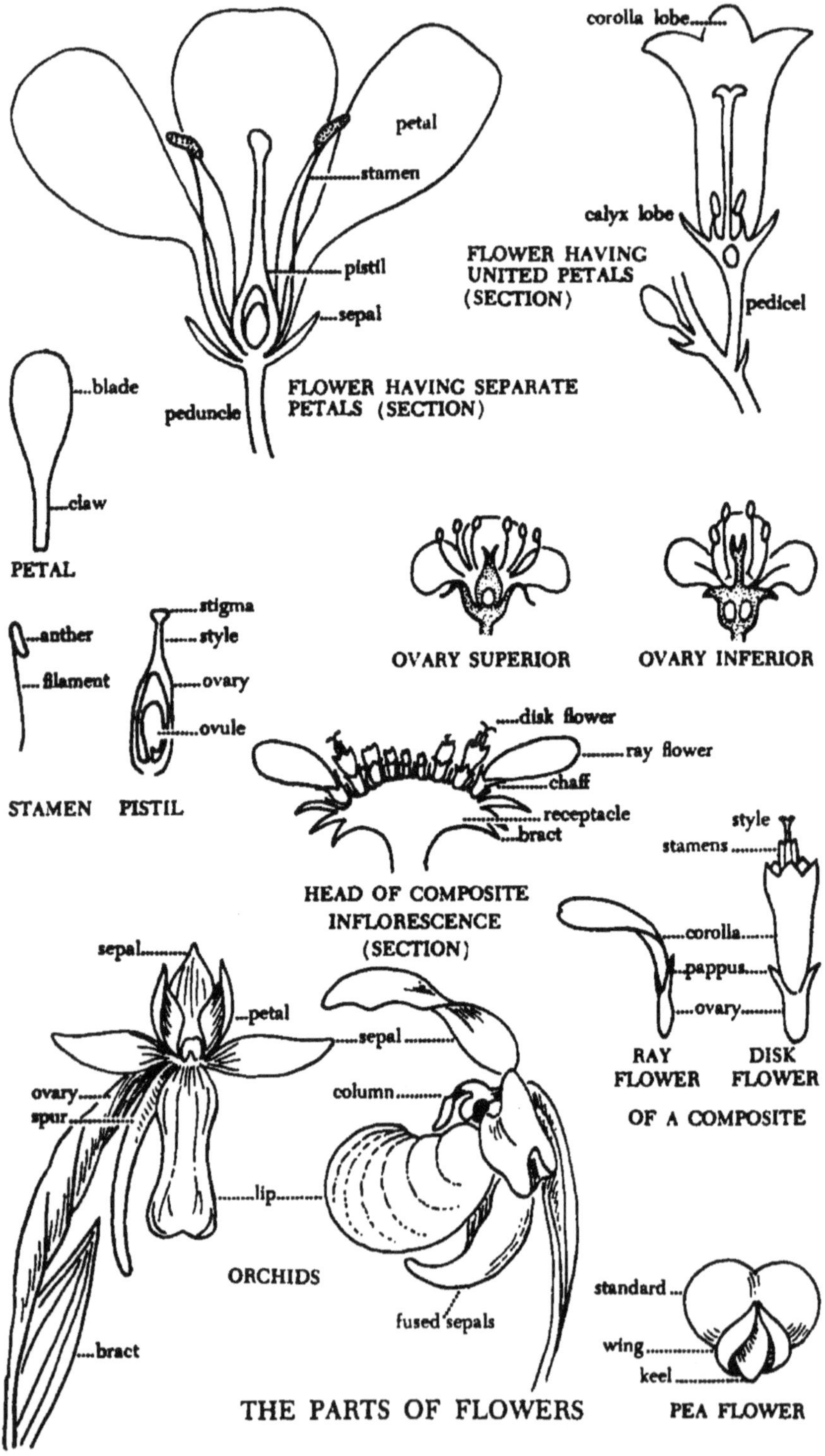
corolla lobe
petal
stamen
pistil
sepal
calyx lobe
FLOWER HAVING UNITED PETALS (SECTION)
pedicel
blade
peduncle
FLOWER HAVING SEPARATE PETALS (SECTION)
claw
PETAL
stigma
anther
style
filament
ovary
ovule
STAMEN
PISTIL
OVARY SUPERIOR
OVARY INFERIOR
disk flower
ray flower
chaff
receptacle
bract
HEAD OF COMPOSITE INFLORESCENCE (SECTION)
style
stamens
corolla
pappus
ovary
RAY FLOWER
DISK FLOWER
OF A COMPOSITE
sepal
petal
sepal
ovary
spur
column
lip
ORCHIDS
fused sepals
bract
standard
wing
keel
PEA FLOWER

THE PARTS OF FLOWERS

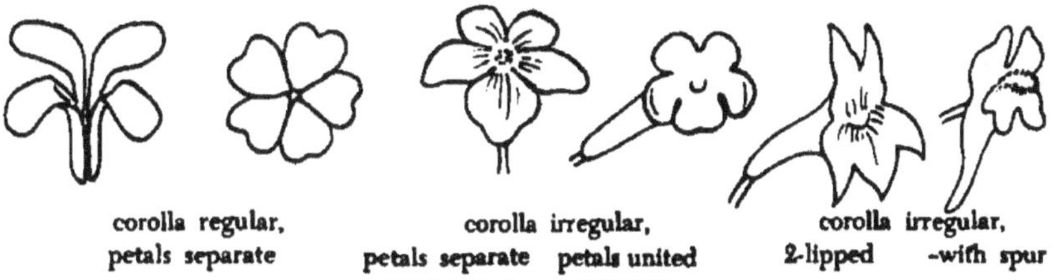

COROLLA TYPES

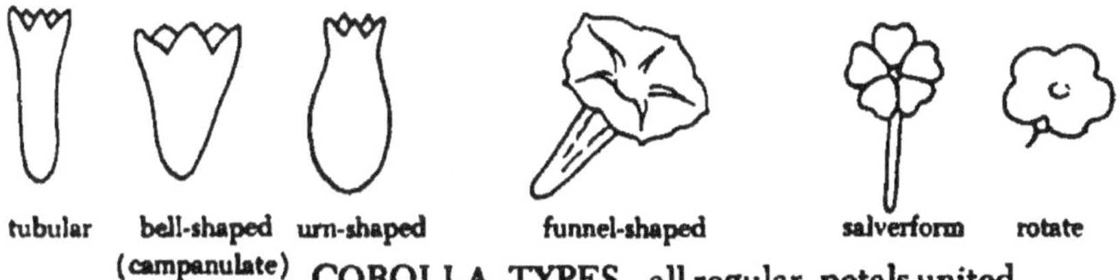

COROLLA TYPES, all regular, petals united

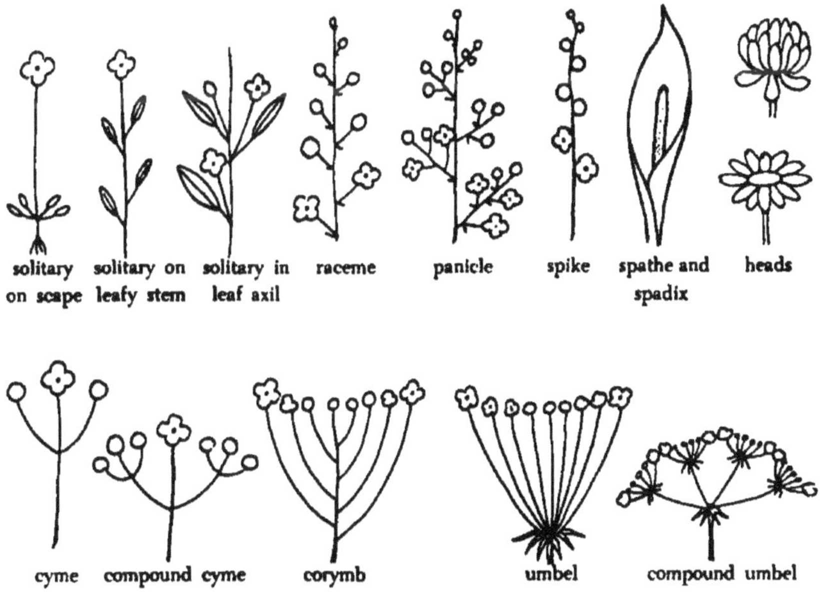

INFLORESCENCE TYPES

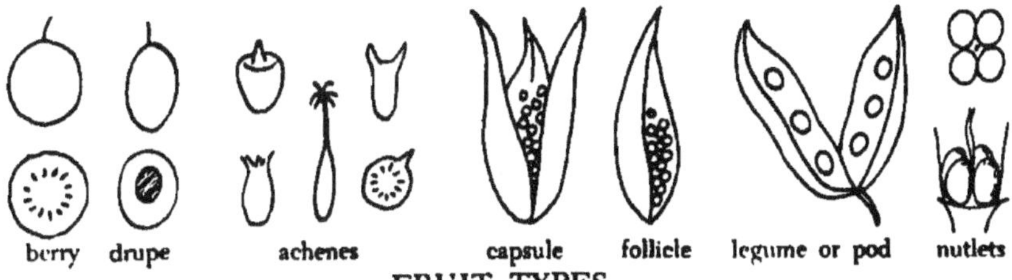

FRUIT TYPES

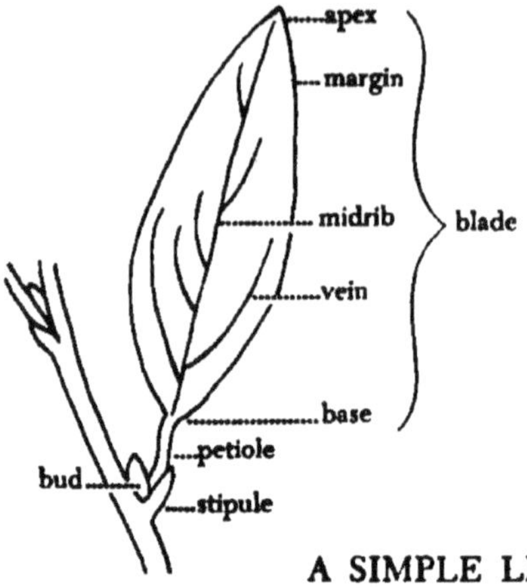

A SIMPLE LEAF

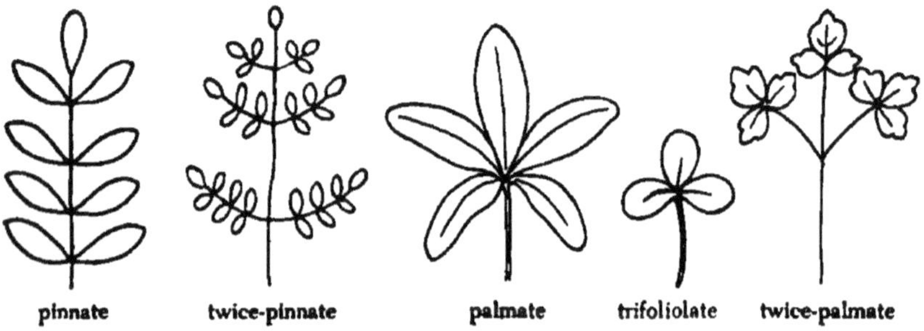

COMPOUND LEAVES

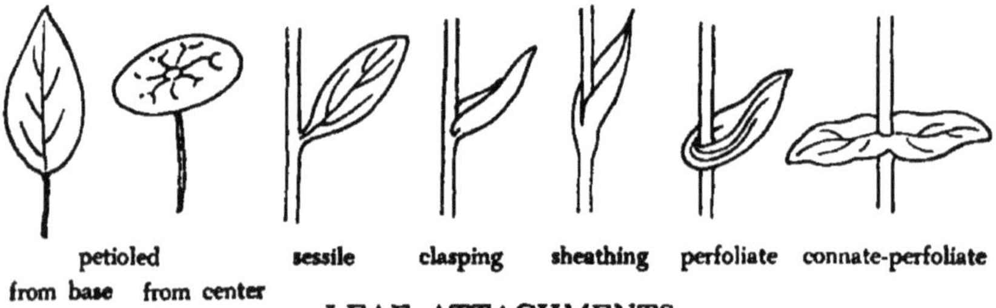

LEAF ATTACHMENTS

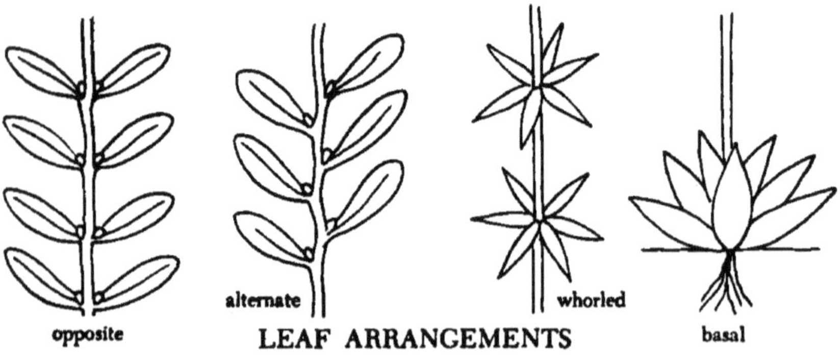

LEAF ARRANGEMENTS

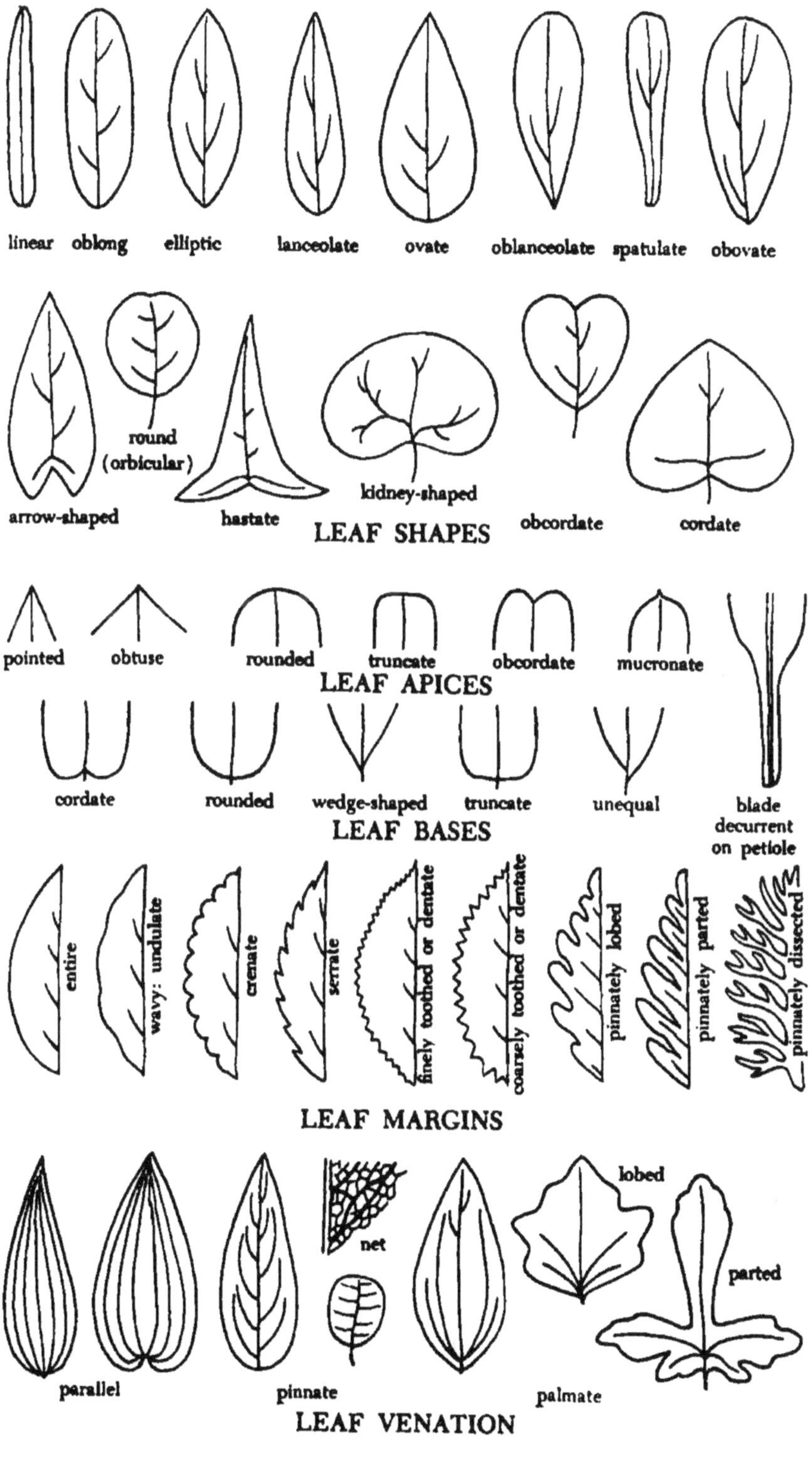
linear
oblong
elliptic
lanceolate
ovate
oblanceolate
spatulate
obovate
arrow-shaped
round
(orbicular)
hastate
kidney-shaped
obcordate
cordate
LEAF SHAPES
pointed
obtuse
rounded
truncate
obcordate
mucronate
LEAF APICES
cordate
rounded
wedge-shaped
truncate
unequal
blade
decurrent
on petiole
LEAF BASES
entire
wavy: undulate
crenate
serrate
finely toothed or dentate
coarsely toothed or dentate
pinnately lobed
pinnately parted
pinnately dissected
LEAF MARGINS
net
lobed
parted
parallel
pinnate
palmate
LEAF VENATION

Books

Chadde, Steve. 2019. *Michigan Flora: Upper Peninsula*. Orchard Innovations. 592 pp.

Chadde, Steve. 2022. *Wetland Plants of the Upper Midwest: A Full Color Field Guide to the Aquatic and Wetland Plants of Michigan, Minnesota and Wisconsin*. Orchard Innovations. 470 pp.

Chadde, Steve & Marjorie T. Bingham. 2022. *Michigan Orchids: An Illustrated Guide to the Wild Orchids of Michigan*. Orchard Innovations. 152 pp.

Voss, E. G. 1972. *Michigan Flora. A Guide to the Identification and Occurrence of the Native and Naturalized Seed-plants of the State. Part I Gymnosperms and Monocots*. Cranbrook Inst. Sci. Bull. 55 & Univ. Michigan Herbarium. xviii + 488 pp.

Voss, E. G. 1985. *Michigan Flora. Part II Dicots (Saururaceae–Cornaceae)*. Cranbrook Inst. Sci. Bull. 59 & Univ. Michigan Herbarium. xx + 724 pp.

Voss, E. G. 1996. *Michigan Flora. Part III Dicots (Pyrolaceae–Compositae)*. Cranbrook Inst. Sci. Bull. 61 & Univ. Michigan Herbarium. xii + 622 pp.

Voss, E. G. & A. A. Reznicek. 2012. *Field Manual of Michigan Flora*. University of Michigan Press, Ann Arbor. xiv + 990 pp.

Online

Michigan Botanical Society. https://michiganbotanicalsociety.org/

Michigan Flora Online. A. A. Reznicek, E. G. Voss, & B. S. Walters. February 2011. University of Michigan. https://www.michiganflora.net

Michigan Natural Features Inventory. https://mnfi.anr.msu.edu/

NOTE Synonyms are listed in *italics*.

Achillea millefolium, 45
Acoraceae, 227
Acorus
- americanus, 227
- calamus, 227
- *calamus var. americanus, 227*

Actaea
- pachypoda, 190
- rubra, 191

Agalinis
- purpurea, 165
- tenuifolia, 166

Ageratina altissima, 45
Agrimonia
- gryposepala, 203
- parviflora, 203

Agrostemma githago, 100
Ajuga
- genevensis, 150
- reptans, 150

Aletris farinosa, 250
Alisma subcordatum, 228
Alismataceae, 228
Alkekengi officinarum, 220
Allium tricoccum, 239
Alyssum
- alyssoides, 87
- *saxatile, 87*

Amaranthaceae, 25
Amaranthus,, 25
Amaryllidaceae, 238
Ambrosia
- artemisiifolia, 46
- *coronopifolia, 46*
- psilostachya, 46
- trifida, 46

Amphicarpaea bracteata, 122
Anaphalis margaritacea, 47
Andersonglossum boreale, 83
Anemonastrum canadense, 191
Anemone
- *canadensis, 191*
- cylindrica, 192
- multifida, 192
- quinquefolia, 193
- virginiana, 192, 193

Anemonella thalictroides, 201
Anemonoides quinquefolia, 193
Angelica
- atropurpurea, 27
- venenosa, 27

Antennaria
- *fallax, 47*
- parlinii, 47

Anthemis
- arvensis, 48
- cotula, 48

Anticlea elegans, 248
Aphyllon uniflorum, 168
Apiaceae, 26
Apios americana, 123
Aplectrum hyemale, 254
Apocynaceae, 34
Apocynum
- androsaemifolium, 35
- cannabinum, 35

Aquilegia
- canadensis, 194
- vulgaris, 194

Arabidopsis lyrata, 87
Arabis
- *canadensis, 88*
- *glabra, 93*
- *lyrata, 87*

Araceae, 229
Aralia
- hispida, 38
- nudicaulis, 39
- racemosa, 39

Araliaceae, 38
Arctium minus, 48
Arctostaphylos uva-ursi, 112
Arenaria stricta, 102
Arethusa bulbosa, 254
Arisaema
- *atrorubens, 230*
- dracontium, 229
- triphyllum, 230

Aristolochiaceae, 41
Arnoglossum atriplicifolium, 48
Artemisia
- *campestris, 49*
- caudata, 49

Asarum canadense, 41
Asclepias
- exaltata, 36
- incarnata, 36
- syriaca, 36
- tuberosa, 37
- viridiflora, 38

Asparagaceae, 239
Asphodelaceae, 242
Aster
cordifolius, 50
ericoides, 50
laevis, 51
lateriflorus, 50
macrophyllus, 49
novae-angliae, 51
puniceus, 51
Asteraceae, 42
Aureolaria
flava, 166
pedicularia, 166
virginica, 166
Aurinia saxatilis, 87

Balsaminaceae, 80
Baptisia
lactea, 124
leucantha, 124
leucophaea, 124
tinctoria, 123
Barbarea vulgaris, 88
Berberidaceae, 81
Berteroa incana, 88
Bidens
cernua, 52
coronata, 52
trichosperma, 52
Blephilia
ciliata, 142
hirsuta, 142
Blitum capitatum, 25
Boraginaceae, 83
Borodinia canadensis, 88
Brasenia schreberi, 159
Brassicaceae, 86
Butomaceae, 232
Butomus umbellatus, 232

Cacalia atriplicifolia, 48
Cactaceae, 93
Cakile edentula, 89
Calla palustris, 230
Calopogon
pulchellus, 255
tuberosus, 255
Caltha palustris, 194
Calypso bulbosa, 255
Calystegia
sepium, 106
spithamaea, 107
Campanula
americana, 95
aparinoides, 97
rotundifolia, 94
Campanulaceae, 94
Campanulastrum americanum, 95
Capnoides sempervirens, 170
Caprifoliaceae, 97
Capsella bursa-pastoris, 89
Cardamine
bulbosa, 89
concatenata, 90
diphylla, 90
douglassii, 90
pennsylvanica, 90
Caryophyllaceae, 100
Cassia fasciculata, 124
Castilleja coccinea, 166
Caulophyllum thalictroides, 81
Celastraceae, 105
Centaurea
cyanus, 53
diffusa, 52
maculosa, 53
stoebe, 53
Cephalanthus occidentalis, 211
Cerastium
arvense, 101
glomeratum, 101
Chamaecrista fasciculata, 124
Chamaenerion angustifolium, 162
Chamaesaracha grandiflora, 220
Chelidonium majus, 172
Chelone glabra, 175
Chenopodium
album, 25
capitatum, 25
Chimaphila
maculata, 113
umbellata, 112
Chrysanthemum leucanthemum, 65
Cichorium
endivia, 53
intybus, 53
Cicuta
bulbifera, 27
maculata, 27
Circaea
alpina, 161
canadensis, 161
quadrisulcata, 161

Cirsium
- arvense, 54
- muticum, 54
- pitcheri, 55
- vulgare, 55

Cistaceae, 106
Claytonia
- caroliniana, 157
- virginica, 157

Clematis
- occidentalis, 195
- *verticillaris, 195*
- virginiana, 195

Clinopodium
- arkansanum, 142
- vulgare, 143

Clintonia borealis, 245
Colchicaceae, 243
Collinsia verna, 175
Collinsonia canadensis, 144
Comandra
- *richardsiana, 214*
- umbellata, 214

Comarum palustre, 203
Commelinaceae, 233
Conium maculatum, 28
Conopholis americana, 167
Convolvulaceae, 106
Convolvulus
- arvensis, 107
- *sepium, 106*
- *spithamaeus, 107*

Coptis
- *groenlandica, 196*
- trifolia, 196

Corallorhiza
- maculata, 256
- odontorhiza, 257
- striata, 257
- trifida, 257

Coreopsis
- lanceolata, 55
- tripteris, 56

Cornaceae, 108
Cornus canadensis, 108
Corydalis
- aurea, 171
- *sempervirens, 170*

Crassulaceae, 108
Crocanthemum canadense, 106
Cryptotaenia canadensis, 29
Cucurbitaceae, 109
Cynoglossum
- *boreale, 83*
- officinale, 83

Cypripedium
- acaule, 258
- arietinum, 259
- *calceolus var. pubescens, 260*
- candidum, 259
- parviflorum, 260
- reginae, 261

Dactylorhiza viridis, 266
Dasiphora fruticosa, 204
Datura stramonium, 219
Daucus carota, 29
Decodon verticillatus, 153
Dentaria
- *diphylla, 90*
- *laciniata, 90*

Desmodium
- canadense, 125
- *nudiflorum, 125*

Dianthus armeria, 101
Dicentra
- canadensis, 171
- cucullaria, 171

Dicotyledons, 25
Dipsacus
- fullonum, 97
- *sylvestris, 97*

Drosera
- anglica, 111
- intermedia, 111
- linearis, 110
- rotundifolia, 110

Droseraceae, 110
Drymocallis arguta, 204

Echinocystis lobata, 109
Echium vulgare, 84
Enemion biternatum, 196
Epifagus virginiana, 167
Epigaea repens, 113
Epilobium
- angustifolium, 162
- ciliatum, 162
- hirsutum, 162
- leptophyllum, 162
- strictum, 162

Ericaceae, 111
Erigenia bulbosa, 29

Erigeron
canadensis, 56
philadelphicus, 57
pulchellus, 57
Erodium cicutarium, 137
Erythranthe
geyeri, 172
michiganensis, 173
moschata, 173
Erythronium
albidum, 246
americanum, 245
Eupatorium
perfoliatum, 57
purpureum, 59
rugosum, 45
Euphorbia
corollata, 120
cyparissias, 120
marginata, 121
Euphorbiaceae, 120
Eurybia macrophylla, 49
Euthamia graminifolia, 58
Eutrochium
maculatum, 58
purpureum, 59

Fabaceae, 121
Fagopyrum
esculentum, 182
sagittatum, 182
Fallopia
ciliinodis, 183
convolvulus, 183
scandens, 183
Fragaria
vesca, 205
virginiana, 204

Galearis spectabilis, 261
Galium
aparine, 212
boreale, 211
lanceolatum, 212
sylvaticum, 212
tinctorium, 212
triflorum, 212
verum, 212
Gaultheria
hispidula, 114
procumbens, 114
Gaura biennis, 164
Gentiana
alba, 133
andrewsii, 134
crinita, 135
flavida, 133
procera, 136
puberula, 134
puberulenta, 134
quinquefolia, 135
rubricaulis, 134
Gentianaceae, 133
Gentianella quinquefolia, 135
Gentianopsis
crinita, 135
virgata, 136
Geraniaceae, 136
Geranium
bicknellii, 138
maculatum, 137
robertianum, 138
Gerardia
flava, 166
paupercula, 165
pedicularia, 166
tenuifolia, 166
virginica, 166
Geum
aleppicum, 207
canadense, 205
fragarioides, 210
laciniatum, 206
macrophyllum, 207
rivale, 206
Glechoma hederacea, 143
Gnaphalium
macounii, 70
obtusifolium, 71
Goodyera
oblongifolia, 262
pubescens, 263
repens, 262
tesselata, 263
Grindelia squarrosa, 59

Habenaria
blephariglottis, 267
clavellata, 267
dilatata, 268
flava, 268
hookeri, 270
hyperborea, 268
lacera, 271

obtusata, 269
orbiculata, 269
psycodes, 270
viridis, 266
Halenia deflexa, 136
Hedeoma pulegioides, 144
Helenium
autumnale, 59
flexuosum, 60
nudiflorum, 60
Helianthemum canadense, 106
Helianthus
annuus, 61
decapetalus, 61
divaricatus, 62
giganteus, 62
maximiliani, 62
occidentalis, 63
tuberosus, 63
Heliopsis helianthoides, 63
Hemerocallis
fulva, 242
lilioasphodelus, 243
Hepatica
acutiloba, 197
americana, 196, 197
Heracleum maximum, 30
Hesperis matronalis, 91
Heuchera richardsonii, 217
Hibiscus
moscheutos, 155
palustris, 155
Hieracium
aurantiacum, 69
florentinum, 70
venosum, 64
Houstonia longifolia, 213
Hydrastis canadensis, 197
Hydrophyllaceae, 138
Hydrophyllum
appendiculatum, 139
canadense, 138
virginianum, 139
Hylodesmum
glutinosum, 125
nudiflorum, 125
Hylotelephium telephium, 108
Hypericaceae, 139
Hypericum
ascyron, 140
fraseri, 139
kalmianum, 140
majus, 140
perforatum, 140
virginicum, 139
Hypopitys americana, 116
Hypoxidaceae, 244
Hypoxis hirsuta, 244

Impatiens
balsamina, 80
capensis, 80
pallida, 80
Inula helenium, 64
Iridaceae, 233
Iris
lacustris, 234
pseudacorus, 235
versicolor, 234
virginica, 235
Isopyrum biternatum, 196

Jeffersonia diphylla, 81
Juncaginaceae, 236

Kalmia polifolia, 115
Krigia virginica, 64
Kummerowia stipulacea, 128

Lactuca
canadensis, 65
sativa, 65
serriola, 65
Lamiaceae, 141
Lathyrus
japonicus, 126
latifolius, 127
ochroleucus, 126
palustris, 127
pratensis, 126
sylvestris, 127
Ledum groenlandicum, 118
Lentibulariaceae, 151
Leonurus cardiaca, 144
Lespedeza
capitata, 127
hirta, 128
violacea, 128
Leucanthemum vulgare, 65
Leucophysalis grandiflora, 220
Liatris
aspera, 66
cylindracea, 66
scariosa, 67
spicata, 67

Liliaceae, 237, 245
Lilium
  michiganense, 246
  philadelphicum, 246
Linaceae, 153
Linaria
  dalmatica, 176
  vulgaris, 175
Linnaea borealis, 98
Linum
  lewisii, 153
  usitatissimum, 153
Liparis loeselii, 263
Listera
  convallarioides, 265
  cordata, 265
Lithospermum
  canescens, 84
  caroliniense, 84
  croceum, 84
Lobelia
  cardinalis, 95
  kalmii, 96
  siphilitica, 96
  spicata, 96
Lobularia maritima, 87
Lupinus
  perennis, 128
  polyphyllus, 129
Lychnis
  alba, 103
  coronaria, 103
Lycopus americanus, 144
Lysimachia
  borealis, 186
  ciliata, 187
  nummularia, 187
  quadrifolia, 188
  terrestris, 188
  thyrsiflora, 188
Lythraceae, 153
Lythrum
  alatum, 154
  salicaria, 154

Maianthemum
  canadense, 239
  racemosum, 240
  stellatum, 240
  trifolium, 241
Malaxis
  brachypoda, 264
  monophyllos, 264
  unifolia, 264
Malva
  moschata, 155
  neglecta, 156
Malvaceae, 154
Matricaria
  discoidea, 67
  matricarioides, 67
Medeola virginiana, 247
Medicago
  lupulina, 129, 130
  sativa, 129
Melampyrum lineare, 168
Melanthiaceae, 248
Melilotus
  albus, 129
  officinalis, 129
Mentha
  arvensis, 145
  x piperita, 145
  spicata, 146
Menyanthaceae, 156
Menyanthes trifoliata, 156
Mertensia virginica, 85
Micranthes pensylvanica, 216
Mimulus
  alatus, 173
  glabratus, 172
  ringens, 173
Mirabilis
  albida, 159
  jalapa, 159
  nyctaginea, 158
Mitchella repens, 213
Mitella
  diphylla, 216
  nuda, 217
Monarda
  citriodora, 147
  didyma, 146
  fistulosa, 146
  punctata, 146
Moneses uniflora, 115
Monocotyledons, 227
Monotropa
  hypopitys, 116
  uniflora, 115
Montiaceae, 157
Myosotis
  laxa, 86
  scorpioides, 85

Nabalus
albus, 68
altissimus, 68
Nanopanax trifolius, 40
Nartheciaceae, 250
Nasturtium officinale, 91
Nelumbo lutea, 158
Nelumbonaceae, 158
Neottia
auriculata, 265
convallarioides, 265
cordata, 265
Nepeta cataria, 147
Nuphar
advena, 159
variegata, 160
Nyctaginaceae, 158
Nymphaea odorata, 160
Nymphaeaceae, 159

Oenothera
biennis, 163
fruticosa, 163
gaura, 164
laciniata, 163
parviflora, 163
perennis, 164
Onagraceae, 161
Oplopanax horridus, 38
Opuntia
cespitosa, 93
humifusa, 93
Orchidaceae, 252
Orchis spectabilis, 261
Orobanchaceae, 165
Orobanche
fasciculata, 168
uniflora, 168
Orthilia secunda, 116
Osmorhiza
claytonii, 30
longistylis, 31
Oxalidaceae, 169
Oxalis
montana, 169
stricta, 170
Oxypolis rigidior, 33

Packera
aurea, 69
paupercula, 69
Palustricodon aparinoides, 97
Panax
quinquefolius, 38, 40
trifolius, 40
Papaveraceae, 170
Parnassia
glauca, 105
parviflora, 105
Pastinaca sativa, 31
Pedicularis
canadensis, 168
lanceolata, 169
Peltandra virginica, 231
Penstemon
calycosus, 176
digitalis, 176
hirsutus, 176
pallidus, 176
Persicaria
amphibia, 183
lapathifolia, 183
maculosa, 184
pensylvanica, 184
sagittata, 184
Phlox
divaricata, 178
paniculata, 178
pilosa, 178
subulata, 178
Phryma leptostachya, 173
Phrymaceae, 172
Physalis
alkekengi, 220
heterophylla, 220
pubescens, 220
Physostegia virginiana, 147
Phytolacca americana, 174
Phytolaccaceae, 174
Pilosella
aurantiaca, 69
piloselloides, 70
Pinguicula vulgaris, 151
Plantaginaceae, 174
Platanthera
blephariglottis, 267
clavellata, 267
dilatata, 268
flava, 268
hookeri, 270
huronensis, 268
lacera, 271
leucophaea, 260
macrophylla, 270

obtusata, 269
orbiculata, 269, 270
psycodes, 270
Podophyllum peltatum, 82
Pogonia ophioglossoides, 271
Polemoniaceae, 178
Polemonium reptans, 179
Polygala
paucifolia, 181
polygama, 180
sanguinea, 180
senega, 180
verticillata, 181
Polygalaceae, 179
Polygaloides paucifolia, 181
Polygonaceae, 182
Polygonatum
biflorum, 242
canaliculatum, 242
pubescens, 241
Polygonella articulata, 185
Polygonum
articulatum, 185
cilinode, 183
lapathifolium, 183
pensylvanicum, 184
persicaria, 184
sagittatum, 184
Pontederia cordata, 274
Pontederiaceae, 274
Potentilla
anserina, 207
argentea, 207
arguta, 204
fruticosa, 204
intermedia, 207
norvegica, 208
palustris, 203
recta, 208
simplex, 208
tridentata, 210
Prenanthes
alba, 68
altissima, 68
Primula
intercedens, 189
mistassinica, 189
Primulaceae, 186
Prunella vulgaris, 148
Pseudognaphalium
macounii, 70
obtusifolium, 71
Pterospora andromedea, 116
Pulsatilla nuttalliana, 193
Pycnanthemum virginianum, 148
Pyrola
americana, 117
asarifolia, 117
chlorantha, 118
elliptica, 117
virens, 118

Ranunculaceae, 189
Ranunculus
abortivus, 198
acris, 198
carolinianus, 199
fascicularis, 199
flabellaris, 200
hispidus, 199
hispidus, 199
longirostris, 200
repens, 199
septentrionalis, 199
Ratibida pinnata, 71
Rhododendron groenlandicum, 118
Rorippa
islandica, 92
palustris, 92
Rosa palustris, 209
Rosaceae, 202
Rubiaceae, 211
Rubus
nutkanus, 209
odoratus, 210
parviflorus, 209
Rudbeckia
hirta, 71
laciniata, 72
triloba, 72
Rumex
acetosella, 185
crispus, 186

Sabulina michauxii, 102
Sagittaria latifolia, 228
Sanguinaria canadensis, 172
Sanicula
canadensis, 31
gregaria, 32
marilandica, 32
odorata, 32
Santalaceae, 214
Saponaria officinalis, 102

Sarracenia purpurea, 214
Sarraceniaceae, 214
Satureja
  arkansana, 142
  hortensis, 143
  vulgaris, 143
Saxifraga pensylvanica, 216
Saxifragaceae, 215
Scrophularia lanceolata, 218
Scrophulariaceae, 218
Scutellaria
  epilobiifolia, 148
  galericulata, 148
  lateriflora, 149
Sedum
  acre, 109
  purpureum, 108
Senecio
  aureus, 69
  pauperculus, 69
Senna hebecarpa, 124
Sibbaldia tridentata, 210
Silene
  antirrhina, 103
  coronaria, 103
  cucubalus, 104
  latifolia, 103
  stellata, 104
  vulgaris, 104
Silphium
  perfoliatum, 72
  terebinthinaceum, 73
Sisymbrium
  altissimum, 92
  officinale, 92
Sisyrinchium
  albidum, 235
  angustifolium, 236
Sium suave, 32
Smilacaceae, 251
Smilacina
  racemosa, 240
  stellata, 240
  trifolia, 241
Smilax
  ecirrhata, 251
  lasioneuron, 251
Solanaceae, 219
Solanum
  dulcamara, 221
  nitidibaccatum, 221
  pseudocapsicum, 221
Solidago
  canadensis, 74
  gillmanii, 74
  graminifolia, 58
  hispida, 74
  juncea, 75
  nemoralis, 75
  racemosa, 74
Sonchus arvensis, 75
Sparganium eurycarpum, 275
Spiranthes
  cernua, 272
  incurva, 272
  lacera, 272
  lucida, 272
  romanzoffiana, 273
Stachys
  hyssopifolia, 150
  palustris, 149
  tenuifolia, 150
Stellaria media, 104
Streptopus
  amplexifolius, 247
  lanceolatus, 247
  roseus, 247
Stylophorum diphyllum, 172
Symphyotrichum
  cordifolium, 50
  ericoides, 50
  laeve, 51
  lanceolatum, 50
  lateriflorum, 50
  novae-angliae, 51
  puniceum, 51
Symplocarpus foetidus, 231

Taenidia integerrima, 33
Tanacetum
  bipinnatum, 76
  huronense, 76
  vulgare, 76
Taraxacum
  erythrospermum, 77
  officinale, 77
Tephrosia virginiana, 124
Teucrium canadense, 150
Thalictrum
  dasycarpum, 201
  dioicum, 200
  thalictroides, 201
Thaspium
  chapmanii, 33
  trifoliatum, 33

Tiarella
cordifolia, 217
stolonifera, 217
Tofieldia
glutinosa, 252
pusilla, 252
Tofieldiaceae, 252
Tradescantia
ohiensis, 233
virginiana, 233
Tragopogon
dubius, 77
major, 77
porrifolius, 78
pratensis, 78
Triantha glutinosa, 252
Trientalis borealis, 186
Trifolium
agrarium, 130
aureum, 130
hybridum, 131
pratense, 130
repens, 131
Triglochin
maritima, 236
palustris, 237
Trillium
cernuum, 249
erectum, 249
flexipes, 249
grandiflorum, 250
Triosteum perfoliatum, 98
Turritis glabra, 93
Typha
angustifolia, 276
latifolia, 275
Typhaceae, 274

Utricularia
cornuta, 151
intermedia, 152
macrorhiza, 152
vulgaris, 152
Uvularia
grandiflora, 243
sessilifolia, 244

Vaccinium
macrocarpon, 118
oxycoccos, 119
Valeriana
officinalis, 99
uliginosa, 99
Verbascum
blattaria, 218
thapsus, 219
Verbena
bracteata, 221
hastata, 222
stricta, 223
Verbenaceae, 221
Vernonia
altissima, 78
gigantea, 78
missurica, 79
Veronica
americana, 177
beccabunga, 177
Veronicastrum virginicum, 177
Vicia
americana, 131
caroliniana, 132
cracca, 132
villosa, 132
Viola
affinis, 223
canadensis, 224
conspersa, 224
cucullata, 224
incognita, 225
labradorica, 224
lanceolata, 225
minuscula, 225
pallens, 225
papilionacea, 223
pedata, 225
pubescens, 226
renifolia, 226
rostrata, 226
sororia, 224
Violaceae, 223

Waldsteinia fragarioides, 210

Xanthium
italicum, 79
spinosum, 79
strumarium, 79
Xyridaceae, 276
Xyris montana, 276

Zigadenus glaucus, 248
Zizia aurea, 34

Adam-and-Eve, 254
Adder's-Mouth, Green, 264
  White, 264
Agrimony, Swamp, 203
  Tall Hairy, 203
Agueweed, 135
Alfalfa, 129
Allegheny Monkey-Flower, 173
Alsike Clover, 131
Alumroot, Richardson's, 217
Alyssum, Hoary, 88
  Pale, 87
  Sweet, 87
Amaranth Family, 25
American Brooklime, 177
American Cow-Wheat, 168
American Dog Violet, 224
American Germander, 150
American Ginseng, 40
American Lopseed, 173
American Lotus, 158
American Pennyroyal, 144
American Pokeweed, 174
American Sea-Rocket, 89
American Squawroot, 167
American Vetch, 131
American Water-Plantain, 228
American White Water-Lily, 160
American Wild Mint, 145
American Wintergreen, 117
Anemone, Meadow, 191
  Rue, 201
  Wood, 193
Angelica, Hairy, 27
  Purple-Stem, 27
Anise-Root, 31
Arbutus, Trailing, 113
Arrow-Arum, 231
Arrow-Grass Family, 236
Arrow-Grass, Marsh, 237
  Seaside, 236
Arrow-Leaf Tearthumb, 184
Arrowhead, Broadleaf, 228
Arum Family, 229
Asparagus Family, 239
Asphodel Family, 250
Asphodel, False, 252
  Sticky False, 252
Aster Family, 42
Aster, Calico, 50
  Large-Leaved, 49
  New England, 51
  Red-Stemmed, 51
  Smooth Blue, 51
  White, 50
Auricled Twayblade, 265
Autumn Coralroot, 257
Autumn-Crocus Family, 243
Avens, Purple, 206
  Rough, 206
  Water, 206
  White, 205

Baby's-breath, 212
Bachelor's Button, 53
Balsam Ragwort, 69
Balsam, 80
Balsam-Apple, 109
Baneberry, Red, 191
  White, 190
Barberry Family, 81
Barren Strawberry, 210
Basil, Wild, 143
Bastard-Toadflax, 214
Bay Forget-Me-Not, 86
Beach Pea, 126
Beach Wormwood, 49
Bearberry, 112
Bearded Iris, 235
Beardtongue, Hairy, 176
Bedstraw, Fragrant, 212
  Northern, 211
  Yellow, 212
Beebalm, Scarlet, 146
  Spotted, 146
Beechdrops, 167
Beggarticks, Nodding, 52
Bellflower, Tall, 95
Bellwort, Large-Flower, 243
  Sessile-Leaf, 244
Bergamot, Wild, 146
Biennial Evening-Primrose, 164
Bigbract Verbena, 221
Bindweed, Black, 183
  Field, 107
  Fringed Black, 183
  Hedge, 106
  Low False, 107
Bird-Foot Violet, 225
Birthwort Family, 41
Bishop's-Cap, Naked, 217
Bittercress, Pennsylvania, 90
Bittersweet Family, 105
Bittersweet Nightshade, 221

Black Bindweed, 183
Black Medick, 129, 130
Black-Eyed Susan, 71
Bladder Campion, 104
Bladderwort Family, 151
Bladderwort, Common, 152
  Horned, 151
  Intermediate, 152
Blazing Star, Rough, 66
  Spiked, 67
Bleeding-Heart, 171
Bloodroot, 172
Blue Bugle, 150
Blue Cohosh, 81
Blue Vervain, 222
Blue Flag, Northern, 234
Blue-Eyed Mary, Spring, 175
Blue-Eyed-Grass, Narrow-Leaf, 236
  White, 235
Blue-Pod Lupine, 129
Bluebead-Lily, Yellow, 245
Bluebell Family, 94
Bluebell, 94, 95
Bluebell, Marsh, 97
Bluebells, Virginia, 85
Bluejacket, 233
Bluet, Long-Leaf Summer, 213
Blunt-Leaf Orchid, 269
Bog Laurel, 115
Bog Willowherb, 162
Bog Yellowcress, 92
Bogbean, 156
Boneset, 57
Borage Family, 83
Bouncing Bet, 102
Bristly Sarsaparilla, 38
Broad-Fruit Bur-Reed, 275
Broad-Leaf Cattail, 275
Broad-Leaf Enchanter's-Nightshade, 161
Broad-Leaf Waterleaf, 138
Broad-Lipped Twayblade, 265
Broadleaf Arrowhead, 228
Brook Lobelia, 96
Brooklime, American, 177
  European, 177
Broom-Rape Family, 165
Broomrape, Clustered, 168
  Naked, 168
Brown-Eyed Susan, 72
Buck-Bean Family, 156
Buck-Bean, 156
Buckwheat Family, 182
Buckwheat, Common, 182
Bugle, Blue, 150
  Carpet, 150
Bugleweed, 144
Bulblet-Bearing Water-Hemlock, 27
Bulbous Cress, 89
Bull Thistle, 55
Bullhead-Lily, 160
Bunchberry, 108
Bur-Reed, Broad-Fruit, 275
Burdock, Lesser, 48
Bush-Clover, Hairy, 128
  Purple, 128
  Round-Head, 127
Butter-and-Eggs, 175
Buttercup Family, 189
Buttercup, Creeping, 199
  Early, 199
  Hairy, 199
  Kidney-Leaf, 198
  Swamp, 199
  Tall, 198
Butterfly-Weed, 37
Butterweed, 56
Butterwort, Common, 151
Buttonbush, 211

Cactus Family, 93
Calamint, Low, 142
Calamus Family, 227
Calico Aster, 50
Calypso, 255
Campion, Starry, 104
  White, 103
Canada Frostweed, 106
Canada Goldenrod, 74
Canada Thistle, 54
Canada-Pea, 132
Canadian Black Snakeroot, 31
Canadian Lousewort, 168
Canadian White Violet, 224
Candy-Flower Family, 157
Cardinal-Flower, 95
Carolina Springbeauty, 157
Carpet Bugle, 150
Carrion-Flower, Midwestern, 251
  Upright, 251
Carrot, Wild, 29
Cat-Tail Family, 274
Catchfly, Sleepy, 103
Catfoot, 71
Catnip, 147

Cattail, Broad-Leaf, 275
Narrow-Leaf, 276
Celandine, 172
Chamomile, Corn, 48
Stinking, 48
Checkered Rattlesnake-Orchid, 263
Chervil, Wild, 29
Chickweed, Common, 104
Field, 101
Chicory, 53
Chinese Lantern-Plant, 220
Cinquefoil, Downy, 207
Marsh, 203
Oldfield, 208
Rough, 208
Shrubby, 204
Silvery, 207
Sulphur, 208
Tall, 56, 204
Three-Toothed, 210
Clammy Ground-Cherry, 220
Clematis, Purple, 195
Climbing False Buckwheat, 183
Closed Bottle Gentian, 134
Clover, Alsike, 131
Large Hop, 130
Red, 130
White, 131
White Sweet, 129
Yellow Sweet, 129
Clustered Black Snakeroot, 32
Coastal Jointweed, 185
Cocklebur, Rough, 79
Cohosh, Blue, 81
Colicroot, White, 250
Columbine, Garden, 194
Red, 194
Comfrey, Northern Wild, 83
Common Bladderwort, 152
Common Blue Violet, 223
Common Buckwheat, 182
Common Butterwort, 151
Common Chickweed, 104
Common Dandelion, 77
Common Evening-Primrose, 163
Common Flax, 153
Common Fleabane, 57
Common Hound's-Tongue, 83
Common Mallow, 156
Common Milkweed, 36
Common Motherwort, 144
Common Mullein, 219
Common Ragweed, 46
Common Self-Heal, 148
Common St. John's-Wort, 140
Common Sunflower, 61
Common Tansy, 76
Common Viper's-Bugloss, 84
Common Yarrow, 45
Coneflower, Cutleaf, 72
Gray-Head, 71
Coralroot, Autumn, 257
Spotted, 256
Striped, 257
Yellow, 257
Coreopsis, Lance-Leaf, 55
Corn Chamomile, 48
Corn Cockle, 100
Cornel, Dwarf, 108
Cornflower, 53
Corydalis, Golden, 171
Cow-Parsnip, 30
Cow-Wheat, American, 168
Cowbane, 33
Cowslip, 194
Cranberry, Large, 118
Small, 119
Crane's-Bill, Spotted, 137
Cream False Indigo, 124
Creeping Buttercup, 199
Creeping Foamflower, 217
Creeping Snowberry, 114
Creeping Wintergreen, 114
Creeping-Jennie, 187
Cress, Bulbous, 89
Cucumber-Root, Indian, 247
Cucumber, Wild, 109
Culver's-Root, 177
Cup-Plant, 72
Curly Dock, 186
Curly-Cup Gumweed, 59
Cut-Leaf Toothwort, 90
Cutleaf Coneflower, 72
Cutleaf Evening-Primrose, 163
Cypress Spurge, 120

Daffodil Family, 238
Daisy Fleabane, 57
Daisy, Ox-Eye, 65
Dalmatian Toadflax, 176
Dame's Rocket, 91
Dandelion, Common, 77
Dwarf, 64
Red-Seeded, 77

Day-Lily, Orange, 242
    Yellow, 243
Deathcamas, Mountain, 248
Dense Gayfeather, 67
Deptford Pink, 101
Devil's-Club, 38
Devil's-Bite, 67
Diffuse Knapweed, 52
Dock-Leaf Smartweed, 183
Dog Violet, American, 224
Dogbane Family, 34
Dogbane, Spreading, 35
Dogwood Family, 108
Downy Cinquefoil, 207
Downy False Foxglove, 166
Downy Gentian, 134
Downy Rattlesnake Orchid, 263
Downy Yellow Violet, 226
Dragon's-Mouth, 254
Duck-Potato, 228
Dune Goldenrod, 74
Dutchman's-Breeches, 171
Dwarf Dandelion, 64
Dwarf Ginseng, 40
Dwarf Lake Iris, 234
Dwarf Rattlesnake-Orchid, 262

Early Buttercup, 199
Early Goldenrod, 75
Early Meadow-Rue, 200
Eastern Prickly-Pear, 93
Elecampane, 64
Enchanter's-Nightshade, Broad-Leaf, 161
    Small, 161
Endive, Garden, 53
European Brooklime, 177
Evening-Primrose Family, 161
Evening-Primrose, Biennial, 164
    Common, 163
    Cutleaf, 163
    Northern, 163
Everlasting, Macoun's, 70

Fairy-Slipper, 255
False Asphodel, 252
False Buckwheat, Climbing, 183
False Hellebore Family, 248
False Indigo, Cream, 124
    Prairie, 124
    White, 124
False Lily-of-the-Valley, 239
False Rue-Anemone, 196
False Solomon's-Seal, 240
Featherling Family, 252
Fen Grass-of-Parnassus, 105
Few-Leaf Sunflower, 63
Field Bindweed, 107
Field Chickweed, 101
Field Sow-Thistle, 75
Field Mint, 145
Figwort Family, 218
Figwort, Lance-Leaf, 218
Fireweed, 162
Flag, Southern Blue, 235
Flat-Topped Goldenrod, 58
Flax Family, 153
Flax, Common, 153
    Prairie, 153
Fleabane, Common, 57
Flowering Spurge, 120
Flowering-Rush, 232
Flowering-Rush Family, 232
Foamflower, Creeping, 217
Forget-Me-Not, Bay, 86
    True, 85
Four-O'clock, 158, 159
Four-O'Clock Family, 158
Foxglove, Downy False, 166
    Purple False, 165
    Slender False, 166
Fragrant Bedstraw, 212
Fringed Black Bindweed, 183
Fringed Loosestrife, 187
Fringed Polygala, 181
Frostweed, Canada, 106

Garden Columbine, 194
Garden Endive, 53
Garden Heliotrope, 99
Garden Lettuce, 65
Garden Phlox, 178
Gayfeather, Dense, 67
    Ontario, 66
    Tall, 66
Gentian Family, 133
Gentian, Closed Bottle, 134
    Downy, 134
    Greater Fringed, 135
    Lesser Fringed, 136
    Red-Stemmed, 134
    Spurred, 136
    Stiff, 135
    Yellow, 133
Geranium Family, 136

Geranium, Wild, 137
Germander, American, 150
Ghost-Pipe, 115
Giant Sunflower, 62
Ginger, Wild, 41
Ginseng Family, 38
Ginseng, American, 40
    Dwarf, 40
Goat's-Beard, Yellow, 78
Goat's-Rue, 124
Golden Alexanders, 34
Golden Corydalis, 171
Golden Ragwort, 69
Golden-Tuft, 87
Goldenrod, Canada, 74
    Dune, 74
    Early, 75
    Flat-Topped, 58
    Gray, 75
    Hairy, 74
Goldenseal, 197
Goldthread, Three-Leaf, 196
Goosefoot, 25
Gourd Family, 109
Grass-of-Parnassus, Fen, 105
    Small-Flowered, 105
Grass-Pink, Tuberous, 255
Gray Goldenrod, 75
Gray-Head Coneflower, 71
Great Blue Lobelia, 96
Great Ragweed, 46
Great Waterleaf, 139
Greater Canadian St. John's-Wort, 140
Greater Fringed Gentian, 135
Greek-Valerian, 179
Greenbrier Family, 251
Green Club-Spur Orchid, 267
Green Adder's-Mouth, 264
Green Comet Milkweed, 38
Green Dragon, 229
Green-Flower Wintergreen, 118
Green Nightshade, 221
Ground-Cherry, Clammy, 220
    Large-Flowered, 220
Ground-Ivy, 143
Groundnut, 123
Ground-Nut, 40
Gumweed, Curly-Cup, 59

Hairy Angelica, 27
Hairy Beardtongue, 176
Hairy Bush-Clover, 128
Hairy Buttercup, 199
Hairy Goldenrod, 74
Hairy Puccoon, 84
Hairy Solomon's-Seal, 241
Hairy Sweet Cicely, 30
Hairy Vetch, Winter Vetch, 132
Hairy Woodmint, 142
Harbinger-Of-Spring, 29
Harebell, 94
Harlequin, Rock, 170
Hawkweed, Orange, 69
    Smooth, 70
Heartleaf Twayblade, 265
Heath Family, 111
Hedge Bindweed, 106
Hedge-Mustard, 92
Hedgenettle, Smooth, 150
Heliotrope, Garden, 99
Hepatica, Round-Lobed, 196
    Sharp-Lobed, 197
Herb-Robert, 138
Hoary Alyssum, 88
Hoary Puccoon, 84
Hoary Vervain, 223
Hog-Peanut, 122
Honewort, 29
Honeysuckle Family, 97
Hooded Ladies'-Tresses, 273
Hooker's Orchid, 270
Horned Bladderwort, 151
Horse-balm, 144
Horsemint, 146
Horseweed, 56
Hound's-Tongue, Common, 83
Husk-Tomato, 220

Indian Cucumber-Root, 247
Indian-Hemp, 35
Indian Paintbrush, 166
Indian-Pipe, 115
Indian-Plantain, Pale, 48
Indigo, Wild, 123
Intermediate Bladderwort, 152
Iris Family, 233
Iris, Bearded, 235
    Dwarf Lake, 234
    Yellow, 235
Ironweed, Missouri, 79
    Tall, 78

Jack-in-the-Pulpit, 230
Jacob's-Ladder, 179

Jerusalem Artichoke, 63
Jerusalem Cherry, 221
Jimsonweed, 219
Joe-Pye Weed, Spotted, 58
    Sweet-Scented, 59
Jointweed, Coastal, 185

Kidney-Leaf Buttercup, 198
Kidney-Leaf White Violet, 226
Kingdevil, 70
Kinnikinick, 112
Knapweed, Diffuse, 52
    Spotted, 53
Korean Lespedeza, 128

Labrador-Tea, 118
Ladies'-Tresses, Hooded, 273
    Shining, 272
    Slender, 272
    Sphinx, 272
Lady's-Slipper, Large Yellow, 260
    Pink, 258
    Ram's-Head, 259
    Showy, 261
    Small White, 259
    Small Yellow, 260
Lady's-Thumb, Spotted, 184
Lake Huron Green Orchid, 268
Lake Huron Tansy, 76
Lamb's-Quarters, 25
Lance-Leaf Coreopsis, 55
Lance-Leaf Figwort, 218
Lance-Leaf Twisted-Stalk, 247
Lantern-Plant, Chinese, 220
Large Cranberry, 118
Large Hop Clover, 130
Large White Trillium,, 250
Large Yellow Lady's-Slipper, 260
Large-Flower Bellwort, 243
Large-Flowered Ground-Cherry, 220
Large-Leaved Aster, 49
Laurel, Bog, 115
Leek, Wild, 239
Lemon-Mint, 147
Lespedeza, Korean, 128
Lesser Burdock, 48
Lesser Fringed Gentian, 136
Lesser Purple Fringed Orchid, 270
Lettuce, Garden, 65
    Prickly, 65
    Wild, 65
Licorice, Wild, 212
Lily Family, 237, 245
Lily-of-the-Valley, False, 239
Lily, Michigan, 246
    Wood, 246
Linear-Leaf Sundew, 110
Live-Forever, 108
Lobelia, Brook, 96
    Great Blue, 96
    Kalm's, 96
    Pale Spike, 96
Loesel's Twayblade, Fen Orchid, 263
Long-Beak Water-Crowfoot, 200
Long-Bract Frog Orchid, 266
Long-Head Thimbleweed, 192
Long-Leaf Summer Bluet, 213
Long-Spur Violet, 226
Loosestrife Family, 153
Loosestrife, Fringed, 187
    Purple, 154
    Swamp, 153
    Tufted, 188
    Whorled, 188
    Winged, 154
Lopseed Family, 172
Lopseed, American, 173
Lotus-Lily Family, 158
Lotus, American, 158
Lousewort, Canadian, 168
Low Calamint, 142
Low False Bindweed, 107
Lupine, Blue-Pod, 129
    Sundial, 128
Lyre-Leaved Rock Cress, 87

Macoun's Everlasting, 70
Mad-Dog Skullcap, 149
Madder Family, 211
Mallow Family, 154
Mallow, Common, 156
Mandrake, 82
Marsh Arrow-Grass, 237
Marsh Bluebell, 97
Marsh Cinquefoil, 203
Marsh Marigold, 194
Marsh Pea, 127
Marsh Skullcap, 148
Marsh St. John's-Wort, 139
Marsh Tickseed, 52
Marsh Violet, 224
Maryland Black Snakeroot, 32
Maximilian Sunflower, 62
May-Apple, 82

Mayweed, 48
Meadow Anemone, 191
Meadow-Parsnip, Yellow, 33
Meadow-Rue, Early, 200
  Purple, 201
Medick, Black, 129, 130
Michigan Lily, 246
Midwestern Carrion-Flower, 251
Milfoil, 45
Milkweed, Common, 36
  Green Comet, 38
  Swamp, 36
  Tall, 36
Milkwort Family, 179
Milkwort, Purple, 180
  Racemed, 180
  Whorled, 181
Mint Family, 141
Mint, American Wild, 145
Missouri Ironweed, 79
Mistassini Primrose, 189
Miterwort, Two-Leaf, 216
Monkey-Flower, Allegheny , 173
  Yellow, 172
Morning-Glory Family, 106
Moss-Pink, 178
Mossy Stonecrop, 109
Moth Mullein, 218
Motherwort, Common, 144
Mountain Deathcamas, 248
Mountain Wood-Sorrel, 169
Mountain-Mint, Virginia, 148
Mullein-Pink, 103
Mullein, Common, 219
  Moth, 218
Musk Mallow, 155
Muskflower, 173
Mustard Family, 86

Naked Bishop's-Cap, 217
Naked Broomrape, 168
Naked-Flower Tick-Trefoil, 125
Narrow-Leaf Blue-Eyed-Grass, 236
Narrow-Leaf Cattail, 276
New England Aster, 51
Nightshade Family, 219
Nightshade, Bittersweet, 221
  Green, 221
Nodding Beggarticks, 52
Nodding Trillium, 249
Northern Bedstraw, 211
Northern Blue Flag, 234
Northern Evening-Primrose, 163
Northern White Violet, 225
Northern Wild Comfrey, 83
Northern Yellow-Eyed-Grass, 276

Obedient-Plant, 147
Oldfield Cinquefoil, 208
One-Flowered Wintergreen, 115
One-Sided Wintergreen, 116
Onionweed Family, 242
Ontario Gayfeather, 66
Orange Day-Lily, 242
Orange Hawkweed, 69
Orchid Family, 252
Orchid, Blunt-Leaf, 269
  Fen, 263
  Green Club-Spur, 267
  Hooker's, 270
  Lake Huron Green, 268
  Lesser Purple Fringed, 270
  Long-Bract Frog, 266
  Pale-Green, 268
  Prairie White Fringed, 260
  Ragged Fringed, 271
  Round-Leaf, 269
  White Bog, 268
  White Fringed, 267
Orchis, Showy, 261
Orpine, 108
Oswego-Tea, 146
Ox-Eye Daisy, 65
Ox-Eye, 63
Oyster-Plant, 78

Painted Cup, 166
Pale Alyssum, 87
Pale Indian-Plantain, 48
Pale Spike Lobelia, 96
Pale Touch-Me-Not, 80
Pale Vetchling, 126
Pale-Green Orchid, 268
Parlin's Pussytoes, 47
Parsley Family, 26
Parsnip, Wild, 31
Partridge-Berry, 213
Partridge-Pea, 124
Pasque-Flower, 193
Pea Family, 121
Pea, Beach, 126
  Marsh, 127
Pearly-Everlasting, 47
Pennsylvania Bittercress, 90

Pennyroyal, American, 144
Pepper-And-Salt, 29
Peppermint, 145
Perennial Ragweed, 46
Phlox Family, 178
Phlox, Garden, 178
    Prairie, 178
    Wild Blue, 178
Pickerelweed, 274
Pickerelweed Family, 274
Pimpernel, Yellow, 33
Pineapple-Weed, 67
Pinedrops, 116
Pinesap, 116
Pink Family, 100
Pink Lady's-Slipper, 258
Pink Wintergreen, 117
Pinkweed, 184
Pipsissewa, 112
Pitcher-Plant Family, 214
Pitcher-Plant, Purple, 214
Pitcher's Thistle, 55
Plantain Family, 174
Pleurisy-Root, 37
Pogonia, Rose, 271
Poison-Hemlock, 28
Pokeweed Family, 174
Pokeweed, American, 174
Polygala, Fringed, 181
Pond-Lily, Yellow, 159
Poppy Family, 170
Prairie False Indigo, 124
Prairie Flax, 153
Prairie Phlox, 178
Prairie White Fringed Orchid, 260
Prairie-Dock, 73
Prickly Lettuce, 65
Prickly-Pear, Eastern, 93
Primrose Family, 186
Primrose, Mistassini, 189
Prince's Pine, 112
Puccoon, Hairy, 84
    Hoary, 84
Purple Avens, 206
Purple Bush-Clover, 128
Purple Clematis, 195
Purple False Foxglove, 165
Purple Loosestrife, 154
Purple Meadow-Rue, 201
Purple Milkwort, 180
Purple Pitcher-Plant, 214
Purple Salsify, 78
Purple-Flowering Thimbleberry, 210
Purple-Head Sneezeweed, 60
Purple-Stem Angelica, 27
Pussytoes, Parlin's, 47
Putty-Root, 254

Queen Anne's Lace, 29

Racemed Milkwort, 180
Ragged Fringed Orchid, 271
Ragweed, Common, 46
    Great, 46
    Perennial, 46
Ragwort, Balsam, 69
    Golden, 69
Ram's-Head Lady's-Slipper, 259
Ramp, 239
Rattlesnake-Orchid, Checkered, 263
    Downy, 263
    Dwarf, 262
    Western, 262
Rattlesnake-Root, Tall, 68
    White, 68
Rattlesnake-Weed, 64
Rattleweed, 123
Red Baneberry, 191
Red Clover, 130
Red Columbine, 194
Red Windflower, 192
Red-Seeded Dandelion, 77
Red-Stemmed Aster, 51
Red-Stemmed Gentian, 134
Red-Stem Stork's-Bill, 137
Richardson's Alumroot, 217
Robin's-Plantain, 57
Rock Cress, Lyre-Leaved, 87
Rock Harlequin, 170
Rock Sandwort, 102
Rocket, Yellow, 88
Rockrose Family, 106
Rose Family, 202
Rose Campion, 103
Rose Mallow, Swamp, 155
Rose Pogonia, 271
Rose, Swamp, 209
Rough Avens, 206
Rough Cinquefoil, 208
Rough Cocklebur, 79
Round-Head Bush-Clover, 127
Round-Leaf Orchid, 269
Round-Leaf Sundew, 110
Round-Lobed Hepatica, 196

Rue Anemone, 201
Rue-Anemone, False, 196

Salsify, Purple, 78
    Yellow, 77
Sandalwood Family, 214
Sandwort, Rock, 102
Sarsaparilla, Bristly, 38
    Wild, 39
Saxifrage Family, 215
Saxifrage, Swamp, 216
Scarlet Beebalm, 146
Scotchmist, 212
Sea-Rocket, American, 89
Seaside Arrow-Grass, 236
Self-Heal, Common, 148
Seneca Snakeroot, 180
Senna, Wild, 124
Sessile-Leaf Bellwort, 244
Sharp-Lobed Hepatica, 197
Sheep Sorrel, 185
Shepherd's-Purse, 89
Shining Ladies'-Tresses, 272
Shinleaf, 117
Showy Lady's-Slipper, 261
Showy Orchis, 261
Showy Tick-Trefoil, 125
Shrubby Cinquefoil, 204
Shrubby St. John's-Wort, 140
Sicklepod, 88
Silkweed, 36
Silverweed, 207
Silvery Cinquefoil, 207
Skullcap, Mad-Dog, 149
    Marsh, 148
Skunk-Cabbage, 231
Sleepy Catchfly, 103
Slender False Foxglove, 166
Slender Ladies'-Tresses, 272
Small Cranberry, 119
Small Enchanter's-Nightshade, 161
Small Sundrops, 164
Small White Lady's-Slipper, 259
Small Yellow Lady's-slipper, 260
Small-Flowered Grass-of-Parnassus, 105
Smartweed, Dock-Leaf, 183
    Water, 183
Smooth Blue Aster, 51
Smooth Hawkweed, 70
Smooth Hedgenettle, 150
Smooth Solomon's-Seal, 242
Snakeroot, Canadian Black, 31
    Clustered Black, 32
    Maryland Black, 32
    Seneca, 180
    White, 45
Sneezeweed, 59
Sneezeweed, Purple-Head, 60
Snow-on-the-Mountain, 121
Snowberry, Creeping, 114
Soapwort, 102
Solomon's-Seal, False, 240
    Hairy, 241
    Smooth, 242
    Starry False, 240
    Three-Leaf False, 241
Sorrel, Sheep, 185
Southern Blue Flag, 235
Sow-Thistle, Field, 75
Spatterdock, 159
Spearmint, 146
Sphinx Ladies'-Tresses, 272
Spiderwort Family, 233
Spiderwort, 233
Spikenard, 39
Spoon-Leaf Sundew, 111
Spotted Beebalm, 146
Spotted Coralroot, 256
Spotted Joe-Pye Weed, 58
Spotted Knapweed, 53
Spotted Lady's-Thumb, 184
Spotted Touch-Me-Not, 80
Spreading Dogbane, 35
Spring Blue-Eyed Mary, 175
Springbeauty, Carolina, 157
    Virginia, 157
Spurge Family, 120
Spurge, Flowering, 120
Spurred Gentian, 136
Squawroot, American, 167
Squirrel-Corn, 171
St. John's-Wort Family, 139
St. John's-Wort, Common, 140
    Greater Canadian, 140
    Marsh, 139
    Shrubby, 140
Star-Grass, Yellow, 244
Starflower, 186
Starry Campion, 104
Starry False Solomon's-Seal, 240
Sticky False Asphodel, 252
Sticky Willy, 212
Stiff Gentian, 135
Stinking-Benjamin, 249

Stonecrop Family, 108
Stonecrop, Mossy, 109
Stork's-Bill, Red-Stem, 137
Strawberry-Blite, 25
Strawberry, Barren, 210
  Virginia, 204
  Woodland, 205
Striped Coralroot, 257
Striped Wintergreen, 113
Sulphur Cinquefoil, 208
Summer Savory, 143
Sundew Family, 110
Sundew, Linear-Leaf, 110
  Round-Leaf, 110
  Spoon-Leaf, 111
Sundial Lupine, 128
Sundrops, 163
Sundrops, Small, 164
Sunflower, Common, 61
  False, 63
  Few-Leaf, 63
  Giant , 62
  Maximilian, 62
  Pale, 61
  Swamp, 59
  Thin-Leaf, 61
  Woodland, 62
Swamp Agrimony, 203
Swamp Buttercup, 199
Swamp-Candles, 188
Swamp Loosestrife, 153
Swamp Milkweed, 36
Swamp Pink, 254
Swamp Rose, 209
Swamp Rose Mallow, 155
Swamp Saxifrage, 216
Swamp Sunflower, 59
Swamp Thistle, 54
Swamp Valerian, 99
Sweet Alyssum, 87
Sweet Cicely, Hairy, 30
Sweet-Scented Joe-Pye Weed, 59
Sweetflag, 227

Tall Bellflower, 95
Tall Buttercup, 198
Tall Cinquefoil, 204
Tall Coreopsis, 56
Tall Gayfeather, 66
Tall Hairy Agrimony, 203
Tall Ironweed, 78
Tall Milkweed, 36
Tall Rattlesnake-Root, 68
Tall Thimbleweed, 193
Tall Tickseed, 56
Tall Tumblemustard, 92
Tansy, Common, 76
  Lake Huron, 76
Tearthumb, Arrow-Leaf, 184
Teasel, Wild, 97
Thimbleberry, 209
Thimbleberry, Purple-Flowering, 210
Thimbleweed, Long-Head, 192
  Tall, 193
Thin-Leaf Sunflower, 61
Thistle, Bull, 55
  Canada, 54
  Pitcher's, 55
  Swamp, 54
Thoroughwort, 57
Three-Leaf False Solomon's-Seal, 241
Three-Leaf Goldthread, 196
Three-Toothed Cinquefoil, 210
Tick-Trefoil, Naked-Flower, 125
  Showy, 125
Tickseed, Marsh, 52
  Tall, 56
Tinker's-Weed, 98
Toadflax, Dalmatian, 176
Toothwort, Cut-Leaf, 90
  Two-Leaf, 90
Touch-Me-Not Family, 80
Touch-Me-Not, Pale, 80
  Spotted, 80
Tower-Mustard, 93
Trailing Arbutus, 113
Trillium, Large White, 250
  Nodding, 249
  Red, 249
Trout-Lily, White, 246
  Yellow, 245
True Forget-Me-Not, 85
Tuberous Grass-Pink, 255
Tuberous White Water-Lily, 160
Tuckahoe, 231
Tufted Loosestrife, 188
Tufted Vetch,, 132
Tumblemustard, Tall, 92
Turtlehead, White, 175
Twayblade, Auricled, 265
  Broad-Lipped, 265
  Heartleaf, 265
  Loesel's, 263
Twinflower, 98

Twinleaf, 81
Twisted-Stalk, Lance-Leaf, 247
Two-Leaf Miterwort, 216
Two-Leaf Toothwort, 90

Umbrella-Wort, 158
Upright Carrion-Flower, 251

Valerian, Swamp, 99
Verbena, Bigbract, 221
Vervain Family, 221
Vervain, Blue, 222
  Hoary, 223
Vetch, American, 131
  Hairy, 132
  Tufted, 132
  Winter, 132
  Wood, 132
Vetchling, Pale, 126
  Yellow, 126
Violet Family, 223
Violet, Bird-Foot, 225
  Canadian White, 224
  Common Blue, 223
  Downy Yellow, 226
  Kidney-Leaf White, 226
  Long-Spur, 226
  Marsh, 224
  Northern White, 225
  White Bog, 225
Viper's-Bugloss, Common, 84
Virgin's-Bower, 195
Virginia Bluebells, 85
Virginia Mountain-Mint, 148
Virginia Springbeauty, 157
Virginia Strawberry, 204
Virginia Waterleaf, 139

Wakerobin, 250
Watercress, 91
Water-Crowfoot, Long-Beak, 200
  Yellow, 200
Water-Dragon, 230
Water-Hemlock, 27
  Bulblet-Bearing, 27
Water-Horehound, 144
Waterleaf Family, 138
Waterleaf, Broad-Leaf, 138
  Great, 139
  Virginia, 139
Water-Lily Family, 159
Water-Lily, American White, 160
  Tuberous White, 160
Water-Parsnip, 32
Water-Plantain Family, 228
Water-Plantain, American, 228
Watershield, 159
Water Smartweed, 183
Western Rattlesnake-Orchid, 262
White Adder's-Mouth, 264
White Aster, 50
White Avens, 205
White Baneberry, 190
White Blue-Eyed-Grass, 235
White Bog Orchid, 268
White Bog Violet, 225
White Campion, 103
White Clover, 131
White Colicroot, 250
White False Indigo, 124
White Fringed Orchid, 267
White Rattlesnake-Root, 68
White Snakeroot, 45
White Sweet Clover, 129
White Trout-Lily, 246
White Turtlehead, 175
Whorled Loosestrife, 188
Whorled Milkwort, 181
Widow's-Frill, 104
Wild Basil, 143
Wild Bergamot, 146
Wild Blue Phlox, 178
Wild Carrot, 29
Wild Chervil, 29
Wild Cucumber, 109
Wild Geranium, 137
Wild Ginger, 41
Wild Indigo, 123
Wild Leek, 239
Wild Lettuce, 65
Wild Licorice, 212
Wild Parsnip, 31
Wild Sarsaparilla, 39
Wild Senna, 124
Wild Teasel, 97
Willowherb, Bog, 162
Windflower, Red, 192
Winged Loosestrife, 154
Wintergreen, American, 117
  Creeping, 114
  Green-Flower, 118
  One-Flowered, 115
  One-Sided, 116

Pink, 117
Striped, 113
Woodland Strawberry, 205
Woodland Sunflower, 62
Woodmint, Hairy, 142
Wood Anemone, 193
Wood Lily, 246
Wood-Poppy, 172
Wood-Sorrel Family, 169
Wood-Sorrel, Mountain, 169
Yellow, 170
Wood Vetch, 132
Wormwood, 168
Beach, 49
Woundwort, 149

Yarrow, Common, 45
Yellow Bedstraw, 212
Yellow Bluebead-Lily, 245
Yellow Coralroot, 257
Yellowcress, Bog, 92
Yellow Day-Lily, 243
Yellow-Eyed-Grass Family, 276
Yellow-Eyed-Grass, Northern, 276
Yellow Gentian, 133
Yellow Goat's-Beard, 78
Yellow Iris, 235
Yellow Meadow-Parsnip, 33
Yellow Monkey-Flower, 172
Yellow Pimpernel, 33
Yellow Pond-Lily, 159
Yellow Rocket, 88
Yellow Salsify, 77
Yellow Star-Grass, 244
Yellow Star-Grass Family, 244
Yellow Sweet Clover, 129
Yellow Trout-Lily, 245
Yellow Vetchling, 126
Yellow Water-Crowfoot, 200
Yellow Wood-Sorrel, 170

0 MM 1 2 3 4 5 6 7 8 9 10 11 12 13 14 15

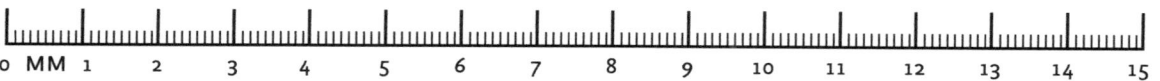

www.ingramcontent.com/pod-product-compliance
Lightning Source LLC
Chambersburg PA
CBHW052109030426
42335CB00025B/2898